Zlatan Stojković, Ph. D.

# COMPUTER-AIDED DESIGN IN POWER ENGINEERING

## Application of Software Tools

Springer

ACADEMIC MIND
University of Belgrade - Faculty of Electrical Engineering

**COMPUTER-AIDED DESIGN IN POWER ENGINEERING**
**Application of Software Tools**

Zlatan Stojković, Ph. D.

Reviewers
Jovan Đorđević, Ph. D.
Dragoslav Perić, Ph. D.
Milan Savić, Ph. D.


ACADEMIC MIND, Belgrade, Serbia
UNIVERSITY OF BELGRADE, Faculty of Electrical Engineering
SPRINGER-VERLAG, Berlin Heidelberg, Germany

Design of cover page
Zorica Marković, Academic Painter

Printed in Germany

ISBN 978-86-7466-433-9
ISBN 978-3-642-30205-3

*For*
*Danica, Tatjana, Dušan and Lea*

# CONTENTS

PREFACE 11

ACKNOWLEDGMENT 17

1. COMPUTER-AIDED MODELING AND SIMULATION 19
1.1. Introduction . . . 19
1.2. Definitions . . . 19
1.3. Categories of approaches to modeling . . . 19
1.4. Phases of development . . . 20
1.5. The role of computers in modeling and simulation . . . 21
1.6. Methodology of computer-aided modeling and simulation 21
1.6.1. Formulation of the problem . . . 21
1.6.2. Analysis of the problem . . . 23
1.6.3. Forming the mathematical model . . . 24
1.6.3.1 General considerations . . . 24
1.6.3.2 General principles of good modeling . . . 36
1.6.4. Selection of computerized methods . . . 49
1.6.5. Forming the program . . . 50
1.6.5.1 Development of the algorithm . . . 51
1.6.5.2 Requirements for resources . . . 54
1.6.5.3 Program characteristics . . . 54
1.6.6. Simulation . . . 86
1.6.6.1 Aspects of performing simulation . . . 86
1.6.6.2 Advantages of simulation . . . 98
1.7. Supplement: GIC – software tool for calculating the grounding grid impulse characteristics . . . 99
1.7.1. Purpose of the program . . . 99
1.7.2. Mathematical model of a grounding grid in an impulse regime . . . 100
1.7.3. Sequence of calculation . . . 100
1.7.4. Description of the software . . . 101
1.7.4.1 Defining the parameters for calculating grounding grid impulse characteristics . . . 102

1.7.4.2 Executing the calculation . . . . . . . . . . . 106
1.7.4.3 Display of calculation results . . . . . . . . 106
1.7.5. Calculation examples . . . . . . . . . . . . . . . . . 108
1.7.6. Input-output values and possible errors . . . . . . . 113
1.7.6.1 Input-output variables . . . . . . . . . . . . 113
1.7.6.2 Possible Errors . . . . . . . . . . . . . . . . 114
1.7.6.3 Messages when working with the software tool GIC . . . . . . . . . . . . . . . . . . . 116
1.7.7. Using windows commands . . . . . . . . . . . . . . 117
1.7.7.1 Opening the program . . . . . . . . . . . . . 117
1.7.7.2 Appearance of screen with windows commands . . . . . . . . . . . . . . . . . . . . 117
1.7.7.3 Description of individual commands using the help menu . . . . . . . . . . . . . . . . 117

**LITERATURE** **130**

**2. APPLICATION OF SOFTWARE TOOLS IN POWER ENGINEERING CALCULATIONS** **135**

2.1. MATLAB® /Simulink® . . . . . . . . . . . . . . . . . . . 135
2.1.1. Introduction . . . . . . . . . . . . . . . . . . . . . 135
2.1.2. Basic groups of blocks in Simulink® software* . . . 138
2.1.3. Additional module SimPowerSystems™* . . . . . . 142
2.1.4. Application of MATLAB® technical computing software in calculation of characteristic values of fault current . . . . . . . . . . . . . . . . . . . . . . . . 160
2.1.4.1 General considerations . . . . . . . . . . . . 160
2.1.4.2 Calculation of characteristic values of the fault current . . . . . . . . . . . . . . . . 160
2.1.4.3 Calculation process using the Simulink® module . . . . . . . . . . . . . . . . . . . . . . 162
2.1.5. SPLCAD software tool for designing medium-voltage overhead lines . . . . . . . . . . . . . . . . . . . . . 167
2.1.5.1 Introduction . . . . . . . . . . . . . . . . . . 167
2.1.5.2 Description of SPLCAD software tool and an example of its application . . . . . . . . 169
2.1.6. Simulink® model of a turbine regulator in the "Kokin Brod" hydroelectric power plant . . . . . . . . . . . 180
2.1.6.1 General data on the hydroelectric power plant 180
2.1.6.2 Modeling the "Kokin Brod" HPP . . . . . . 180
2.1.6.3 Display of the "Kokin Brod" HPP turbine regulator . . . . . . . . . . . . . . . . . . . . 184

2.1.6.4 Simulink® model of turbine regulator . . . . 185
2.1.6.5 Comparative analysis of numerically and experimentally obtained results . . . . . . . . 187
2.2. EMTP / ATP . . . . . . . . . . . . . . . . . . . . . . . 188
2.2.1. Introduction . . . . . . . . . . . . . . . . . . . . . . 188
2.2.2. Examples of calculations of overvoltages caused by switching operations of the disconnector in metal-enclosed $SF_6$ gas insulated switchgears . . . . . . . 192
2.2.2.1 General considerations . . . . . . . . . . . . 192
2.2.2.2 Electromagnetic transient processes in secondary circuits of measurement transformers in GIS . . . . . . . . . . . . . . . . . . . 197
2.2.2.3 Calculation of growth of potential of the metal enclosure of an $SF_6$ gas insulated switchgear caused by switching operations of the disconnector . . . . . . . . . . . . . 207
2.3. MS Excel / MS Access . . . . . . . . . . . . . . . . . 216
2.3.1. Introduction . . . . . . . . . . . . . . . . . . . . . . 216
2.3.2. Application of databases in designing high-voltage substations . . . . . . . . . . . . . . . . . . . . . 219
2.3.2.1 Criteria for selection of high-voltage equipment . . . . . . . . . . . . . . . . . . . . . . 219
2.3.2.2 Calculation of characteristic values of fault current . . . . . . . . . . . . . . . . . . . . 221
2.3.3. Application of databases in selection of high-voltage equipment . . . . . . . . . . . . . . . . . . . . . 226
2.3.3.1 Definition of database types . . . . . . . . . 226
2.3.3.2 Example of the application of MS Excel . . 228
2.3.3.3 Example of the application of MS Access . . 232
2.3.4. Application of MS Excel macros for design in power engineering . . . . . . . . . . . . . . . . . . . . 238
2.3.4.1 Developing macros in MS Excel . . . . . . . 238
2.3.4.2 Examples of the application of macros . . . 241
2.3.5. Application of MS Excel in designing the power supply of telecommunications equipment . . . . . . 246
2.3.5.1 Technical description of power supply with direct voltage . . . . . . . . . . . . . . . . 246
2.3.5.2 Technical calculations . . . . . . . . . . . 248
2.4. AutoCAD . . . . . . . . . . . . . . . . . . . . . . . . 255
2.4.1. Introduction . . . . . . . . . . . . . . . . . . . . . . 255
2.4.2. Spatial (3D) model of a turbogenerator . . . . . . . 255

2.4.3. Calculation of the moment of inertia of busbars . . 260
2.4.4. Designing lightning protection for general and special purpose structures . . . . . . . . . . . . . . . . 263
2.4.4.1 General considerations . . . . . . . . . . . . 263
2.4.4.2 Theoretical assumptions for calculation of the protection level and protected zone . . . 264
2.4.4.3 Examples of calculations of the protected zone for a general purpose structure . . . . 271
2.4.4.4 Example of calculation of protected zone for a special purpose structure . . . . . . . . 274
2.4.5. Designing lightning protection for substations . . . 281
2.4.5.1 General considerations . . . . . . . . . . . . 281
2.4.5.2 Overview of the method for evaluating the protected zone of substation as a special purpose structure . . . . . . . . . . . . . . . 281
2.4.5.3 Example of calculation of protected zone for substation . . . . . . . . . . . . . . . . 290
2.4.6. Designing lightning protection for overhead lines . 293
2.4.6.1 General considerations . . . . . . . . . . . . 293
2.4.6.2 Review of the method for calculating the protected zone of overhead lines as structures with limited danger . . . . . . . . . . 295
2.4.6.3 Program organization . . . . . . . . . . . . 297
2.4.6.4 Examples of calculation of overhead line shielding zone . . . . . . . . . . . . . . . . 300
2.4.7. Conclusions . . . . . . . . . . . . . . . . . . . . . . 303

**LITERATURE** **305**

**3. APPLICATION OF SOFTWARE TOOLS IN DRAFTING GRAPHICAL DOCUMENTATION** **312**
3.1. Introduction . . . . . . . . . . . . . . . . . . . . . . . . 312
3.2. Graphical documentation . . . . . . . . . . . . . . . . . 313
3.2.1. Electrotechnical graphic symbols . . . . . . . . . . 313
3.2.2. Types of electrical diagrams . . . . . . . . . . . . . 315
3.2.3. Marking of devices and connections in diagrams . . 321
3.3. EPLAN . . . . . . . . . . . . . . . . . . . . . . . . . . 334
3.3.1. Introduction . . . . . . . . . . . . . . . . . . . . . . 334
3.3.2. EPLAN concept . . . . . . . . . . . . . . . . . . . . 334
3.3.3. Main functions of the CAE tool EPLAN Electric P8 . . . . . . . . . . . . . . . . . . . . . . . . 336
3.3.4. Example of the application of EPLAN . . . . . . . 337

3.4. Application of the program SIMARIS DESIGN and SIMARIS SIVACON for design and installation of low-voltage switchgear . . . 339
3.4.1. Introduction . . . 339
3.4.2. Application of the program SIMARIS DESIGN . . 340
3.4.3. Application of the program SIMARIS SIVACON . . 359

**LITERATURE** **367**

**4. APPLICATION OF SOFTWARE TOOLS IN POWER ENGINEERING PROJECT MANAGEMENT** **369**

4.1. About projects . . . 369
4.1.1. Definition of a project . . . 369
4.1.2. Types of designs . . . 369
4.1.3. Participants in realization of the project . . . 371
4.1.4. Content of main design . . . 377
4.1.5. Phases within the realization of the project from the perspective of the investor . . . 386
4.1.6. Types of contracts . . . 388
4.1.7. Example contract . . . 392
4.2. Basics of project management . . . 395
4.2.1. Project cycle . . . 396
4.2.2. Phases of project management . . . 397
4.3. Examples of project management . . . 398
4.3.1. Application of the program MS Excel . . . 398
4.3.1.1 Forming dynamic Gantt charts . . . 399
4.3.1.2 Description of activities . . . 400
4.3.2. Application of the program MS Project . . . 403
4.3.2.1 MS Project basics . . . 403
4.3.2.2 Application of MS Project in the drafting a complete project . . . 405
4.3.2.3 Closing considerations . . . 425

**LITERATURE** **427**

**LIST OF ABBREVIATIONS** **430**

**INDEX OF KEY WORDS** **434**

# PREFACE

A design represents each process by which a goal or group of goals is reached. The realization of a design is inconceivable without the use of a computer. Computer-Aided Design in Power Engineering represents a vital engineering discipline which has been studied at the Faculty of Electrical Engineering at the University of Belgrade for nearly two decades. The use of computers in power engineering is elaborated upon through the aspects of calculation, development of technical documentation and project management.

Within the reformed studies at the Faculty of Electrical Engineering at the University of Belgrade, the aspects of design and application of software tools are taught within the basic studies of the Power Systems Department in the courses *Computer-Aided Design in Power Engineering*, *Workshop for Software Tools in Power Engineering* and *Computer-Aided Design in Power Engineering Project*, or within the power systems doctoral studies program in the course, *The Application of Software Tools in Power Engineering.*

A continuation of the previously realized monographic work – "Computer-Aided Design in Power Engineering – Software Tools" is the monographic work "Computer-Aided Design in Power Engineering – Application of Software Tools". This work relates to the application of existing and independently developed software tools in solving a series of problems from the field of designing power engineering structures and systems. The material is presented in four chapters.

The first chapter relates to computer-aided modeling and simulation in power engineering. The complex problems from the mentioned area are displayed in an original and illustrative manner. By working through eleven original examples, the reader is led through all the phases necessary in the procedures of computer-aided modeling and simulation. The concepts of computer-aided modeling and simulation are explained in the practical design examples along with the methods for model development, the principles of carrying out computer-aided simulation and the possible limitations. The entire procedure is shown in a concise way, step by step, from the formulation and analysis of the engineering problem, to the formation of the mathematical model, selection of computer methods, formation of the program and execution of the simulation. Special attention is given to the reasons for possible errors in calculations.

With consideration of fact that engineers from the field of power systems often develop software tools in practice, the required documentation which a software tool must contain has been displayed along with the procedure for its testing. This chapter displays a software tool for the analysis of grounding systems during the effects of impulse current as an example of a software tool with all the necessary characteristics to make it user friendly. This software tool was created using the MATLAB® technical computing software and the program Visual C++.

Calculations represent an integral part of the design and are necessary for the selection and verification of the characteristics of power engineering equipment. The application of software tools in power engineering calculations is elaborated upon in the second chapter. The most significant programs used today in power engineering calculations have been singled out from the multitude of what is available. These programs are organized into four sections: MATLAB®, EMTP/ATP, MS Excel & MS Access and AutoCAD.

The first section elaborates upon MATLAB® which represents a software tool for solving mathematical problems, analysis of data and visualizations. This technical computing software dominates in university centers, scientific research institutions and design firms throughout the entire world and is useful in solving a multitude of engineering problems. The application of this software tool is illustrated through three examples from the engineering practice. In the first example, the characteristic values of the fault current, necessary for equipment selection, have been calculated using the additional Simulink® software. The application of this software is displayed on a part of an equivalent circuit of a power system. The obtained results are discussed and compared with the results obtained from the application of the classic procedure of calculation. The second example displays the independently developed software tool SPLCAD (Software Power Line Computer-Aided Design) for designing medium-voltage overhead lines. The tool was developed using the MATLAB® technical computing software as the development platform for creating the user interface. Creating and working with databases was achieved using software tools MS Access and MS Excel. The third example presents a Simulink® model of the turbine regulator within the "Kokin Brod" hydroelectric power plant. A section of the comparative analysis of the numerically and experimentally determined results for the elaborated turbine regulator is presented.

Problems during the designing of power systems and problems related to the planning and exploitation of power systems are often solved through the application of the software tool EMTP (Electromagnetic Transient Program) / ATP (Alternative Transient Program), which is described in the second section of the second chapter. The use of the software tool ATP is illustrated in the examples of calculation of transient phenomena in metal-enclosed $SF_6$ gas insulated switchgears (GIS). The first example relates to the numerically and experimentally determined electromagnetic transient processes in the secondary circuits of the measurement transformers of the 123 kV three-phase enclosed, $SF_6$ GIS Karlsruhe-Oberwald. The second example illustrates the procedure for calculating the increase of potential of the metal enclosure of the 420 kV single-phase enclosed, $SF_6$ GIS within the "Višegrad" hydroelectric power plant.

The third section of the second chapter explains the use of the software tools MS Excel and MS Access and contains three parts. In the first part of the third section, some possibilities for the use of databases in the design of high-voltage substations are displayed. The characteristic values of the fault current, as an integral part of the criteria for the selection of high-voltage equipment, have been calculated using the programming languages Visual Basic for Applications (VBA) and Visual Basic (VB). Different types of databases with elements of high-voltage equipment have been developed using the software tools MS Excel and MS Access. The basic principles of the architecture of databases and formation of filtering criteria are discussed based on the use of the program MS Query and VBA. The main elements for the proper creation of relational databases in MS Access are also illustrated. The use of databases is presented in the examples of selection of high-voltage circuit breakers and disconnectors.

Modern design in power engineering entails the automation of calculations through the use of macros. In the second part of the third section, the procedure for forming a macro as a VBA procedure is presented. The use of macros is illustrated in the examples of calculation of the total electricity and thermal impulse of injected impulse current into the tested structure and the automation of work with databases. In the third part of the third section, the use of the Microsoft Office software package in the design of a power supply for the telecommunications equipment of a control-commutation center is demonstrated. The technical calculations and corresponding selection of the aforementioned devices were completely automated using MS Excel and VB, which is illustrated in the examples of the selection of accumulator batteries and rectifiers.

The fourth section of the second chapter is dedicated to AutoCAD as a powerful software tool for computer-aided design. This section presents some of the aspects of advanced use of AutoCAD in computer-aided design of power system substations and structures. The first example illustrates an effective technique of 3D modeling of some of the most complex structures, such as a turbogenerator. The calculation of the physical characteristics of a 3D model is executed in the example of the inertia moment of a U-profile busbar. The developed software tool for designing lightning protection for general and special purpose structures was realized using the programs AutoLISP, Visual LISP and VBA. By using this software tool, calculations were made for the zone of protection from lightning discharge for a residential structure, special purpose structure, substation and overhead line.

The graphical documentation represents an integral part of the design. The application of software tools in the development of graphical documentation is the subject of the third chapter. The most frequently used types of electrical diagrams are described along with examples of how they are formed. Electrotechnical graphic symbols and markings of devices and connections in installations are presented. The application of the software tool AutoCAD, as the basic tool for development of graphical documentation in power engineering, is clearly presented in the examples of the different types of electrical diagrams of a concrete transformer station. The elements of the program EPLAN, which enables the efficient formation of wiring diagrams and connection diagrams, are also considered. Application of the version R.6.1 of the program SIMARIS SIVACON for design and installation of 8PT SIEMENS SIVACON low-voltage switchgear is displayed in the example of a transformer station with a single-pole diagram created using the program SIMARIS DESIGN.

In that way the collection of software tools which enable the complete automation of the development of graphical documentation of a power engineering design is presented, through which efficiency in work is greatly increased.

In the fourth chapter, the application of software tools within project management in power engineering is discussed. In the first section, the basic elements of a design and the participants in its realization are analyzed. In the second section, the basics of project management are presented, while in the third section examples are presented which illustrate the management of a concrete design of technological development through the application of the software tools MS Excel and MS Project.

Literature is provided in each chapter. A list of abbreviations and index of key words are provided at the end of the monographic work.

This monographic work has an educative, engineering and scientific aspect. Considering that it contains a series of examples from the direct application of software tools in the practice of design in the field of power engineering, the monographic work may be of some benefit to experts who are involved in power engineering design. The educative and scientific aspects have been confirmed by the publishing of a large number of works in journals from various categories, as well as at conferences. For that reason, the material is also intended as an educational resource in the area of design of power systems and structures for all levels of study.



Dr. Zlatan Stojković, Full Professor

# ACKNOWLEDGMENT

I would like to acknowledge the help of all of those involved in the development and review process of the monograph, without whose support this project could not have been successfully completed.

I would also like to give special thanks to the Alexander von Humboldt Foundation, Bonn, Germany and the Ministry for Science and Technological Development of Serbia for the support provided during scientific research work.

The monograph includes the current versions of software tools thanks to the donations and consent for using the programs from the following companies:

- The MathWorks, Inc., USA for technical computing software MATLAB®, Simulink® software and SimPowerSystems™ module;
- Norwegian University of Science and Technology, Norway (Dr. Hans Kristian Høidalen) for consent for using the program ATPDraw™ for Windows;
- OSA Engineering, Belgrade, Serbia, official representative of Autodesk for a trial version of the program AutoCAD 2011;
- VESIMPEX, Inc., Belgrade, Serbia, general representative of the company EPLAN for the software EPLAN;
- SIEMENS, Inc., Belgrade, Serbia for the software tools SIMARIS DESIGN - Version R.6.1 and SIMARIS SIVACON, as well as the design firm MONTPROJEKT, Belgrade, Serbia for their cooperation;
- The ELMS for MSDNAA Software Centre, E-Academy Inc. for the software tool Project Professional 2007.

I would like to take this opportunity to thank Manchester University Press as the publisher of the International Journal of Electrical Engineering Education (IJEEE) for their consent to republish the pictures and tables from reference 84 in Chapter 2.

Special thanks go to the translator Mr. Anthony Bock and Mr. Željko Hrček, who performed the language and technical elaboration of the monographic work with great skill.

In closing, I want to express the gratitude to my family, to whom I dedicate this monograph, for their support and encouragement.

Zlatan Stojković

# 1. COMPUTER-AIDED MODELING AND SIMULATION

## 1.1. Introduction

The purpose of this chapter is to familiarize the reader with the following:

- The definition of the concept of computer-aided modeling and simulation;
- A review of the method for model development;
- The principles of carrying out computer-aided simulation and its limitations;
- A review of an example for initialization of computer-aided modeling and simulation.

## 1.2. Definitions

A model, for the purposes of this chapter, is a system of hypotheses, data and conclusions. This comes in the form of a mathematical system of equations and logical rules, or a program for solving a system of equations and logical rules.

Modeling is the process through which a suitable system of hypotheses, data and conclusions is reached. Simulation is the execution of a model and the forming of certain results. In order to properly understand the presented material, it is important to recognize the difference between the concepts of modeling and simulation.

## 1.3. Categories of approaches to modeling

The systems can be classified in various ways, of which some refer to the mathematical form of the equations. Table 1.1 displays one of the possible ways of classifying the systems.

Z. Stojković, *Computer-Aided Design in Power Engineering*,
DOI: 10.1007/978-3-642-30206-0_1, © Springer-Verlag Berlin Heidelberg 2012

Table 1.1 – Classifications of systems

| Category | Opposite category |
|---|---|
| Deterministic | Stochastic |
| Continual | Discreet |
| Dynamic | Static |
| Time varying | Stationary |
| Linear | Nonlinear |
| Real-time system | Batch processing system |

## 1.4. Phases of development

Development of a computerized model and simulation are carried out in several steps [1, 2]:

– transitioning from an actual system into a mathematical presentation,
– forming a mathematical model in written form,
– determining the appropriate computerized method for solving the mathematical model,
– creating a program,
– verifying the program.

None of the aforementioned phases may be omitted, and in order for them to be realized one must possess various knowledge and skill sets [3-6]. Very few individuals have the ability to successfully complete all of the aforementioned phases of development. Because of this, developing a model and solving an engineering problem most often requires a team to be formed. Within this team it is important to establish proper coordination and communication between the team members.

The analytical and numerical solving of formed systems of equations requires the proper knowledge from theoretical mathematics and computerized methods [7-11]. After determining the technique for solving the problem comes implementation on a computer which requires the proper skills in programming [12, 13].

When creating the program it is necessary to ensure that it is verified with the proper mathematical model. Along with that it is necessary to conceptualize a model producing results which agree with those produced in actual systems.

## 1.5. The role of computers in modeling and simulation

The role of computers is explicitly included in the phases of creating and verifying the program. Development of the mathematical model implicitly includes the role of computers. Individuals with experience in modeling and programming automatically gravitate towards the mathematical models which are most easily and effectively solved using a computer.

The models and programming methods formed by various authors are always different. From this we can derive the rule that $n$ different individuals or teams will form $n$ different models.

## 1.6. Methodology of computer-aided modeling and simulation

Solving any engineering problem using a computer requires that the proper methodology be followed. This consists of the following steps:

- formulation of the problem,
- analysis of the problem,
- formation of the mathematical model,
- selection of the computerized method,
- formation of the program,
- verification of the link between the model and program,
- verification of the validity of the model and program.

A structural diagram with the aforementioned steps is provided in Fig. 1.1.

A detailed description of the aforementioned steps is provided below.

### 1.6.1. Formulation of the problem

Solving engineering problems requires investing a certain amount of effort and employing the necessary resources. There are five fundamental groups of questions, in literature known as the "five basic W's", which must be answered before beginning the process of computer-aided modeling [1, 2]:

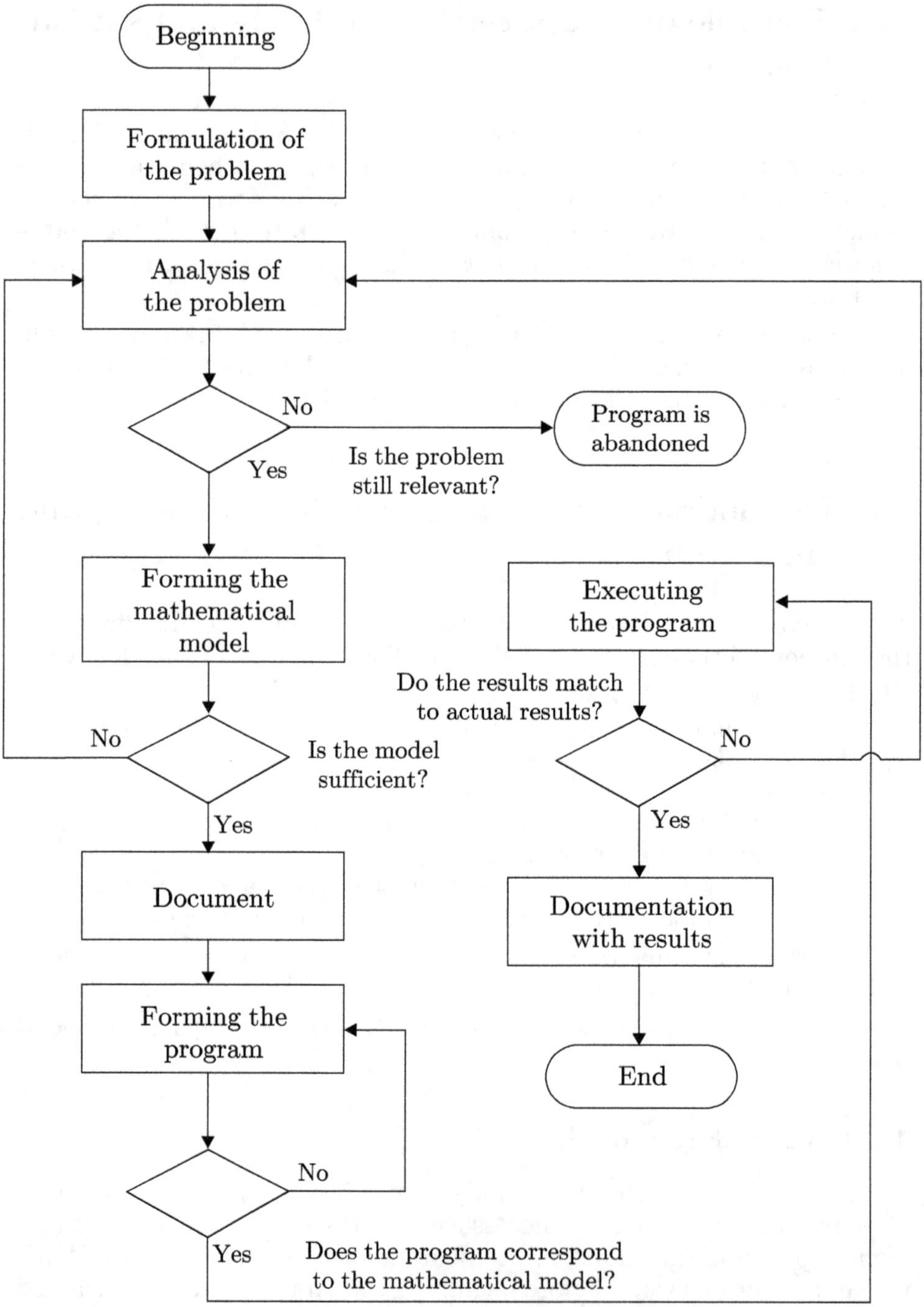


Fig. 1.1 – Steps in computer-aided modeling and simulation

1. Why is it necessary to create a model and simulation?
2. What is the essential problem?
3. What resources are required for solving the problem?
4. What programming language should be used?
5. What characteristics of the computer are necessary for solving the stated problem?
6. What are the expected results?
7. Who will form the mathematical model?
8. Who will realize the program?
9. Who will use the program?
10. Where are the resources necessary for realization of the model?
11. Where will the simulation be carried out?
12. When will the project be completed?

### 1.6.2. Analysis of the problem

Formulation of the problem and analysis of the problem are connected by an iterative process. As it is displayed in the structural diagram in Fig. 1.1, after formulating the problem, it is necessary to answer the question regarding whether or not it is appropriate to continue the project. The decision made regarding this issue is contained in the strategy of the entire project, which is defined on the basis of the drafted project documentation and studies which are conducted in this phase, such as: the opportunity study, the prefeasibility study and the feasibility study [14].

One of the most important factors in analyzing the problem relates to the complexity of the system. This complexity occurs as a result of the following two factors:

- there is a large number of mutually connected components within the system,
- the nature of the components is not well understood.

Eliminating the complexity of the system caused by a large number of mutually connected components is most often done through the process of decomposition [9]. In accordance with this, the appropriate software tools are also created for solving complex technical problems, during which the user, in addition to the possibilities of independent programming, is provided with the corresponding additional modules for solving the most common mathematical problems.

## 1.6.3. Forming the mathematical model

### 1.6.3.1 General considerations

A system, defined as a device or process which has one or more inputs or one or more outputs, may be modeled using various procedures. A mathematical model contains a number of advantages in comparison to other models. This type of model makes it possible to:

- clearly define variables,
- explicitly express hypotheses,
- determine the output variables based on the complexity of the relation of the model.

Generally, a linear model is easier to solve than a nonlinear model, a regular differential equation is easier to solve than a partially differential equation, and an analytical function is easier to solve than an infinite series, Bessel function, etc. Additionally, deterministic models are easier to solve than stochastic models.

The general methodology of modeling consists of four steps:

- defining the system,
- determining the input values of the system,
- determining the output values of the system,
- determining the relation between the input values, output values and parameters of the system.

In the first step it is necessary to clearly define the considered system, the relation between the components of the system (connection of the parts of the system is not always exceptionally important for the problem which is being solved), the desired goal, etc.

The process of modeling requires clearly defined input and output values. For both groups of values the following must be done:

- identify them,
- when possible measure them,
- through the iterative process of analysis and modeling, determine what the significant values are and, finally, what is their relation.

Some of the input values may be controllable and desirable while others may be uncontrollable and undesirable. Weather conditions such as temperature, humidity and wind speed are examples of uncontrollable input values when forming some models. Serious difficulties in the process of modeling may result when external uncontrollable factors have a large impact on the analyzed occurrence or process.

Determining the relation between input values, output values and the parameters of the system is the most difficult step when forming a model. The essence of good modeling is in the determination of significant variables and forming the relations which describe how they are mutually connected.

The general methodology of modeling will be illustrated in the example of grounding grid impulse characteristics.

**Example 1.1.** This example displays the procedure of modeling the impulse characteristics of a grounding grid. The behavior of a grounding grid during an electrical discharge of lightning defines its impulse characteristics [15-17]. Injecting an impulse current into a grounding grid leads to the occurrence of time variable voltages and currents along the grounding grid.

The calculation of grounding grid impulse characteristics is a complex problem due to the large number of various factors which affect them. The constructive parameters of the grounding grid, electrical characteristics of the soil, as well as the shape, amplitude and location of injection of the impulse current have the most important influence on these characteristics [18-25].

Generally speaking, each grounding grid can be divided into a certain number of linear elements. Each linear element can be modeled through concentrated parameters using the corresponding equivalent circuit ($\Gamma$, $\Pi$, $T$ – circuit). Fig. 1.2a shows a grounding grid with its $i$ element provided by end nodes $k$ and $l$. This element is modeled using a $\Gamma$ circuit (Fig. 1.2b), $\Pi$ circuit (Fig. 1.2c) and $T$ circuit (Fig. 1.2d).

The symbols on Figs. 1.2b, 1.2c and 1.2d have the following meanings: $R_i$ – series resistance of the element, $L_s$ – self-inductance of the element, $G_i$ – parallel conductance of the element, $C_i$ – parallel capacitance of the element.

*a) Identification of input and output values*

The discharge impulse current which is injected into the grounding grid represents the input value. The amplitude and form of the discharge impulse current in reality have a stochastic character and are subject to corresponding laws of distribution [26]. Measurement of these values was done in a series of experimental studies on actual grounding grids, during which the amplitude and form were adjusted through the selection of the parameters of the impulse current generator [16, 19, 20].

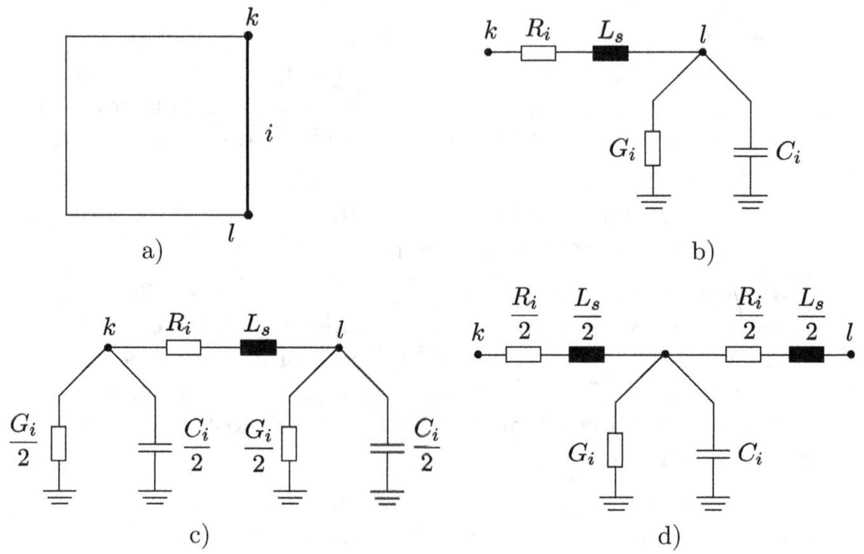


Fig. 1.2 –
a) Disposition of the grounding grid with marked element $i$; b) Reversed Γ circuit of the $i$ element of the grounding grid; c) Π circuit of the $i$ element of the grounding grid; d) $T$ circuit of the $i$ element of the grounding grid

There are two groups of output values. The first group is represented by independent output values, those being the voltages of the grounding grid nodes and the currents of the grounding grid inductive branches. The second group refers to the voltages of the grounding grid branches and the currents through capacitive and conductive branches of the grounding grid. Those are dependent output values which are determined through an effective technique of mathematical model generation on the basis of independent output values. The process of measuring independent output values is described in detail in [16, 19, 20].

*b) Grounding grid parameters*

The parameters of an equivalent circuit of a grounding grid can be modeled as frequency independent or frequency dependent parameters [16]. Additionally, depending on the structure of the soil in which the grounding grid is placed, it is necessary to encompass cases of uniform and two-layer soil. A multiple-layer soil structure can be reduced to a two-layer soil structure and still maintain sufficient accuracy.

Fig. 1.3 shows a horizontal square-shaped grounding grid with its elements modeled in a reversed Γ circuit.

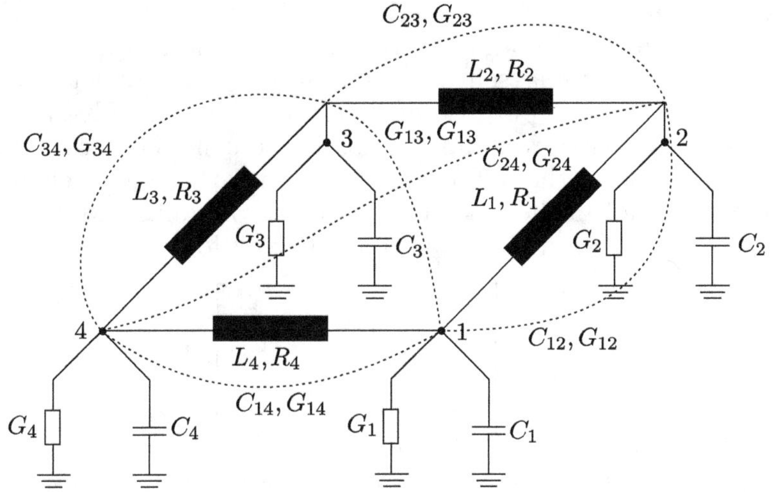


Fig. 1.3 - Application of a reversed Γ circuit for modeling the elements of a grounding grid

The symbols on the circuit in Fig. 1.3 have the following meanings: $i$, $j$ – symbols of elements ($i, j = 1, \ldots, 4$), $G_i$ – parallel conductance of the $i$ element, $C_i$ – parallel capacitance of the $i$ element, $R_i$ – series resistance of the $i$ element, $L_s$ – self-inductance of the $i$ element, $G_{ij}$ – mutual conductance of the $i$ and $j$ elements, $C_{ij}$ – mutual capacitance of the $i$ and $j$ elements.

The process of determining these parameters according to a frequency independent model and frequency dependent model is described in detail in [16, 17].

The series resistance of a conductor of an arbitrary shape of the cross-section during the flow of direct current is determined using the following expression:

$$R_i = \frac{l_i}{\sigma_i \cdot S_i} \tag{1.1}$$

where: $\sigma_i$ – specific electrical conductance of the material of the conductor ($1/\Omega$m), $S_i$ – area of the cross-section ($\mathrm{m}^2$), $l_i$ – length of the conductor (m), $R_i$ – series resistance of the conductor ($\Omega$).

The value of the series resistance of an element of the grounding grid does not depend on the structure or parameters of the soil.

Self and mutual parallel conductances of the elements of a grounding grid have an important role in determining the steady-state characteristics of complex grounding systems. The numerical and analytical methods for determining self and mutual parallel conductances have been developed for the purpose of determining the steady-state characteristics of complex grounding grids. These conductances are determined on the basis of the self and mutual resistances of the elements of the grounding grid. In the case of a complex grounding grid, the matrix of self and mutual resistances has the following form:

$$[r] = \begin{bmatrix} r_{11} & r_{12} & \cdots & r_{1n} \\ r_{21} & r_{22} & \cdots & r_{2n} \\ & & \vdots & \\ r_{n1} & r_{n2} & \cdots & r_{nn} \end{bmatrix} \tag{1.2}$$

where: $r_{ij}$ – mutual resistance of conductors $i$ and $j$, $r_{ii}$ – self-resistance of conductor $i$, $n$ – total number of grounding grid elements.

Matrix $[r]$ is a square non-singular symmetrical matrix of the $n^{\text{th}}$ order. Due to the symmetry of the problem, the condition that $r_{ij} = r_{ji}$ is always fulfilled. In more complex cases, the calculation of the matrix $[r]$ is performed on the basis of the procedure displayed in [27-29]. The method is applicable for the case of a grounding grid of an arbitrary shape in uniform and two-layer soil.

The value of power frequency grounding resistance $R_s$ can be determined from the expression:

$$R_s = \frac{1}{[I]^t [r]^{-1} [I]} \tag{1.3}$$

where $[I]$ is the $n$-dimensional vector of units.

The matrix of self and mutual conductances $[G]$ is determined from the following expression:

$$[G] = [r]^{-1} \tag{1.4}$$

The matrix of self and mutual capacitances of the grounding grid elements $[C]$ in uniform soil is determined according to the following expression [30]:

$$[C] = \varepsilon \cdot \rho \cdot [G] \tag{1.5}$$

where: $\epsilon$ – dielectric constant of soil, $\rho$ – soil resistivity.

Matrix $[G]$ is defined through expression (1.4). Expression (1.5) is derived from the analogy of electrostatic and stationary current fields. Based on this expression, it is possible to approximately determine the corresponding matrix of self and mutual capacitances of the grounding grid elements in two-layer soil.

The dielectric constant of soil is defined by the expression:

$$\varepsilon = \varepsilon_0 \cdot \varepsilon_r = 8,85 \cdot 10^{-12} \cdot \varepsilon_r \quad (\mathrm{F/m}) \tag{1.6}$$

where: $\epsilon_r$ – relative dielectric constant of soil. Its value depends on the type of soil and falls in the range of 1-100 [31]. For a normal soil moisture content, this constant has a value of 10.

The mutual inductance of conductors $i$ and $j$, marked with the symbol $L_{ij}$, is determined using the following expression:

$$L_{ij} = \frac{\mu \cdot l_i \cdot l_j \cdot \alpha_{ij} \cdot \cos\theta_{ij}}{\rho} \tag{1.7}$$

where: $\mu$ – magnetic permeability of soil (H/m), $\rho$ – soil resistivity ($\Omega$m), $l_i$, $l_j$ – length of conductors $i$ and $j$, respectively (m), $\alpha_{ij}$ – mutual resistance of conductors $i$ and $j$, in unlimited homogenous soil of a soil resistivity $\rho$ ($\Omega$), $\theta_{ij}$ – angle between conductors $i$ and $j$.

The self-inductance of the linear conductor $L_s$ can be determined using the following expression:

$$L_s = L_i + L_e \tag{1.8}$$

where: $L_i$ – internal inductance of the conductor, $L_e$ – external inductance of the conductor.

The internal inductance of a cylindrical conductor with a round cross-section under direct current is:

$$L_i = \frac{\mu_m \cdot l_i}{8 \cdot \pi} \tag{1.9}$$

where: $\mu_m$ – magnetic permeability of the material of the conductor (H/m), $l_i$ – length of the conductor (m).

The external inductance of the conductor is obtained from the expression (1.7), for $i = j$. Expression (1.7) is applicable for both uniform and two-layer soil structures. The consequence of magnetic homogeneity of a multi-layer soil structure is that the self and mutual resistance is always determined for the case of unlimited homogenous soil with a soil resistivity $\rho$.

It can be concluded that the self and mutual inductances of the grounding grid elements do not depend on the structure of the soil.

*c) Determining the relation between the input values, output values and parameters of the system*

Based on the circuit displayed in Fig. 1.3, it can be concluded that the grounding in the form of a grid represents a complex electrical circuit which can be solved using various techniques [32]. One of the most effective techniques for solving this electrical circuit is based on the forming of a normalized system of differential equations.

The term normalized system of differential equations implies a system of first-order differential equations written so that all derivatives are variables on the left side and variables on the right side. Variables with derivatives which are figured in differential equations are called state variables and they correspond to independent output values. The remaining variables are called algebraic variables and they correspond to dependent output values.

By solving the state equation, all variables in the circuit are obtained, which differs from the classic approach where the entire circuit is solved one variable at a time. The state equation is a system of $n$ first-order differential equations, while the classic approach is a system reduced to one $n$-order equation. Along with simplicity, the advantage of using a computer is that several state variables can be calculated simultaneously.

Generally, the system of differential equations in the normal form appears as follows:

$$[x'] = [M] \cdot [x] + [F(t)] \tag{1.10}$$

$$[y] = [N] \cdot [x] + [G(t)] \tag{1.11}$$

where: $[x]$, $[x']$ – vectors of state variable and their derivatives according to time, $[y]$ – vector of algebraic variables with derivatives which are not figured in differential equations, $[M]$, $[N]$ – square matrices of the coefficients, $[F(t)]$, $[G(t)]$ – vectors of the source functions.

The process of forming the state equation using the theory of graphs has been illustrated for the considered grounding grid. The following concepts are used in the explanation of the electrical circuit state:

- The graph of the electrical circuit is composed of a collection of nodes and oriented branches. The ends of each of the branches make up the nodes;
- The subgraph of the electrical circuit represents a subcollection of the branches and nodes of that circuit. Each node of the graph can belong to one or more branches;
- The degree of the node represents the number of branches which collect into that node;
- The graph is connected if each node can be reached by only going through the branches of the graph;
- The bundle represents a collection of branches of the graph which must be removed (terminated) in order for the graph to be divided into two unconnected subgraphs. They are always oriented towards a selected branch of the bundle, from one subgraph towards another;
- The contour of the graph represents a subgraph with the characteristic that all of its nodes are of the 2nd degree. They are always oriented towards a selected branch of the contour;
- The tree of a connected graph represents a subgraph which has the characteristics of being connected and encompassing all nodes of the graph while not forming a contour;
- The co-tree represents a subgraph composed of elements which do not belong to the tree (complimentary tree). The branches of the co-tree are called connectors.

The forming of a normal tree represents the first phase in forming the state equation. The term normal tree implies a tree which contains the maximum number of capacitive branches and the minimum number of inductive branches. The goal of forming a normal tree is to ensure that only voltages on the condensers which belong to the tree and currents on the inductors which belong to the co- tree are figured into the state equation. In this way a minimal form for the state equation is achieved.

The symbols for trees of individual subgraphs are: $T_{\overline{el_1},el_2}$ – a tree in a circuit composed of an $el_2$ element, while $el_1$ elements are short connected and the rest are removed. Forming a normal tree is performed in four steps:

**Step I.** Forming the tree $T_C$ (a tree in a circuit which contains only capacitive branches, while all other branches are terminated);

**Step II.** Forming the tree $T_{\overline{C},G}$ (a tree in a circuit which contains only conductances, while capacitances are short connected and the remaining branches are terminated). The bundle of branches of this tree represents an empty set because these elements are always in capacitive contours which in this case are short connected;

**Step III.** Forming the tree $T_{\overline{C},\overline{G},R}$ (a tree in a circuit composed of resistivities, while capacitances and conductances are short connected and the remaining elements are terminated);

**Step IV.** Forming the tree $T_{\overline{C},\overline{G},\overline{R},L}$ (tree in a circuit in which all elements except for inductances are short connected).

A normal tree is represented by the union of these trees:

$$T = T_C \cup T_{\overline{C},G} \cup T_{\overline{C},\overline{G},R} \cup T_{\overline{C},\overline{G},\overline{R},L} \tag{1.12}$$

The process of forming a normal tree according to the aforementioned steps has been performed for the equivalent circuit of the grounding grid displayed in Fig. 1.3. In the first step a normal tree $T_C$ is formed which contains only capacitive branches. It is formed in such a way that it encompasses all capacitances according to the reference point. Because the tree $T_C$ encompasses all nodes of the grid, it at the same time represents a normal tree of the considered grid.

Fig. 1.4a displays tree $T_C$, and Fig. 1.4b illustrates a complete graph with the tree marked by thick lines. The symbol for the element which the considered branch of the graph refers to is provided alongside each branch of the graph. The branches of the normal tree are marked by full lines. The branches of the complimentary normal tree are marked by dashed lines. The orientation of the graph is completely arbitrary, being that the branches which correspond to a normal tree are oriented toward the reference node.

The complete information about the connection of the electrical circuit is obtained from the basic incident matrices. The basic incident matrices are matrices which contain information about the graph and they are of a minimal form. The rank of basic incident matrices equals the number of types.

By applying the displayed process, the following normalized system of differential equations is obtained for a grounding grid in an

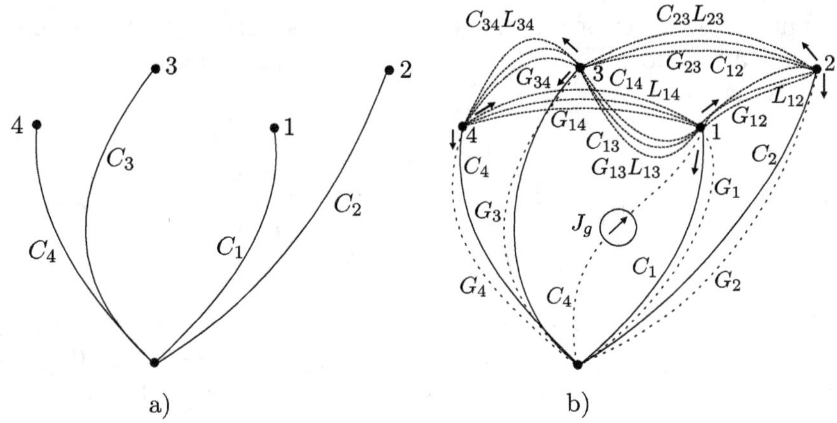


Fig. 1.4 - Tree $T_C$ composed of capacitive branches (a) and complete graph of the circuit (b)

impulse period:

$$\begin{bmatrix} \dfrac{du_1}{dt} \\ \dfrac{di_2}{dt} \end{bmatrix} = -\begin{bmatrix} C & 0 \\ 0 & L \end{bmatrix}^{-1} \cdot \left( \begin{bmatrix} G & K \\ -K^t & R \end{bmatrix} \cdot \begin{bmatrix} u_1 \\ i_2 \end{bmatrix} - \begin{bmatrix} I_C \\ e_L \end{bmatrix} \right) \tag{1.13}$$

The symbols in the expression above have the following meanings: $[C]$, $[L]$, $[G]$, $[R]$ – matrices of capacitance, inductance, conductance and resistance, respectively, $[K]$ – incident matrix, $[K]^t$ – transposited incident matrix, $[0]$ – zero matrix, $[u_1]$ – voltage vector of grounding grid nodes, $[i_2]$ – current vector of inductive branches of the grounding grid, $[I_C]$ – current vector of equivalent current sources connected to grounding grid nodes, $[e_L]$ – electromotor force vector of equivalent voltage sources in the grounding grid branches.

For the displayed square grounding grid, matrices in expression (1.13) have the following form:

$$[r] = \begin{bmatrix} r_{11} & r_{12} & r_{13} & r_{14} \\ r_{21} & r_{22} & r_{23} & r_{24} \\ r_{31} & r_{32} & r_{33} & r_{34} \\ r_{41} & r_{42} & r_{43} & r_{44} \end{bmatrix} \tag{1.14}$$

The calculation of matrix $[G]$ and $[C]$ is performed using expressions (1.4) and (1.5). The submatrix of self and mutual inductances has the following form:

$$[L] = \begin{bmatrix} L_{11} & L_{12} & L_{13} & L_{14} \\ L_{21} & L_{22} & L_{23} & L_{24} \\ L_{31} & L_{32} & L_{33} & L_{34} \\ L_{41} & L_{42} & L_{43} & L_{44} \end{bmatrix} \tag{1.15}$$

For a normally placed conductor, the mutual inductance equals zero, or $L_{12} = L_{21} = L_{23} = L_{32} = L_{34} = L_{43} = L_{14} = L_{41} = 0$. Due to symmetry $L_{13} = L_{31} = L_{24} = L_{42}$ is valid, so matrix $[L]$ has the form:

$$[L] = \begin{bmatrix} L_{11} & 0 & L_{13} & 0 \\ 0 & L_{22} & 0 & L_{24} \\ L_{31} & 0 & L_{33} & 0 \\ 0 & L_{42} & 0 & L_{44} \end{bmatrix} \tag{1.16}$$

The matrix of series resistances of the grounding grid elements has the following form:

$$[R] = \begin{bmatrix} R_1 & 0 & 0 & 0 \\ 0 & R_2 & 0 & 0 \\ 0 & 0 & R_3 & 0 \\ 0 & 0 & 0 & R_4 \end{bmatrix} \tag{1.17}$$

while the series resistance of the $i$ element is determined using expression (1.1).

Matrix $K$ represents the matrix of connections between the electrical values of bundles and contours. Within these matrices, the columns correspond to individual branches, and the types correspond to individual bundles. The elements of the matrix are $+1$ if a branch belongs to a bundle and their orientations are identical, $-1$ if the branch belongs to a bundle and it is oriented oppositely, and 0 if the branch does not belong to a bundle. For the graph displayed in Fig. 1.4b, matrix $K$ has the following form:

$$[K] = \begin{bmatrix} 1 & 0 & 0 & -1 \\ -1 & 1 & 0 & 0 \\ 0 & -1 & 1 & 0 \\ 0 & 0 & -1 & 1 \end{bmatrix} \tag{1.18}$$

The vector of current sources in capacitive bundles is defined by the relation:

$$[I_C] = \begin{bmatrix} J_g(t) \\ 0 \\ 0 \\ 0 \end{bmatrix} \tag{1.19}$$

The waveshape of a discharge impulse current $J_g(t)$ is displayed in the form of a double-exponential function:

$$J_g(t) = I\left(e^{-\alpha t} - e^{-\beta t}\right) \tag{1.20}$$

The discharge impulse current is normally assigned in the sense of the maximum value and nominal rise time of the impulse current front and nominal time to half-value of impulse current wave. The connection between the standard characteristics of the wave and the value from expression (1.20) is determined using the following relations [26]:

$$\beta = \frac{2.746}{T_1} \tag{1.21}$$

$$\alpha = \frac{0.693}{T_2} \tag{1.22}$$

$$I = I_m \left[\left(\frac{\alpha}{\beta}\right)^{-\frac{\alpha}{\alpha-\beta}} - \left(\frac{\alpha}{\beta}\right)^{-\frac{\beta}{\alpha-\beta}}\right]^{-1} \tag{1.23}$$

where: $T_1$ – the nominal rise time of the impulse current front, $T_2$ – the nominal time to half-value of impulse current wave, $I_m$ – the maximum value of the current.

The electromotor force vector of voltage sources according to the contours is:

$$[e_L] = \begin{bmatrix} 0 \\ 0 \\ 0 \\ 0 \end{bmatrix} \tag{1.24}$$

In order to solve the system of differential equations provided by expression (1.13), it is necessary to know the initial conditions. The initial conditions of the state variables are defined by the expression:

$$\begin{bmatrix} u_1(t_0) \\ i_2(t_0) \end{bmatrix} = \begin{bmatrix} u_{10} \\ i_{20} \end{bmatrix} \tag{1.25}$$

where: $t_0$ – the moment when observation of the occurrence begins, $u_{10}$ – the vector of initial values of the voltage of grounding grid nodes, $i_{20}$ – the vector of initial values of the current of inductive branches of the grounding grid.

Within the analysis of the impulse characteristics of the grounding grid, $t_0$ represents the zero moment. Considering that the grounding grid does not contain accumulated electrostatic and electromagnetic energy, expression (1.25) becomes:

$$\begin{bmatrix} u_1(t_0) \\ i_2(t_0) \end{bmatrix} = \begin{bmatrix} 0 \\ 0 \end{bmatrix} \tag{1.26}$$

The system of state differential equations is completely defined by expressions (1.13) and (1.26).

### 1.6.3.2 General principles of good modeling

The general principles of good modeling are:

- begin modeling with that which is known,
- develop complex models modularly,
- use the iterative process,
- model only those elements which are necessary,
- form assumptions,
- state the limitations,
- form equivalents suitable for engineering practice.

Short descriptions of the fundamental characteristics of the aforementioned principles are provided below.

*a) Begin modeling with that which is known*

Evidently, the mathematical model of a grounding grid in an impulse period is possible to form on the basis of the relation of elements of the grounding grid according to Kirchhoff's laws for current and voltage. By applying the methodology displayed in Section 1.6.3.1, a minimal form of a system of state differential equations is generated. By carefully considering the matrix relations (1.13), it is noticed that the first and second type in this relation actually present equations which correspond to Kirchhoff's law for current and Kirchhoff's law for voltage, respectively.

*b) Develop complex models modularly*

A modular approach implies the step-by-step expansion of a model. This process enables the transition from simple configurations to complex configurations with the most effective elimination of shortcomings which go hand in hand with the modeling of complex systems.

This approach is illustrated in the example of a complex grounding in the form of a grid. Because this grounding grid is composed of a series of meshes, and each mesh is composed of several linear elements, in the first step it is necessary to develop a model of a linear grounding grid (Fig. 1.5a). In the second step, a grounding grid model is formed which is composed of just one mesh (Fig. 1.5b). After testing this model, it is generalized for a complex grounding grid composed of $m$ meshes (Fig 1.5c).

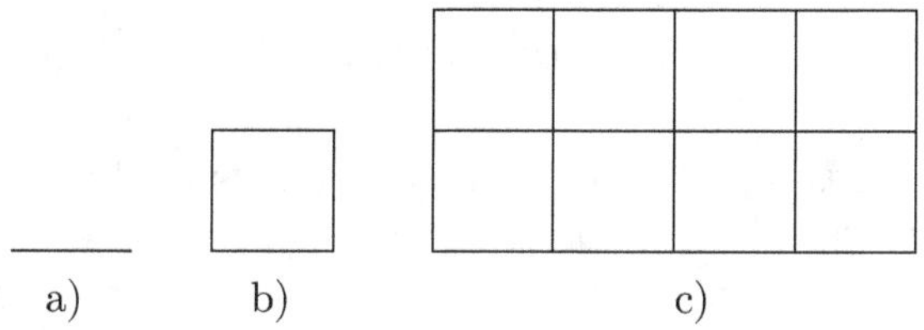

Fig. 1.5 – Modular approach to modeling a grounding grid;
a) Linear grounding grid, b) Grounding grid composed of one mesh, c) Complex grounding grid composed of $m$ meshes

*c) Use the iterative process*

Iterative process implies making repeated attempts following certain modifications. The structural diagram shown in Fig 1.1 clearly shows the important loops in the process of modification. Within each displayed block there is the possibility for smaller iterative loops.

Computer-aided simulation contains the following phases [1, 2, 33]:

- forming calculation results,
- comparing these results with expected results for a known case,
- improving the model until the appropriate accuracy is reached.

It is exceptionally rare for one to obtain results on the first attempt which are completely satisfactory and do not require improvement. Even authors of the most successful software tools state that in the case of repeated modeling of a certain problem, it should be done differently, in an even better way.

Modeling is learning. The iterative process of improving a model is the essence of the modeling process.

*d) Model only those elements which are necessary*

Forming equations which properly describe the behavior of a system is often a real art form. The complete concrete modeling of all the elements of a system often leads to problems, considering the large number of additional details related to input and output variables, as well as the extended time of simulation. Due to this fact, in modeling it is mandatory to possess extensive knowledge of engineering for the purpose of including the necessary elements while simultaneously eliminating the unnecessary elements.

Modeling just the necessary characteristics is of great significance for two reasons:

- this process leads to the forming of suitable, when possible, simpler relations which present the considered system with relative accuracy,
- forming a proper model is directly related to finding an effective and robust technique for solving that model.

In addition to this, within modeling it is exceptionally important to accurately define:

- which parameters are encompassed and in which way,
- which parameters have a dominant influence on the calculation results.

The complexity of the modeling elements of a power system is reflected in the various natures of its parameters. According to the first division, parameters are divided into frequency independent, frequency dependent or current dependent, and according to the second division, they are divided into linear or nonlinear. The nature of the parameters directly affects the method for solving the corresponding mathematical model. In power systems, for example, the presence of nonlinear parameters of a system is common. A system of nonlinear equations in that case cannot be solved with a linear algebra tool which is otherwise very powerful, but rather one must apply some of the general methods for solving nonlinear systems [7, 9, 13].

Defining the parameters which have a dominant influence on the calculation results can be done by conducting a sensitivity analysis.

This analysis is conducted by varying just one parameter of a completely formed model. Based on the sensitivity analysis, it is possible to simplify a complete model, which is of great importance in the practical consideration of problems in power systems.

An illustration of this process is provided in the following example.

**Example 1.2.** Fig. 1.6 displays the disposition of a simple square horizontal grounding grid placed at a depth of $h = 0.5$ m in uniform soil with a soil resistivity of $\rho = 100$ Ωm. The grounding grid is made from copper wire with a cross section of $S_{cu} = 50$ mm$^2$. The sides of the square are 10 m in length. It is assumed that the current is injected into the grounding grid at one of the corners of the square. The current of a discharge amounts to $I = 5$ kA, nominal rise time of impulse current front $T_1 = 1.5$ $\mu$s, and the nominal time to half-value of impulse current wave $T_2 = 30$ $\mu$s.

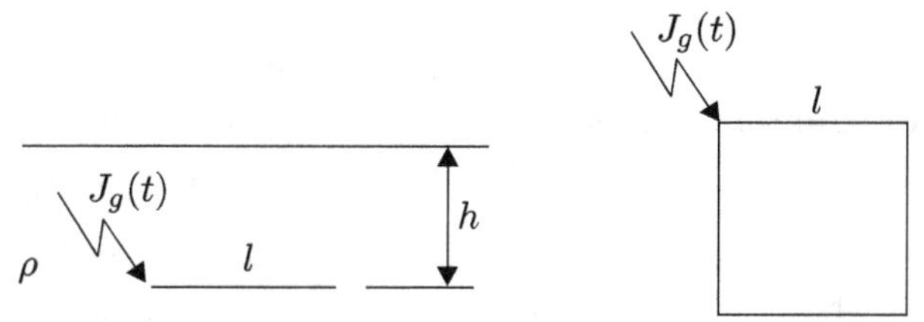


Fig. 1.6 – Disposition of analyzed grounding grid

The purpose of the sensitivity analysis is to show the effect of series resistances of the grounding grid elements on its impulse characteristics. The sensitivity analysis is performed using the following steps.

**Step I.** The calculation of impulse characteristics of the considered grounding grid is performed on the basis of the complete mathematical model.

**Step II.** The calculation of impulse characteristics of the grounding grid is repeated in the case of neglected values of series resistances of the grounding grid elements. This means that within the system of state differential equations by expression (1.13) the matrix of series resistances $R$ is reduced to the zero matrix.

Fig. 1.7 displays the voltage response at the injection location for a complete model (a) and a simplified model without series resistances (b).

Based on the results displayed on Figs. 1.7a and 1.7b, one can notice a complete lack of influence of the series resistances of ground-

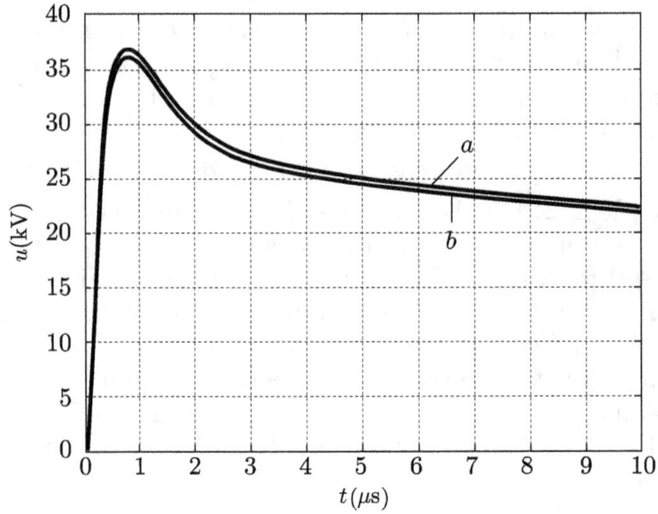


Fig. 1.7 - Voltage response at the injection location for a complete model (a) and simplified model without series resistances (b)

ing grid elements on the voltage values at the injection location. Identical calculation results were also obtained in the calculations of impulse characteristics for grounding grids of a differing shape in soil with various characteristics. This result is the consequence of significantly smaller values of series resistances of grounding grid elements when compared to the corresponding reactances of the elements in the case of lightning discharge. Large reactance values for the elements are the consequence of high frequency which corresponds to the lightning discharge wave front [26].

It can be concluded that in the equivalent circuit, the series resistances of the grounding grid elements is justifiably ignored, by which a simplified corresponding mathematical model is achieved while maintaining calculation accuracy.

*e) Form assumptions*

The formation of a mathematical model is significantly easier for a system with understandable and predictable behavior. However, the behavior of actual systems in certain segments is normally hard to predict, so forming a corresponding mathematical model requires certain assumptions to be made.

The introduction of certain assumptions in the modeling of the nonlinear effect of soil ionization around the grounding grid is illustrated in the following example.

**Example 1.3.** This example illustrates the assumptions under which the mathematical model is formed for calculating grounding grid impulse characteristics when soil ionization is present. According to the results of theoretical and experimental research of soil ionization around a grounding grid, modeling can be done with a fictitious increase of the radius of the grounding grid elements [21-25, 31, 34, 35].

Alteration of the grounding grid parameters with the radius leads to their current dependency and a corresponding nonlinear system of differential equations.

The nonlinear grounding grid model which was implemented in the software tool GIC [17] is based on the calculation of effective radii of the grounding grid elements. According to this procedure, determining the effective radii of linear elements begins with the assumption that the discharge current into the soil along the considered element is fairly constant.

The surface density of the current is determined by the following equation:

$$J_i = \frac{I_{mi}}{2 \cdot \pi \cdot a_i \cdot l_i} \tag{1.27}$$

where: $J_i$ – surface density of current on the $i$ element, $I_{mi}$ – maximum value of discharge current from the $i$ element, $a_i$ – actual radius of the $i$ element, $l_i$ – length of the $i$ element.

The criteria for the occurrence of ionization are defined by the following expression:

$$E_i = \frac{\rho \cdot I_{mi}}{2 \cdot \pi \cdot a_i \cdot l_i} \geqslant E_k \tag{1.28}$$

where $E_k$ – critical strength of the electrical field, which has a typical value of 400 kV/m [17].

The effective radius of the $i$ element is determined by the following expression:

$$a_{ei} = \frac{\rho \cdot I_{mi}}{2 \cdot \pi \cdot l_i \cdot E_k} \tag{1.29}$$

Fig. 1.8 displays the change of effective radii of the elements of a horizontally placed linear grounding for an impulse current injected at its end. The length of each element amounts to 10 m, which leads to an acceptable level of precision and calculation duration [21, 22].

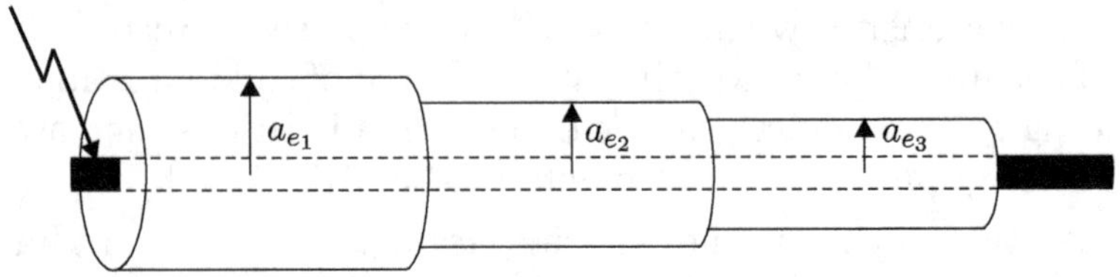


Fig. 1.8 - Change of effective radii of the elements of a horizontally placed linear grounding for an impulse current injected at its end

In addition to the change of the effective radius, for a horizontally placed grounding, there is also a fictitious change to the placement depth. However, for a regular grounding placement depth ($h = 0.5 - 1$ m), this change is not practically significant for analyzing the impulse characteristics of the grounding [16, 21, 22].

In the exhibited process, it is assumed that the grounding grid is approximately equipotential. This assumption is nearly fulfilled in the following cases:

- short grounding grids (grounding grids for overhead line towers),
- long grounding grids which are located in soil of a large resistivity.

An analysis of the influence of soil ionization around long grounding grids placed in soil of a small resistivity is of no practical significance. The ionization zone in this case is not developed, which can be concluded from expression (1.29).

*f) State the limitations*

When forming the mathematical model of an actual system, it is necessary to state the limitations under which it is valid. These limitations can be physical or operational. In practice, limitations are often omitted or are not clearly provided. For the purpose of avoiding these problems, when modeling a system it is necessary to provide answers to the following questions:

- Under which conditions is the displayed mathematical model valid, meaning where is the boundary of validity of the proposed model?
- How far outside of the boundary of validity is the error in calculation for using the proposed mathematical model.

An illustration of this process is provided in the following example.

**Example 1.4.** The limitations of the mathematical model of the grounding grid displayed in Section 1.6.3.1 relate to its inability to encompass the influence of soil ionization on the impulse characteristics of the grounding grid. Considering that this influence appears when a large amplitude impulse current is injected into a grounding grid located in poorly conductive soil, the application of the unmodified mathematical model under those conditions leads to an error in the calculation of the impulse characteristics of the grounding grid. The extent of this error is evaluated in the example of a simple square grounding grid, the disposition of which is displayed in Fig 1.6.

The application of the software tool for calculating the impulse characteristics of a grounding grid GIC [17] has determined the impulse impedance of the considered grounding grid according to a frequency independent model (Fig 1.9a) and a nonlinear model (Fig. 1.9b), for a resistivity of soil $\rho = 100$ Ωm and an amplitude of lightning discharge current $I = 100$ kA. The nominal rise time of the impulse current front and the nominal time to half-value of the impulse current wave are $T_1 = 1.5$ $\mu$s and $T_2 = 30$ $\mu$s, respectively.

Based on the calculation results displayed in Fig. 1.9, it can be concluded that using a frequency independent model outside of the boundary of validity leads to an error in calculation, which in the considered case amounts to $\Delta z = 58.4\%$. This error is determined on the basis of the relation:

$$\Delta z(\%) = \frac{z_a - z_b}{z_b} \cdot 100 \tag{1.30}$$

where: $z_a$, $z_b$ – impulse impedances of the considered grounding grid determined according to a frequency independent model and nonlinear model, respectively.

*g) Form equivalents suitable for engineering practice*

When modeling complex systems, it is very important to form corresponding equivalents which are suitable for the engineering prac-

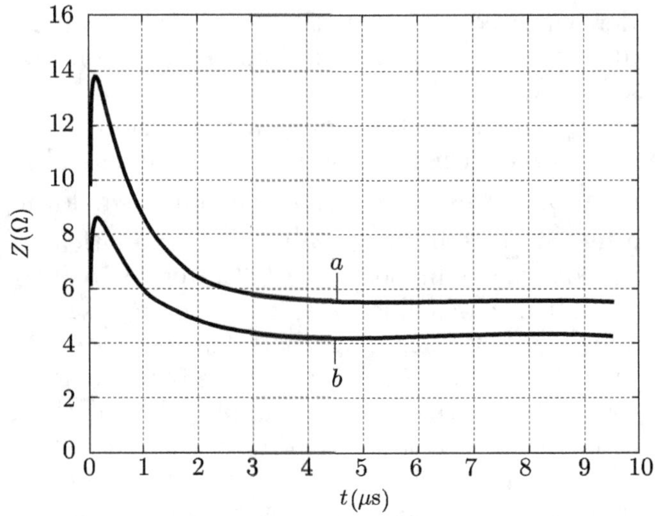


Fig. 1.9 - Time changes of the impulse impedance of the grounding grid determined according to a frequency independent model (a) and nonlinear model (b) for $\rho = 100$ Ωm and $I = 100$ kA

tice. The following example illustrates the process of reducing a complex model of a grounding grid in an impulse period into an equivalent suitable for the engineering practice.

**Example 1.5.** The process of determining the parameters of an equivalent grounding grid is based on the analysis of the voltage response of the grounding grid as a result of injection of an impulse current in the form of the Heaviside function. It is shown below that a complex equivalent circuit of an arbitrarily shaped grounding grid (Fig. 1.10a) can be reduced to an equivalent circuit which contains just three parameters (Fig. 1.10b). The parameters $R_1$, $R_2$ and $L$ are functions of the parameters of an actual grounding grid and are determined on the basis of equal voltage responses of the equivalent circuits in Figs. 1.10a and 1.10b due to the Heaviside function.

Fig. 1.11 displays the typical wave shape of the voltage response of an actual grounding grid as a result of the Heaviside function [16, 18, 26].

Two areas can be seen on the diagram displayed in Fig. 1.11. The first area corresponds to the increase of voltage from zero to the maximum value. This indicates that the maximum value of voltage on the grounding grid is most often reached after 20 ns to 50 ns from

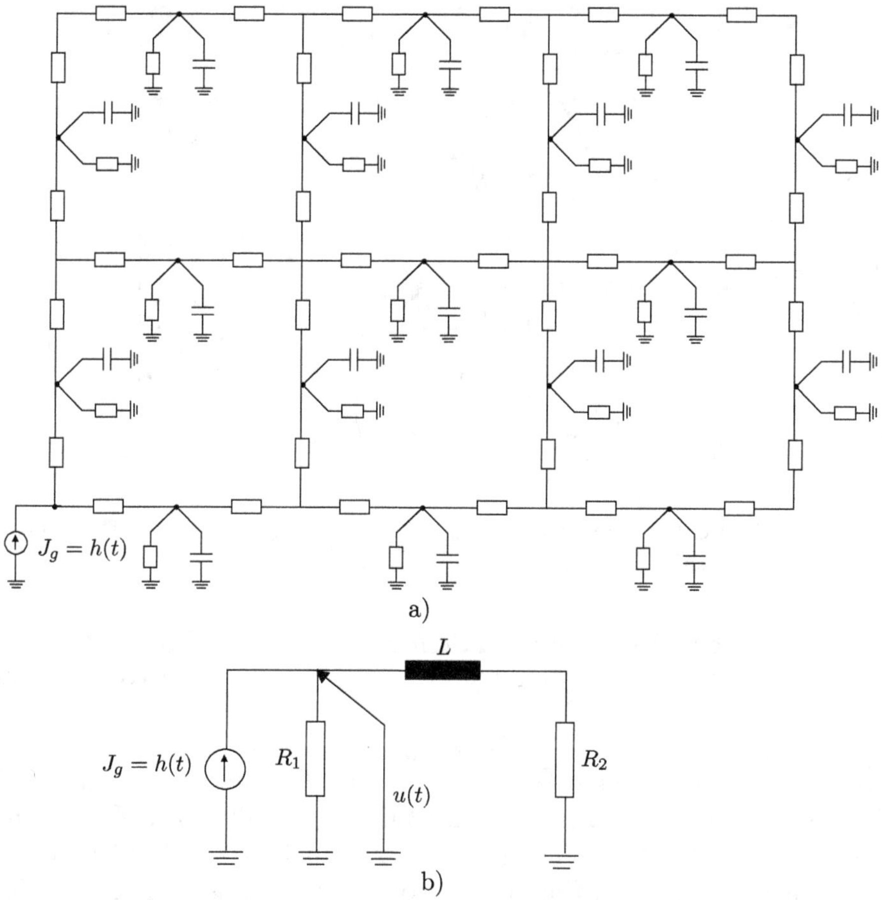


Fig. 1.10 - a) Equivalent circuit of a complex grounding grid, b) *RL* equivalent circuit from Fig 1.10a

the moment of injection of the Heaviside function. The second area is characterized by a slower decrease of voltage from the maximum value to the corresponding stationary value.

The effect of the wave propagation through the grounding grid represents one of the more important influences on the waveshape of the grounding grid response. The speed of the wave propagation in the soil can be determined from the expression:

$$v = \frac{c_0}{\sqrt{\varepsilon_r}} \tag{1.31}$$

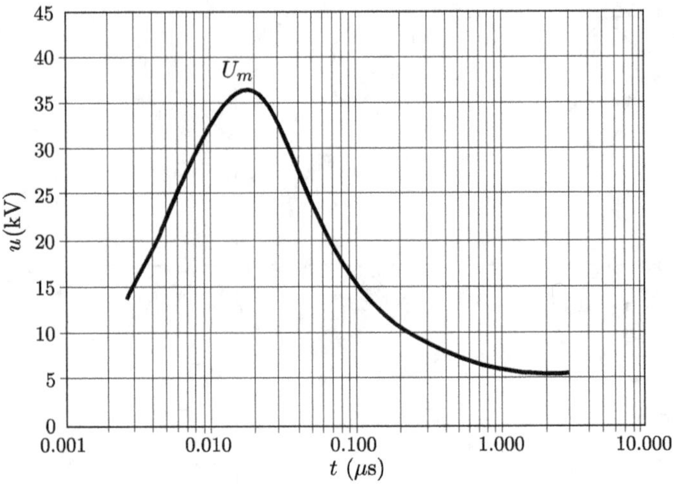


Fig. 1.11 - Typical wave shape of the voltage response of an actual grounding grid as a result of the Heaviside function

where: $v$ – speed of wave propagation in soil, $c_0$ – speed of wave propagation in a vacuum, which amounts to 300 m/$\mu$s, $\epsilon_r$ – relative dielectric constant of soil.

The relative dielectric constant of soil depends on the type of soil and falls in the range of 4 to 81. For a typical value of $\epsilon_r = 9$, the speed of the wave propagation in soil amounts to $v = 100$ m/$\mu$s. For a time period of 20 ns to 50 ns, the wave travels a distance of 2 m to 5 m.

During analysis of the line flashover rate, it was determined that overvoltages on the insulation sufficient enough to cause a flashover result a few $\mu$s after the first occurrence of an lightning discharge current on the grounding grid. At that time it can be considered that the ascending section of the line of the grounding grid response has already passed a value of $U_m$ and that it does not affect the calculation results. Due to this, the voltage response from Fig. 1.11 can be approximated to the voltage response displayed in Fig. 1.12.

The symbols in Fig. 1.12 have the following meanings: $U_m$ – maximum value of voltage, $R_s$ – power frequency grounding resistance, which also numerically represents the stationary value of the voltage response at the injection location (kV) for the impulse current in the form of the Heaviside function (kA), $T$ – circuit time constant.

Determining the parameters of the equivalent grounding grid was done on the basis of the approximated form of the voltage response of

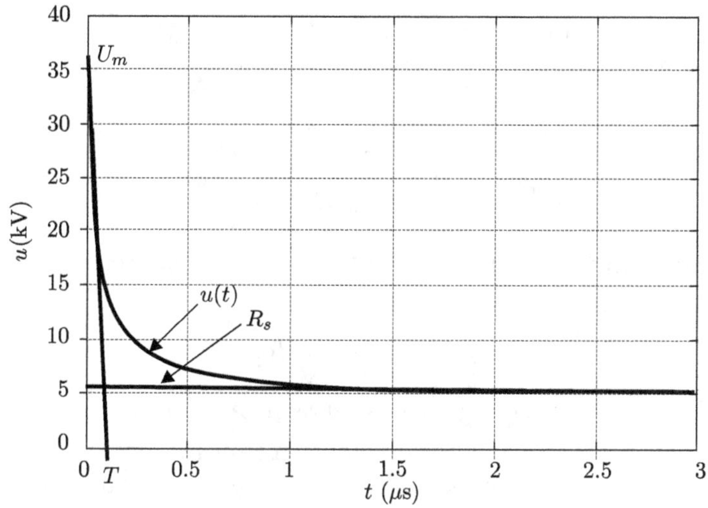


Fig. 1.12 - Approximated wave shape of voltage response of an actual grounding grid as a result of the Heaviside function

an actual grounding grid according to Fig. 1.12. This kind of voltage response can be achieved by injecting a impulse current in the form of the Heaviside function into the RL circuit provided in Fig. 1.10b.

The voltage $u(t)$ in the Laplace domain on the resistor $R_1$, which represents the voltage at the location of current injection, amounts to:

$$U(p) = \frac{R_1 \cdot (R_2 + L \cdot p) \cdot I_m}{(R_1 + R_2 + L \cdot p) \cdot p} \tag{1.32}$$

where $I_m = 1$ kA represents the amplitude of the impulse current in the form of the Heaviside function.

By determining the inverse of the Laplace transform of the expression (1.32), the following is obtained:

$$u(t) = \left( \frac{R_1 \cdot R_2}{R_1 + R_2} + \frac{R_1^2}{R_1 + R_2} \cdot e^{-\frac{R_1 + R_2}{L} \cdot t} \right) \cdot I_m \tag{1.33}$$

Resistances $R_1$ and $R_2$ are determined from the following bounding conditions, with $I_m = 1$ kA:

$$u(t)\big|_{t=0} = U_m \tag{1.34}$$

$$\lim_{t \to \infty} u(t) = R_s \tag{1.35}$$

On the basis of the relations (1.33)-(1.35), the following expressions are obtained for resistances $R_1$ and $R_2$:

$$R_1 = U_m \tag{1.36}$$

$$R_2 = \frac{U_m \cdot R_s}{U_m - R_s} \tag{1.37}$$

The equivalent grounding inductance can be determined from the following expression:

$$L = T \cdot (R_1 + R_2) \tag{1.38}$$

where $T$ is the circuit time constant shown in the diagram in Fig. 1.12.

All parameters of the equivalent grounding grid are calculated on the basis of the maximum value of voltage $U_m$, stationary value of voltage (meaning the power frequency grounding resistance $R_s$) and the circuit time constant $T$ according to the diagram in Fig. 1.12. Parameters of the equivalent grounding grid which are determined in this manner can be applied for an arbitrary waveshape of injected lightning discharge current. Generally, the voltage response of the grounding grid $U_G(p)$ in the Laplace domain on an arbitrary impulse current $I_G(p)$ is defined by the expression:

$$U_G(p) = U(p) \cdot I_G(p) \tag{1.39}$$

where $U(p)$ is the grounding grid response to the impulse current in the form of the Heaviside function.

The voltage response of the grounding grid in the domain of time is obtained through the inverse Laplace transform of the expression (1.39):

$$u_G(t) = u(t) \cdot i_G(0) + \int_0^t u'(t-z) \cdot i_G(z) dz \tag{1.40}$$

Instead of applying the convolution integral for calculating the voltage response of the grounding grid to an arbitrary injection of current, calculation can be done in the time domain by directly injecting the lightning discharge current into the equivalent circuit of the grounding grid and by determining the response function.

The aforementioned process of determining the parameters of the equivalent grounding grid is applicable for cases of moderate values of soil resistivity (up to $\rho = 300\ \Omega\text{m}$).

The software tool GIC contains a completely automated process of determining the parameters of the equivalent circuit [17]. This process is illustrated in Section 1.7.5. For the higher soil resistivity values the ionization effect influence is dominant. This effect leads to the decrease of grounding impulse impedance, as well as the line flashover rate. In these cases, the power frequency grounding model yields the results on the safety side [18].

### 1.6.4. Selection of computerized methods

If a physical phenomenon is interpreted mathematically, most often an idealized model is selected with a solution which is approximately the same as that which actually occurs. When solving mathematical models it is necessary to carry out the following:

- proper selection of computerized methods,
- analysis of errors, considering that errors accompany every physical or technical problem with input information in the form of numerical data.

The fundamental mathematical problems in power systems and the techniques for solving them are provided below:

- Solving a system of linear algebraic equations:
  - Direct methods: Gaussian elimination, Triple-diagonal systems.
  - Iterative methods: Jacobi method, Gauss-Seidel method, Gradient method.
- Solving nonlinear equations:
  - Newton-Raphson method, secant method, interval bisection method, gradient method.
- Solving system of ordinary differential equations:
  - quasinumeric methods (Taylor method, Picard method, small parameter method), Runge-Kutta methods, multistep methods (predictor-corrector formulas).
- Integration:
  - Newton-Cotes formulas, explicite formulas, Romberg method, Hermite integration, Chebyshev method, summary formulas.
- Interpolation:
  - polinomial interpolation, cubic spline, multidimensional interpolation.

- Monte Carlo method.
- Discrete Fourier transform.
- Statistical data description.
- Solving partial differential equations:
    - finite-difference method, finite elements method.

A detailed display of the computerized methods in power systems exceeds the scope of this monograph. For this reason the reader is directed towards specialized literature from this field [7-13, 36].

Generally speaking, the very expansive and complex set of problems in power systems can be divided into two basic fields: 1. Power systems 2. Electromechanical converters. The application of computerized methods in power systems is covered most thoroughly in literature [9], in which the following table is presented.

Table 1.2 - Requirements for computerized methods depending on the field of power systems

| | Power Systems | | Electromechanical converters |
|---|---|---|---|
| | (Stationary Processes) | (Dynamic Processes) | (Dynamic Processes) |
| 1.1 Systems of linear algebraic equations | x | x | |
| 1.2 Self-values and self-vectors | | x | x |
| 1.3 Systems of nonlinear algebraic equations | x | x | |
| 2.1 Systems of linear differential equations | | x | x |
| 2.2 Numerically solving systems of differential equations | | x | x |
| 2.3 Stability of systems of differential equations | | x | x |

### 1.6.5. Forming the program

The purpose of the computer is to produce the essence and not the numbers. When forming a program it is not the amount of effort put into its development that is most important, but rather the correct forming of results when the program is used. Additionally, an excellent presentation of the model by a developed program does not mean much if the model has shortcomings or is not suitable.

The aspects of forming a program are discussed below.

### 1.6.5.1 Development of the algorithm

The first step is to look at the algorithm in the sense of a function or process. Then it is necessary to form a collection of subprograms for each function or process. These special modules can be tested individually.

After successfully testing the considered subprogram, it can be added to the main algorithm. Considering that the programming team is composed of several members, it is necessary to have proper planning for the purpose of working efficiently. This, among other things, means that before beginning development of the model, it is necessary to adopt certain standards and procedures for nomenclature and structures.

On the example of the grounding grid, it is necessary for the team to come to an agreement on the following questions:

- Is the model based on frequency independent parameters or frequency dependent parameters?
- What kind of grounding grid structure should the developed model represent?
- Should nonlinear effects like ionization be encompassed? If they are encompassed, which techniques should be applied for solving the now nonlinear system of equations?

Development of the algorithm is the process by which something is being executed. The algorithm indicates what must be done and the sequence in which it must be done. The most difficult part of any computer modeling or development of a simulation is definitely the development of a proper algorithm. Specifically, it is not enough to write a system of equations, but rather one must define the steps and the sequence in which those steps are executed, as well as the procedure for solving the equations.

Even for the simplest model, it is still necessary to develop an efficient and logical algorithm. A majority of problems which require the development of a corresponding program are not contained within the description of the equations, but rather in the logic, meaning within the development of such an algorithm which encompasses all cases completely and consistently. Essential problems are in the logic and not in mathematical relations.

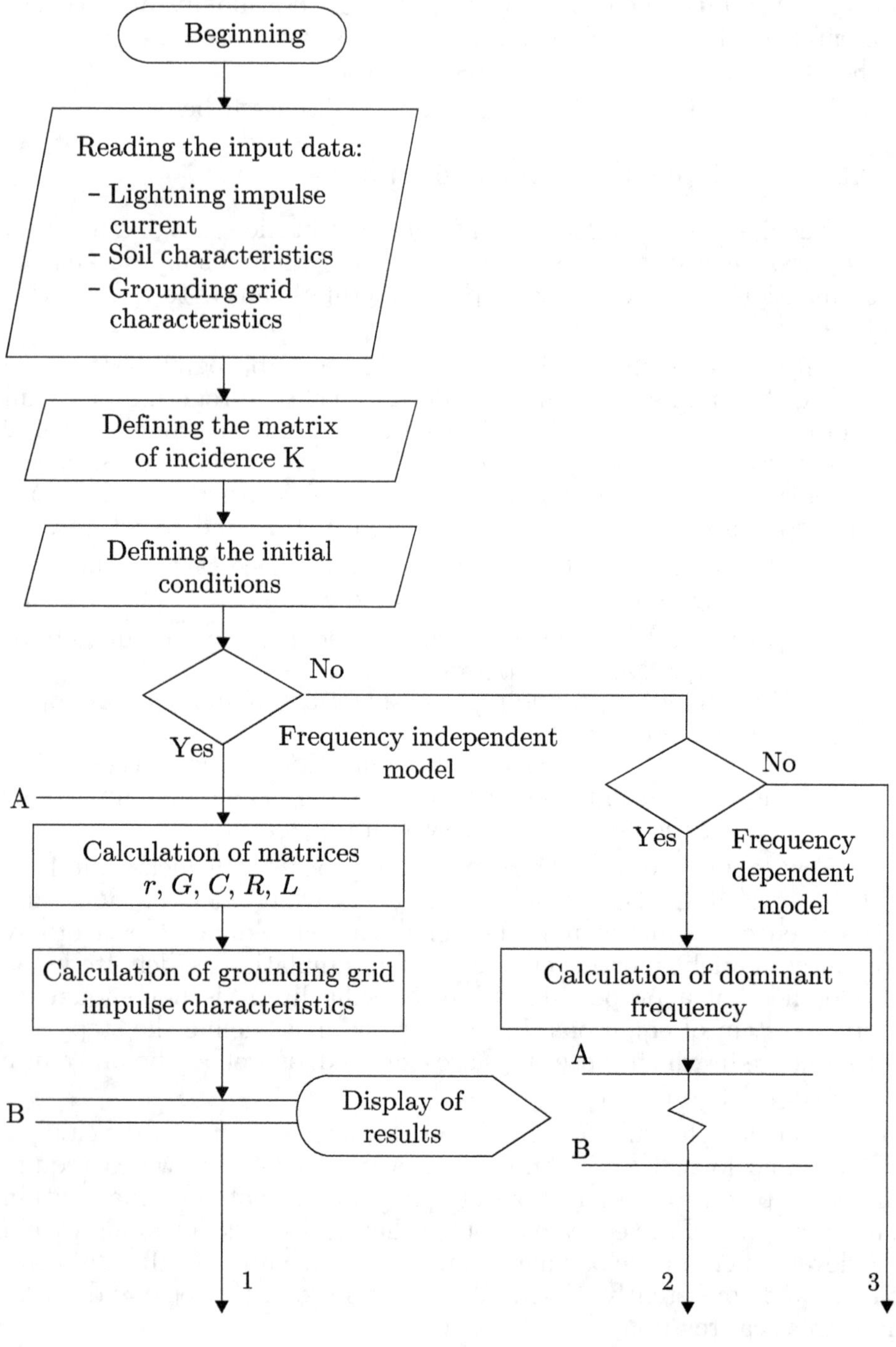
Beginning
Reading the input data:
- Lightning impulse current
- Soil characteristics
- Grounding grid characteristics
Defining the matrix of incidence K
Defining the initial conditions
No
Yes
Frequency independent model
A
Calculation of matrices r, G, C, R, L
Calculation of grounding grid impulse characteristics
B
Display of results
No
Yes
Frequency dependent model
Calculation of dominant frequency
A
B
1
2
3

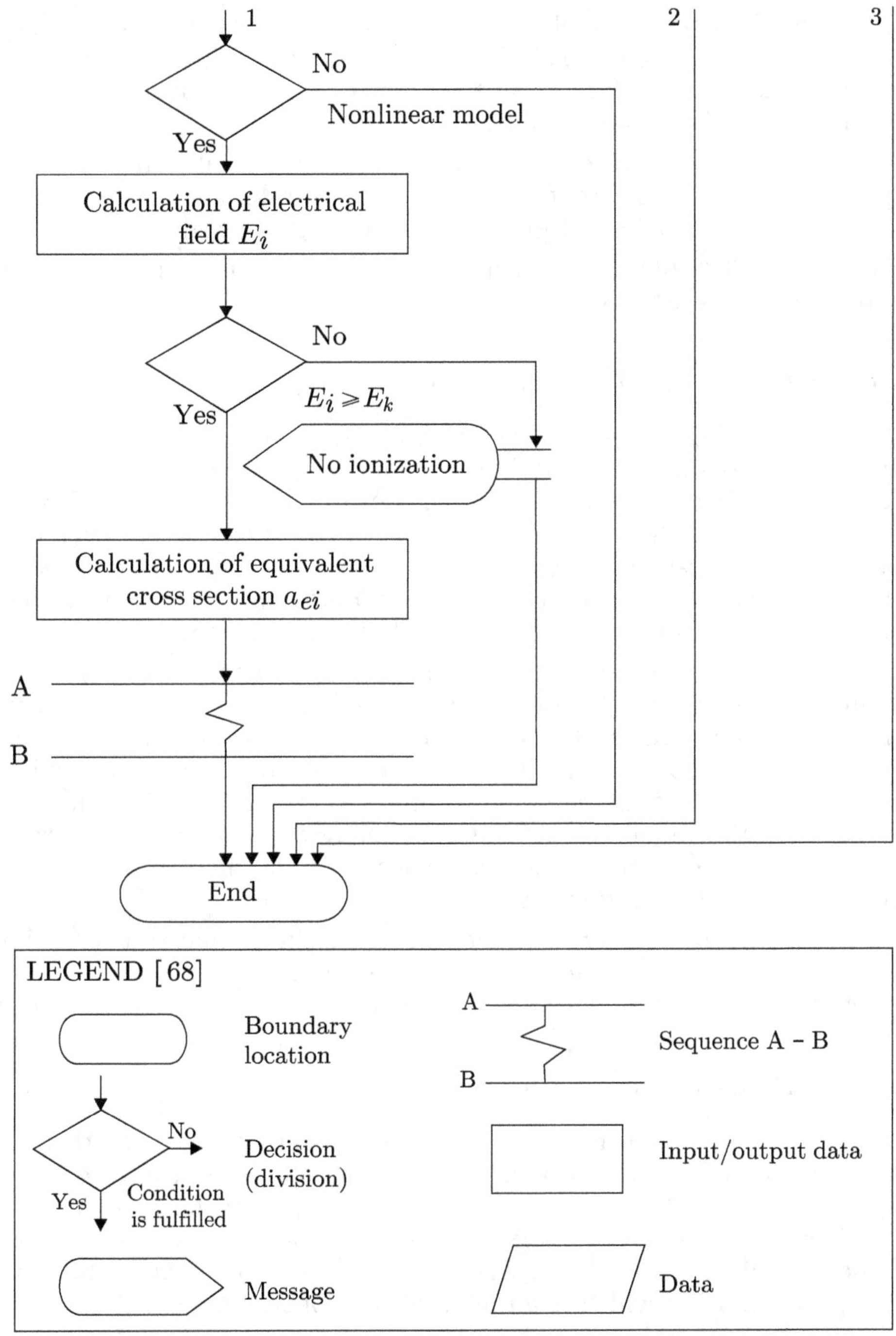


Fig. 1.13 - Algorithm for calculating the impulse characteristics of a grounding grid

The algorithm should not contain too many calculation details, but rather information about what must be done and in what sequence to do it. That which is important to note in every well formed algorithm relates to the places of decisions.

Forming a complete algorithm is the most difficult part of developing a program, but this effort will pay off several fold. The algorithm should be formed before beginning scripting of the program.

Fig. 1.13 displays an example of an algorithm for calculating the impulse characteristics of a grounding grid.

### 1.6.5.2 Requirements for resources

Considering the size of the program (measured in the number of lines in the program, as well as in the number of variables which must be determined), it is necessary to properly evaluate the relation between size and speed of the computer hardware and the size of the software. Basically, one must select hardware/software which can fulfill the needs of the project. The advancement of computers continues to make this problem less of an issue in the sense of both size and speed.

However, there are still problems for which analysis and solving require a complex mathematical apparatus and a significant amount of time for the program to be executed. A typical example of this is using the finite elements method or the finite differences method, which are commonly used for modeling complex physical processes such as processes within breaking elements, modeling cable, modeling a series of problems in electromagnetics, etc. These calculations are mostly used in the process of testing and verification of new products.

Aside from the characteristics of the hardware and software, the degree of complexity of the model and the work regime of the user also influence the execution speed of the program.

### 1.6.5.3 Program characteristics

One of the basic measures of a good program is if the program functions well. However, a program can function well and at the same time be developed in a complex or simple manner. When considering the characteristics of a program, the discussion should not focus just on modeling and simulation, but also on the program as a whole. Finally, the documentation which is an integral part of the program is also an important characteristic.

*a) Program structure*

Every well conceptualized program is modular. Non-modular programs cause a series of problems such as:

- difficulties in upgrading and modernization,
- difficulties in maintenance,
- difficulties in making modifications,
- difficulties in transferring to another computer.

It is necessary to structure the program so that it contains the following three components:

- preprocessor,
- main program,
- postprocessor.

In the preprocessor phase, the user conducts a complete verification of the input file and eliminates possible errors. In this component it is necessary to:

- verify whether the requested data is within the proper range,
- display a warning message for each suspicious case,
- explicitly display errors,
- disable the execution of the main program if any fatal errors are detected.

The main program must be modularly formatted. The basic characteristics of the main program are:

- efficient coding,
- detects errors, so that the problem is always controlled,
- forms technically correct results,
- robust,
- moderate demands for input values,
- automatic values wherever possible,
- appropriate output with moderate details and format.

The postprocessor enables the user to conduct various analyses and generate reports based on the results obtained using the main program. These outputs can come in the form of textual reports, tables or drawings. For simulation programs, which are executed several times, it is necessary to enable a clear separation of the output results for various cases.

*b) Transferability of the program*

Transferability measures the capability of a program which was developed on one computer to be executed on another computer. This is especially important if the program must be widely distributed. A program which possesses this characteristic is significantly easier to maintain.

Programs written in assembly language are in principle non-transferable. In contrast to these programs, programs written using higher level programming languages can be considered transferable. This means that if a version of the program which uses a standard version of a programming language is installed on computers $A$ and $B$, then a program which functions on computer $A$, will, without any changes, also function on computer $B$.

*c) Input/output*

The normal expression which describes the characteristics for the connection between the user and the computer is user orientation. A program which aspires to be user oriented should contain the following characteristics:

- well documented program,
- logical sequence of calculation accompanied by an suitable algorithm,
- description of variables,
- moderate number of inputs and outputs,
- moderate response time,
- interactiveness,
- modularity,
- system of menus,
- user should know what should be done next,
- well thought out help system,
- formulates procedures for eliminating errors.

*d) Suitable programming language*

When making the decision on the selection of a suitable programming language, one must pay attention to the elements displayed in Table 1.3.

Some aspects of the three programming languages which are often used in power systems are briefly covered below.

*d.1 Programming language used in the software tool EMTP/ATP*

The advent of the programming language FORTRAN has made it possible to numerically solve several mathematical and engineering

Table 1.3 - Important elements when selecting a suitable programming language

| Desirable for the Programmer | Desirable for the User |
|---|---|
| Executes the necessary operations in an appropriate manner | Suitable input / output (including graphs) |
| Modularity | Fast calculation |
| Documentation | |
| Transferability | |
| Easy maintenance and upgrading | |

problems. Teams of experts made up of mathematicians, engineers and experts from the field of numerical analysis have developed an entire library of corresponding subprograms for the purpose of making program writing easier and more uniform, as well as achieving a higher level accuracy. The division into the main program which, among other things, contains the corresponding subprograms, has enabled straightforward connection to a family of ready-made subprograms. Even the most complex engineering problems have been programmed using this language and have, in lesser or greater measure, been effectively solved.

The programming language FORTRAN was applied in the original version of the program for simulation of electromagnetic and electromechanical transient processes, as well as transient processes in automated control systems. Subsequently, through certain modifications, one of the most popular software tools emerged called EMTP (Electromagnetic Transients Program), which also has a non-commercial version called ATP (Alternative Transients Program).

Use of the software tool EMTP/ATP is geared towards solving two groups of problems [17, 37, 38]:

- Problems related to the design of power systems which include selection of equipment, coordination of insulation, specifications of protection devices, design of automated control systems, etc.;
- Problems related to the planning and exploitation of power systems.

The ATP program predicts variables of interest within electric power networks as functions of time, typically initiated by some disturbances. Basically, trapezoidal rule of integration is used to solve the differential equations of system components in the time domain.

Non-zero initial conditions can be determined either automatically by a steady-state phasor solution or they can be entered by the user for simpler components.

ATP has many models including rotating machines, transformers, surge arresters, transmission lines and cables. Interfacing capability to the program modules TACS (Transient Analysis of Control Systems) and MODELS (a simulation language) enables modeling of control systems and components with nonlinear characteristics such as arcs and corona. Dynamic systems without any electrical network can also be simulated using TACS and MODELS control system modeling.

Symmetrical or unsymmetrical disturbances are allowed, such as faults, lightning surges and several kind of switching operations including commutation of valves. Frequency-domain harmonic analysis using harmonic current injection method (HARMONIC FREQUENCY SCAN) and calculation of the frequency response of phasor networks using FREQUENCY SCAN feature is also supported.

A partial review of the typical studies which can be solved using this software tool is displayed in Table 1.4.

Table 1.4 - Partial review of the typical studies which can be solved using the software tool EMTP/ATP

| **Typical EMTP-ATP studies** |
|---|
| Lightning overvoltage studies |
| Switching transients and faults |
| Statistical and systematic overvoltage studies |
| Very fast transients in GIS and groundings |
| Machine modeling |
| Transient stability, motor startup |
| Shaft torsional oscillations |
| Transformer and shunt reactor/capacitor switching |
| Ferroresonance |
| Power electronic applications |
| Circuit breaker duty (electric arc), current chopping |
| FACTS devices: STATCOM, SVC, UPFC, TCSC modeling |
| Harmonic analysis, network resonances |

These are just a few of the possibilities of the software tool EMTP/ATP. One of the most important advantages of this tool is its flexibility in modeling systems which enables the user to apply the program for solving several problems within the field of power systems.

The application of the software tool EMTP/ATP is analyzed in Section 2.2.

The model-library of ATP at present consists of the following components:

- uncoupled and coupled linear, lumped R, L, C elements,
- transmission lines and cables with distributed and frequency-dependent parameters,
- nonlinear resistances and inductances, hysteretic inductor, time-varying resistance, TACS/MODELS controlled resistance,
- components with nonlinearities: transformers including saturation and hysteresis, surge arresters (gapless and with gap), arcs,
- ordinary switches, time-dependent and voltage-dependent switches, statistical switching (Monte-Carlo studies),
- valves (diodes, thyristors, triacs), TACS/MODELS controlled switches,
- analytical sources: step, ramp, sinusoidal, exponential surge functions, TACS/MODELS defined sources,
- rotating machines: 3-phase synchronous machine, universal machine model,
- user-defined electrical components that include MODELS interaction.

The original version of the program for numerically solving electromagnetic transient processes was based on the programming language FORTRAN. One of the fundamental shortcomings of this programming language is its relative complexity in forming a program (subprograms, scripting method, indexing method, capability to monitor errors, etc.), as well as the lack of a suitable graphical user interface. For this reason, the application of the program was significantly more difficult.

In the next phase of development, the operation of the software tool ATP was based on the execution of specially formed input files. It was necessary to draw the circuit, mark the nodes and define the parameters of the elements. The input file was then defined using a text editor and the program was practically the same as the programming language because it was composed of commands with specific formats (Fig. 1.14). For this reason, working with the program required special training, and when creating files it was easy to make mistakes. It was painstaking work and there was also the problem of not being able to visualize the files. Due to the aforementioned reasons, in order to use this software tool, one was required to attend advanced training, which was a privilege enjoyed by a relatively small number of engineers.

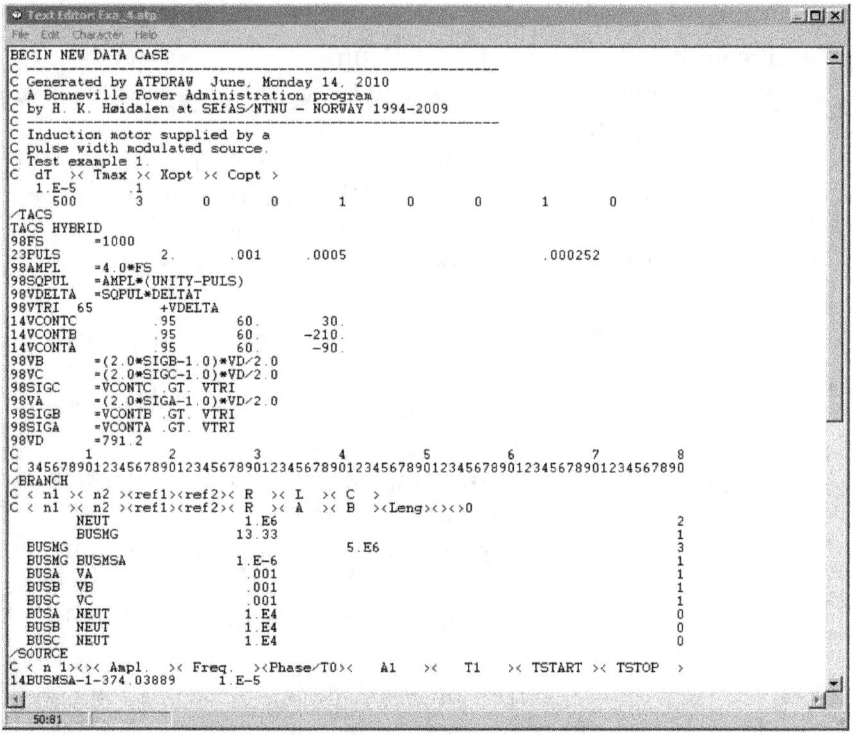

```
BEGIN NEW DATA CASE
C --------------------------------------------------------
C Generated by ATPDRAW  June, Monday 14, 2010
C A Bonneville Power Administration program
C by H. K. Høidalen at SEfAS/NTNU - NORWAY 1994-2009
C --------------------------------------------------------
C Induction motor supplied by a
C pulse width modulated source.
C Test example 1.
C  dT  >< Tmax >< Xopt >< Copt >
   1.E-5       .1
     500        3       0       0       1       0       0       1       0
/TACS
TACS HYBRID
98FS       =1000
23PULS              2.      .001     .0005                          .000252
98AMPL     =4.0*FS
98SQPUL    =AMPL*(UNITY-PULS)
98VDELTA   =SQPUL*DELTAT
98VTRI  65         +VDELTA
14VCONTC            .95        60.        30.
14VCONTB            .95        60.      -210.
14VCONTA            .95        60.       -90.
98VB       =(2.0*SIGB-1.0)*VD/2.0
98VC       =(2.0*SIGC-1.0)*VD/2.0
98SIGC     =VCONTC .GT. VTRI
98VA       =(2.0*SIGA-1.0)*VD/2.0
98SIGB     =VCONTB .GT. VTRI
98SIGA     =VCONTA .GT. VTRI
98VD       =791.2
C        1         2         3         4         5         6         7         8
C 345678901234567890123456789012345678901234567890123456789012345678901234567890
/BRANCH
C < n1 >< n2 ><ref1><ref2>< R  >< L  >< C  >
C < n1 >< n2 ><ref1><ref2>< R  >< A  >< B  ><Leng><><>0
        NEUT                  1.E6                                            2
        BUSMG                13.33                                            1
  BUSMG                                   5.E6                                3
  BUSMG BUSMSA                1.E-6                                           1
  BUSA  VA                     .001                                           1
  BUSB  VB                     .001                                           1
  BUSC  VC                     .001                                           1
  BUSA  NEUT                  1.E4                                            0
  BUSB  NEUT                  1.E4                                            0
  BUSC  NEUT                  1.E4                                            0
/SOURCE
C < n 1><>< Ampl.  >< Freq.  ><Phase/T0><   A1   ><   T1   >< TSTART >< TSTOP  >
14BUSMSA-1-374.03889     1.E-5
```

Fig. 1.14 – Example of an input file for ATP

However, the formation of the graphical preprocessor ATPDraw™ for Windows led to a revolution in the use of this software tool. The user was given the ability to create and edit electrical circuit models, as well as simulation through the ATP software, without the mandatory knowledge of the corresponding programming language [39].

Most of the standard components of ATP as well as TACS are supported, and in addition the user can create new objects based on MODELS or Include (Data Base Module). Line/Cable modeling is also included in ATPDraw™ where the user specifies the geometry and material data and has the option to view the cross section graphically and verify the model in the frequency domain.

ATPDraw$^{TM}$ for Windows is most valuable to new users of ATP-EMTP and is an excellent tool for educational purposes. However, the possibility of multi-layer modeling makes ATPDraw$^{TM}$ for Windows a powerful front-end processor for professionals in analysis of electric power system transients, as well.

Fig. 1.15 displays an equivalent circuit formed using the program ATPDraw$^{TM}$ for Windows.

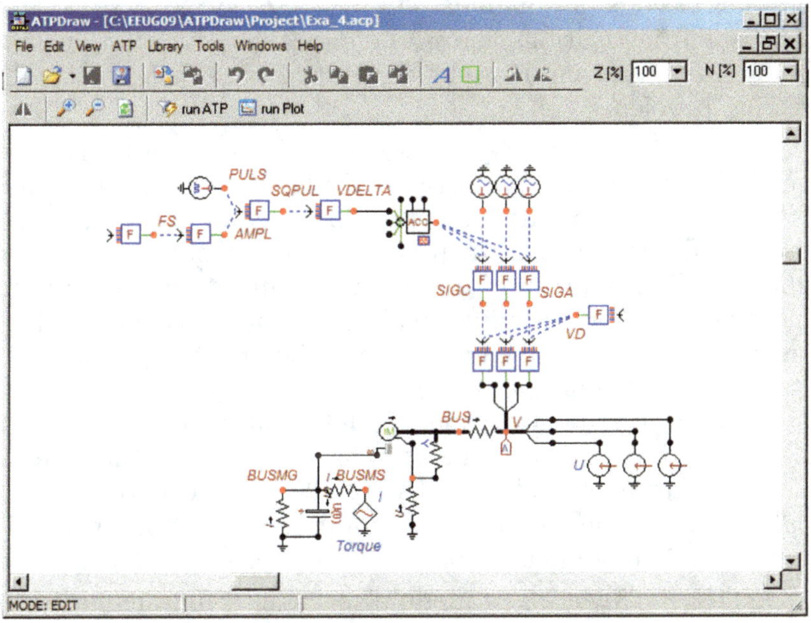


Fig. 1.15 – An equivalent circuit formed using ATPDraw$^{TM}$ for Windows

## *d.2 MATLAB® programming language*

MATLAB® technical computing software is a software tool for solving mathematical problems, analyzing data and visualization [17, 40, 41]. This tool integrates numerical analysis, matrix calculation, data processing and graphical display. It is characterized by its ability to solve all mathematical problems. The advantage of this software tool is in its simple expression of mathematical problems and solutions as they are written in mathematics, by which traditional programming is avoided.

MATLAB® technical computing software is also represented by a programming language which makes many mathematical problems easier to solve making earlier programming languages such as FORTRAN, BASIC or C obsolete. As a result of the simple approach in programming, MATLAB® technical computing software makes it possible to solve mathematical and engineering problems in a significantly shorter amount of time, which is exceptionally important in the field of engineering and science. For this reason MATLAB® technical computing software has become a dominant software tool in universities and scientific research institutions across the entire world. Additionally, it is also being used more and more by design firms in order to solve several engineering problems.

The unbelievable success of this software tool is also a consequence of its conceptual organization. MATLAB® technical computing software has the capability of being upgraded with modules. These additional modules are called Toolboxes. The Control System Toolbox and Simulink® software are two of the most commonly used additional modules. These modules make it possible to simulate dynamic systems within a graphical environment. The family of additional software tools contains functions which are useful for several mathematical and engineering disciplines. Along with the aforementioned modules, MATLAB® technical computing software is also constantly being updated with other new modules. MATLAB® technical computing software also enables the user to form additional functions which is yet another advantage of open-source systems such as this software tool.

Simulink® software is a module specialized for simulation of dynamic systems in a graphical environment. Using Simulink® software enables the analysis of linear, nonlinear, time-continuous or discrete multivariable systems with concentrated parameters [42]. Realization of simulation is achieved by forming a Simulink® model and using the Simulink® function which solves systems of first-order ordinary differential equations. The fundamental advantage of Simulink® software is the simplicity of operation which is reflected in the display of the Simulink® model in the shape of a block diagram. The advantage of this type of display is in the illustrativeness and simplicity of presenting the problem which is required to be solved.

A block diagram is one form of a mathematical model of a system which illustrates the dynamic characteristics of the system, the main variables of the system and the link between those variables. The functional relation between the parts of the system can be viewed in the displayed block diagram. By displaying every element of the system, the principle of the "black box" approach is represented. This approach is based on every element being displayed by a specific block which contains the mathematical relations between the input and output variables of that element. Several blocks are mutually connected by orientational lines which indicate the flow of the signal from one block to another. Additionally, the signals indicate the input and output variables.

Each Simulink® model corresponds to an m-file which is a series of MATLAB® and Simulink® commands and functions which create Simulink® models.

SimPowerSystems™ is a component of the additional Simulink® module. The additional module itself consists of several blocks:

- Electrical sources,
- Elements,
- Power electronic,
- Machines,
- Connectors,
- Measurements,
- Extras,
- Demos,
- Discrete system,
- Power GUI.

As previously mentioned, this module can be used for calculation in several fields within power systems. The applicable fields for this module are:

- analysis of power system networks,
- calculation of load flows of a network,
- analysis of transient processes in a network,
- analysis of a network with nonlinear elements, such as surge arresters,
- analysis of statistical and dynamic stability of synchronized machines,
- analysis of a circuit for vector management of asynchronous drives,
- analysis of the operation of DC machines,
- calculation of excitation circuits of synchronized generators,
- analysis and calculation of parameters for steam and hydro-turbines.

Within the Demos library there are examples which illustrate some of the possible applications of this additional module. These examples are especially useful for beginners and even some more experienced users when mastering the technique of properly creating a mathematical model or block diagram, as well as conducting the simulation of the selected physical system and its behavior.

Examples of the application of MATLAB® technical computing software and the additional Simulink® and SimPowerSystems™ modules are illustrated in Section 2.1.

The simplicity of programming in the MATLAB® technical computing environment is illustrated in the following example.

**Example 1.6.** For the grounding grid displayed in **Example 1.2**, form a program for calculating the impulse characteristics of the grounding grid according to a frequency independent model. Calculate the impulse impedance of the grounding grid in the function of time, the impulse coefficient of the grounding grid in the function of time, the conventional impedance of the grounding grid and the conventional coefficient of the grounding grid. Divide the screen into four section and draw the following: in the first section the time change of the node voltages, in the second section the time change of the current of inductive branches, in the third section the time changes of the impulse impedance of the grounding grid and the impulse coefficient of the grounding grid, in the fourth section the time changes of the voltages at the injection location, injecting current and impulse impedance of the grounding grid.

The program for calculating the impulse characteristics of the grounding grid is created by forming two m-files titled *input* and *output.*

The *m*-files are a specific component of MATLAB® technical computing software and are ASCII files which contain regular text and have a ".*m*" file extension. Both types of files can be transferred between various computer systems and architectures.

There are two types of m-files: command and function files. Both types of files are formed using a text editing or text processing software. Additionally, it is possible for one *m*-file to reference another *m*-file.

The command file is a series of MATLAB® expressions and commands which are executed when the file is referenced. All variables defined in the MATLAB® technical computing environment are referenced in the command file and vice versa.

Function files enable the user to form new functions. In contrast to command files, function files must contain a function in the first line which has the following form:

$$\text{function}[y1, y2, \ldots] = \text{functionname}(x1, x2, \ldots) \tag{1.41}$$

Following this expression comes a series of MATLAB® commands and expressions. $y_1, y_2, \ldots$ represent the input arguments of the function which are obtained when the function is referenced, functionname represents the name of the function or function file, and $x_1, x_2, \ldots$ are the output arguments. Only variables which are input arguments of the function can be referenced from the workspace in the function file.

Writing the program is carried out in the typical way. Selecting the option File>New>M-file opens the MATLAB® text editor through which the $m$-file can be created. This file can also be created in other text editors.

An example of input data is provided in Fig 1.16.

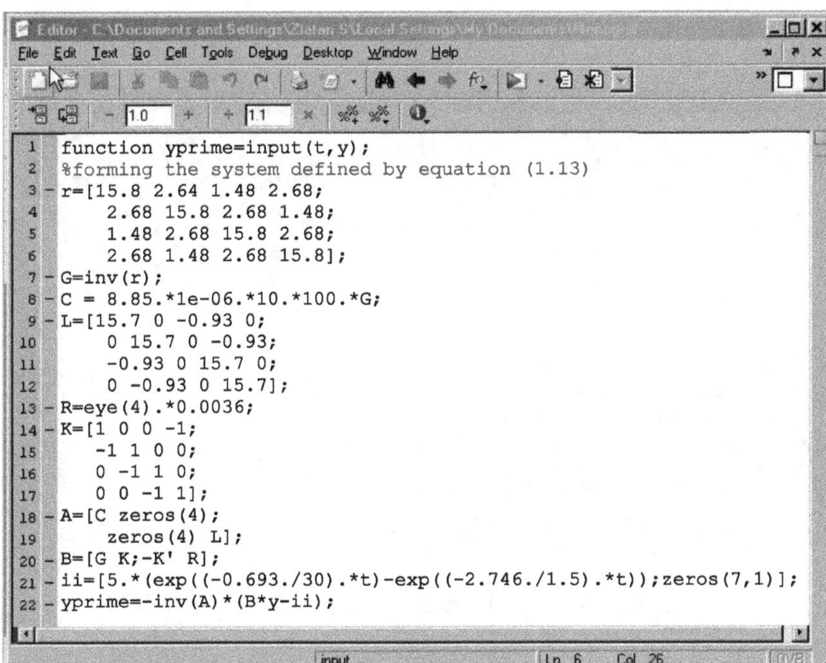


Fig. 1.16 – Appearance of the input file *input.m*

The following provides an explanation of the input file. The command in the first line indicates a function subprogram, while the designation *yprime* represents the derivative $dy/dt$, where $t$ is an independent variable and $y$ is a dependent variable (in this case $y$ represents the voltage vector of the nodes and the current vector of the inductive branches).

A comment is entered into the second line. Comment lines characteristically begin with the % symbol. Comments may also be entered into the program later for the purpose of simplified monitoring.

Next comes the assignment of matrix $r$ (lines 3 to 6) and calculation of matrix $G$, $C$ and $L$ (lines 7, 8 and 9 to 12, respectively). The matrix of series resistances of elements $R$ (line 13), as well as the matrix of incidence K (line 14 to 17) are defined below. Then comes the calculation of matrix $A$ (lines 18 and 19) and matrix $B$ (line 20). Line 21 defines the vector of current and voltage sources ii. Finally, in line 22, based on relation (1.13), a system of differential equations in a normalized form is defined.

The output file output.$m$ is displayed in Fig 1.17.

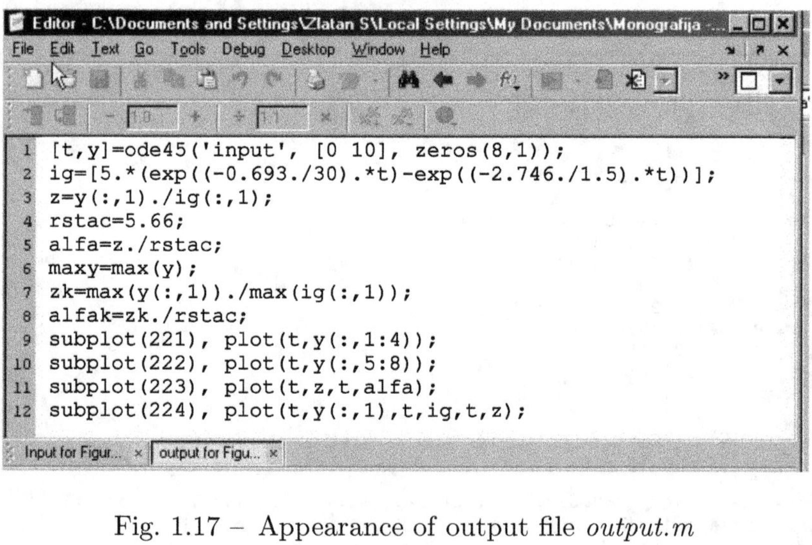


Fig. 1.17 – Appearance of output file *output.m*

The following provides an explanation of the output file. Line 1 contains a command with a syntax defined by the subprogram ode45. The syntax of the subprogram ode45 can be viewed by using the help command:

>>help ode45

The ode45 function is used for numerically solving first-order regular differential equations (or systems of first-order differential equations) which have the following form:

$$\frac{dy}{dt} = f(t, y) \tag{1.42}$$

where: $t$ – independent variable. The solution of the previous differential equation is function $y$ which fulfills the previous equation. Function ode45 has the following syntax:

$$[t, y] = \text{ode45}(\text{'fun\_name'}, \text{t\_range}, y0) \tag{1.43}$$

where fun_name is the name of the local function or function file which calculates the function $f(t, y)$ for the given values $t$ and $y$, t_range is the vector which assigns the interval of the solution and y0 is the initial value for $y$ (the value of $y$ in the initial point of the interval t_range). Vectors $t$ and $y$ are obtained as the result of the previous command.

The output file references the file *input*. The transient regime is calculated from the initial time of 0 to the end time of 10 $\mu$s, which is defined by the vector [0 10]. A square grounding grid has eight independent variables: voltages of four nodes and currents of four inductive branches. Zeros (8,1) represent the initial conditions for solving a system of differential equations, without which the calculation cannot be initiated. Calculation of vector $y$ is conducted for time $t$ and the result is located in the matrix $[t, y]$.

Line 2 defines the impulse current which is injected into the grounding grid. On line 3 the impulse impedance of the grounding grid is calculated which is defined as the relation of voltage at the injection point and the injection current. Line 4 provides the value of stationary power frequency grounding resistance of the grounding grid. Line 5 is the calculation of the impulse coefficient of the grounding grid which is defined as the relation of the impulse impedance of the grounding grid and the power frequency grounding resistance of the grounding grid. The maximum values of independent variables are determined in line 6. The conventional impedance of the grounding grid and conventional coefficient of the grounding grid are determined in lines 7 and 8. Line 9 to 12 relate to the proper division of the screen and drawing the dependencies defined in **Example 1.6**. The screen is divided into four sections and the following is drawn: in the first section the time change of the node voltages (line 9), in the second section the time change of the current of inductive branches (line 10), in the third section the time changes of the impulse impedance of the grounding grid and the impulse coefficient of the grounding grid, in the fourth section the time changes of the voltages of the injection location, injecting current and impulse impedance of the grounding grid.

After creating and saving the program comes the phase of executing the program which can be achieved in two ways:

- Assigning the name of the program without the extension. This assignment is done directly from the MATLAB® command line. In this example that is achieved through the following command:
  >>output
- by selecting the option *File>Run Script.*

After selecting this option it is necessary to assign the name of the output $m$-file or select it from the list of provided $m$-files. Then the program is executed the same as in the previous case.

If the program has been properly formed, the desired graphs will appear on the screen (Fig. 1.18). In the event that there are errors, the user will be provided with information about the errors. In this case it is necessary to edit the program, enter in the necessary changes, save the new version and repeat the initiation of the program. It is especially important to save the changed version. This is one of the most common mistakes made by beginners. All of the effort put into changing the program is in vain if the changes are not saved. Execution of the program in the case of unsaved changes would lead to the same error information appearing as in the previous case.

It is important to note that linking to another m-file may also be done within the program just by entering in its name. Following execution of the referenced $m$-file, all variables remain in the workspace and subsequent commands can use them as known values.

It is necessary to mention that there are several subprograms available for calculation of grounding grid impulse characteristics which are based on the Runge-Kutta method, which is discussed further along in this chapter.

### *d.3 Programming language C/C++*

The programming language C is a general purpose medium level language. This program contains structured types of data and control structures which is characteristic of several programming languages

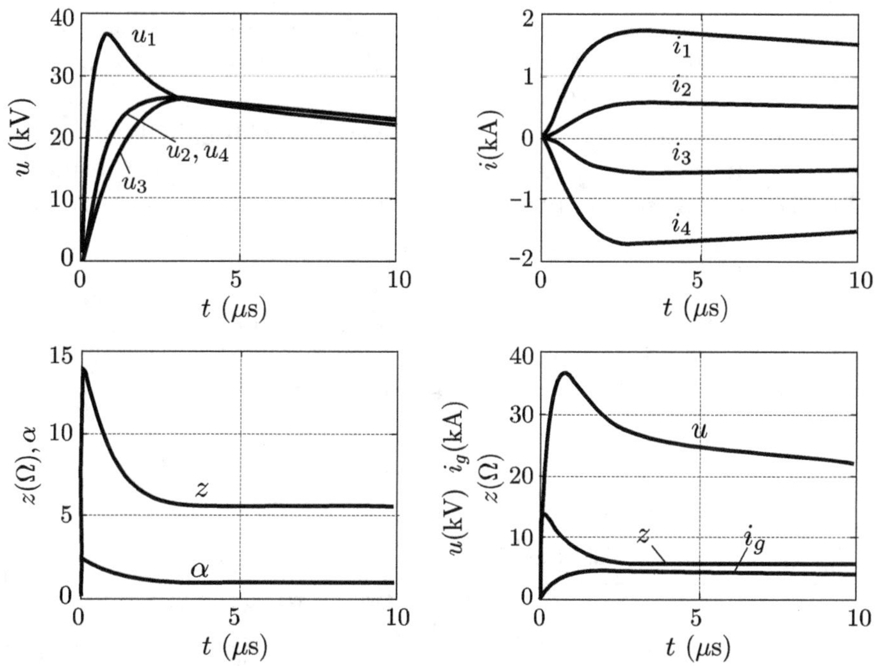


Fig. 1.18 - Calculation results for impulse characteristics of the grounding grid according to **Example 1.2**

[43]. Furthermore, it supports bit manipulation, use of processor registers, access to help address data and is operator-oriented towards the hardware of the computer. The aforementioned traits are characteristic of programming languages like symbolic assembly languages. The program C contains a syntax which enables concise expression and scripting of structured programs.

The basic version of the program C was developed in Bell Laboratories for use with the Unix operating system. This program is similar in characteristics to assembly language, but is much simpler to learn. Following further expansion in 1983/84, the programming language C was finally given its current name C++. Today, this language is one of the most powerful languages for object-oriented programming.

The processing of a program in the language C++ is displayed in detail in [43, 44] and is comprised of the following four steps:

- entering the source text of the program into a file on a disk,
- translating the source text of the program,

– connecting the translated form of the program with the necessary user and system subprograms in feasible form,
– executing the program.

Section 1.7 displays GIC [17], a software tool for calculating grounding grid impulse characteristics, which was created using the modern MATLAB® technical computing software and Visual C++ [40, 45].

*e) Documentation*

Documentation of the program deserves special attention. Table 1.5 displays the minimum characteristics of a well documented program. The first eight characteristics are necessary for both users and programmers.

Table 1.5 – Minimum characteristics of a well documented program

| Line No. | User | Programmer | Documentacion |
|---|---|---|---|
| 1 | x | x | Description of what the program does, methods |
| 2 | x | x | Example of input/output |
| 3 | x | x | Description of input/output |
| 4 | x | x | Possible errors in input data and messages |
| 5 | x | x | Organization of data, files and appearance |
| 6 | x | x | References |
| 7 | x | x | Mathematical model on which the program is based |
| 8 | x | x | Program development notes, repeat check-up |
| 9 | | x | Names of all encompassed subprograms |
| 10 | | x | Program listing |
| 11 | | x | Interpreting the variable concepts |
| 12 | | x | Algorithm of calculation |
| 13 | | x | Written commentary related to the program |
| 14 | x | x | Test example with complete input/output |

Brief commentaries on individual characteristics of program documentation are provided below.

*Characteristic 1.* The description of what the program does should be done in such a way that the potential user has enough elements to decide whether or not its application can resolve the desired problem.

*Characteristic 2.* The input and output represent the actual connection between the user and computer. Possible complex requirements for input data, too much data, or inadequate form for usage (e.g. mixed standard and metric units), may turn away potential users. Similarly, output should be of an appropriate format and size.

*Characteristic 3.* The requirements for input units should be described in great detail. Very often different cases require information organized in different ways.

*Characteristic 4.* Possible limitations in the size of various parameters or specificities which may cause problems in the functioning of the program, as well as units for input values or output values, should be clearly highlighted.

*Characteristic 5.* For an application which encompasses a large collection of data, it is necessary to describe not just the data in question, but also how that data is organized. Most often these large amounts of data must be displayed in the sense of corresponding hierarchal structures.

*Characteristic 6.* When forming a program it is necessary to note the references used. Further development of the program without noted references would be difficult or practically impossible.

*Characteristic 7.* Various approaches and mathematical models are possible for the given system. The exact mathematical model on which development of the program is based represents an integral part of the documentation.

*Characteristic 8.* All phases of program development must be recorded. This record will make often unavoidable subsequent modifications to the program possible.

*Characteristic 9.* The list of names of all subprograms represents an integral part of the documentation.

*Characteristic 10.* The program listing is very useful when making changes or amendments to the program. A significant amount of time can be saved if an analysis of possible errors is performed on the printed version of the program along with entering in the corresponding changes. This analysis is much easier to perform on the printed version of the program than on the version of the program displayed on screen.

*Characteristic 11.* Interpreting the variable concepts is very useful when the program is developed especially for users with inadequate experience in the given field. This interpretation must also contain explicit definitions which will remove the possibility of improper interpretation of the used variables.

*Characteristic 12.* Algorithms of calculation are necessary even when solving the simplest tasks. A logically consistent and complete algorithm makes development of a suitable program possible and also enables it to be easily maintained and upgraded. As mentioned in Section 1.6.5.1, the level of detail of the algorithm should be sufficient enough to show the logic and sequence of execution. For exceptionally complex sections of an algorithm, it is necessary to introduce two levels: the general diagram and diagram with details.

*Characteristic 13.* Written commentaries are important in the phase of program development as well as in the phase of noticed problems and the methods for solving those problems. This kind of documentation makes the process of modernizing the existing program much easier.

*Characteristic 14.* Test examples which make use of all options in the program and, most importantly, contain concrete answers, are very useful and necessary when mastering a new program. These examples make the following possible:

- verification of the validity of the program,
- comparison of the results obtained using different methods,
- comparison of the times for executing the calculation.

As a final point, it is necessary to note that without well-formed corresponding documentation, even the best program for solving a certain program is not practically usable.

*f) Development of software tools*

Existing software tools can be used when they are suitable and applicable for solving the class of problems which is required to be solved. However, it often occurs that software tools in their existing form do not make it possible to directly solve the considered problem. In this case the following should be done:

- new modules should be formed within an existing software tool, or
- a new software tool should be developed.

Practical experience points towards the constant requirement to expand an existing software tool with appropriate modules. This requirement results from the fact that engineering problems exist which must be very carefully encompassed in a way which the performances of existing software tools do not allow. An example of such a case is the software tool EMTP/ATP which in principal solves all problems related to the calculation of transients in power systems. In any case, these types of software tools are developed by extremely specialized teams, which should be respected and appreciated. However, that still doesn't mean that the offered solutions are the best or even adequate, especially when it concerns universal tools which claim to effectively solve a wide spectrum of engineering problems.

One form of a modification to the software tool EMTP/ATP is displayed in the example of calculating electromagnetic transient processes in the secondary circuits of the transformer in the $SF_6$ substation [46-48]. A complete example is illustrated in Section 2.2.

A new software tool should be developed if an existing software tool is unavailable due to commercial reasons or if creating an add-on for an existing software tool is too complex. In any case, an extremely specialized software tool for solving one class of problems may be more efficient and user-oriented when compared to a generic software tool which claims to effectively solve a wide spectrum of engineering problems.

Efficient organization is necessary for development of a well-conceptualized software tool. The team leader is required to select the proper team of programmers for the purpose of handling specific requirements within the development of the program. Each team member works on a section of the program and uses standardized nomenclature and structures for the purpose of easily combining the various, independently written segments into a coherent software tool.

This kind of approach requires careful planning in:

- the development of the structure of the software tool,
- the distribution of responsibilities among the programming teams,
- the linking and testing of the final product.

Development is always an iterative process. Sections which are developed should be periodically shown to potential users for the following reasons:

- potential users can see which phase in the development of the software tool has been completed,

- the exchange of information between the development team and users can help with consideration of the requests and possible corrections from future users in a timely manner,
- contact between users and the programmers will help to motivate members of the programming team.

*g) Testing the program*

Testing the program is the process of verifying the program or system with the goal, revealing and correcting errors, and is an integral part of development of the program. It is applied in each phase of the development cycle [49] and encompasses more than 50% of the time needed for development of the program.

The goals of testing the program are [50, 51]:

- verification and validation of the program conforming with the specifications of the request, which doesn't always mean the program is technically accurate, reliable and safe,
- improvement of the quality of the program,
- estimation of reliability.

The methodology of testing the program implies the cycle consisting of the phases displayed in Fig 1.19.

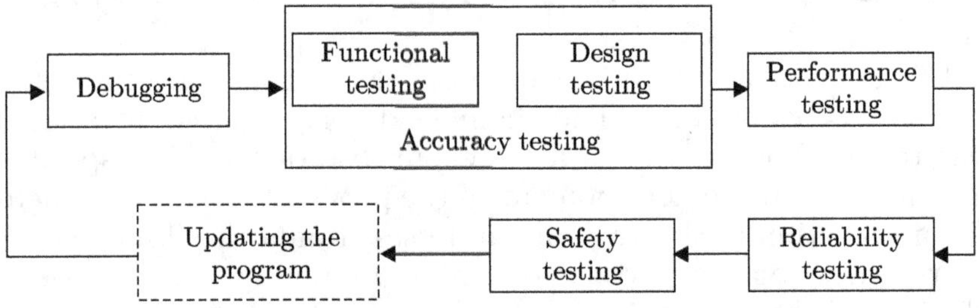


Fig. 1.19 – Phases of testing a software product

The phases of testing a software product are described below [51].

*Debugging* implies the removal of syntax and logical errors during development of the application.

*Testing accuracy* encompasses the following methods:

a) Functional testing ("Black box" method). Only the performance of specific functions is analyzed along with verification of input and output data. The accuracy of output data is verified on the basis of the specifications of the request for the

program. Analysis of the source code is not conducted during these tests. A problem with functional testing may occur in the case of ambiguous requests or an inability to describe all methods for using the program.
Nearly 30% of all code errors are the consequence of incomplete or non-specific specifications [49].

b) Testing program design ("White box" method)
This testing verifies and analyzes the source code and requires good knowledge of programming, the corresponding programming language and the design of the actual software product. The testing plan is determined on the basis of the elements of program implementation, such as the programming language, logic and styles. Tests are conducted on the basis of the program structure. With this method there is the possibility of verifying almost the entire code, for example verifying whether each line of code is executed at least once, verifying all functions or verifying all possible combinations of various programming commands. Specific tests can be used to verify the existence of endless loops or code which is never executed.

*Testing of performance* encompasses locating and removing problems which degrade the program's performance. Testing of performance most often encompasses exploitation of processor resources, data flow and response time. The typical resources which are verified are the capacity range, processor speed, memory usage and consumption of hard disk space.

*Testing reliability* determines the probability of the program to function without any errors or delays. The reliability of a program is an important characteristic of quality. According to [52], the reliability of a program is represented by the ability of the system or components of the system to carry out the requested functions under certain conditions within a certain period of time. Testing robustness and testing the load are examples of reliability tests which attempt to trigger delays in the program in certain situations. An evaluation of the reliability of a program is performed using various measures:

a) Final product measures
The size of a program indicates its complexity. The basic method of determining the size of a program is determining the number of lines of code;
b) Project management measures

Greater reliability is reached by using a better development process, better risk management process, better configuration management process, etc.;

c) Program development measures
Based on the assumption that the quality of a product is a direct function of the process of development, these measures are used for evaluating, monitoring and increasing the reliability and quality of the program;

d) Error and delay measures
Normally these parameters are based on user feedback regarding errors which they have discovered. Based on the data collected about delays and errors, the mean time before failure (MTBF) can be calculated along with other parameters which determine or envisage the reliability of a program;

e) Request for reliability measures
For the purpose of developing a reliable program, the specifications of the request are not allowed to contain ambiguous expressions, optional requests or requests with multiple meanings. The programmer should not be left with the decision on whether a request should be implemented or not.

*Testing security* refers to identifying and removing errors which could potentially endanger security, as well as to validating the effectiveness of protective measures.

The main activities in testing the software are:

- Verifying that the developed program actually presents the proper mathematical model;
- Validating the calculation results with the results determined in actual systems.

Verification of the program is most effective if each module is tested individually. Using a program for discovering errors is mandatory and can be used effectively during the verification phase.

The validity of the results of the program in relation to results obtained from actual systems requires data from an actual facility. This kind of procedure is not customary when collection of the data is difficult or expensive. Data from an actual facility requires an organized, statistical procedure of data collection.

When verifying the validity of the program, one must consider the possible sources of differences between the results obtained through calculations and those obtained from an actual facility. Some of the sources of possible differences are:

- A pessimistic approach to calculation. The effect of this type of approach is illustrated in **Example 1.7**.
- Quality of the model in the sense of limitations in encompassing certain phenomenon which occur under actual conditions. In other words, the model itself is normally approximated and encompasses assumptions and some average values of individual parameters, etc. This approximation of the model can best be noticed when comparing numerically and experimentally obtained results. Experimental results are especially useful under the condition that they are precisely determined. These results initiate steps for making further improvements in the quality of the model which enables more accurate results to be obtained through calculation [16, 20, 46, 47].
- If the mathematical model contains differential equations or integral equations, they will be replaced by approximations in the sense of discrete values (e.g. in the finite differences method or in the numerical method used for solving integral equations). The selection of appropriate subprograms and the steps of integration have a significant influence on the accuracy of calculation, which is described in detail in **Example 1.8**.
- The method of forming input data. Input data can be:
  - imprecisely determined;
  - concentrated; e.g. the average value of the length of inductance of the grounding grid elements 1 $\mu$H/m is adopted and used to perform the calculation failing to take the actual situation into consideration. This inductance is located within the range of 0.8 to 1.5 $\mu$H/m [53];
  - determined on the basis of insufficiently known assumptions.

**Example 1.7.** During the analysis of the lightning protection of the transformer substation, it is necessary to evaluate the risk of failure to the insulation of the power system equipment. For this evaluation it is necessary to repeat a large number of simulations of the transient process due to lightning discharges in order to realize the statistical effect of the random parameters of lightning, the locations of lightning strikes, instant values of operating voltage, configuration of the substation, etc.

During analysis of lightning overvoltages located within the transformer substation on connected lines, it is necessary to properly model the individual elements of the substation on which the overvoltage wave is located [26]. [54-56] display the process of forming an optimal equivalent circuit of the substation by which one can avoid working with complicated circuits with a large number of elements.

The selection of the number, placement location and type of protective device is based on the estimated values of risk of failure to the insulation of the equipment, meaning its reciprocal value called the mean time between failure (MTBF). According to existing criteria, it is necessary that the MTBF be at least 400 years for less important equipment and at least 800 years for very important equipment.

The question is asked, for which transformer substation configuration does one conduct an evaluation of risk of failure to the insulation of individual elements of the substation. It is known that the configuration of a substation changes during the operation of the substation. The most critical circuit with a configuration composed of just one operational transmission line and one transformer during one year is used as the standard in the design practice. However, an analysis conducted for the 400 kV TS 400/110 kV "Novi Sad 3" transformer substation indicates that use of this circuit can lead to unrealistic results.

Fig. 1.20 displays the basic electric circuit of the 400 kV TS 400/110 kV "Novi Sad 3" transformer substation.

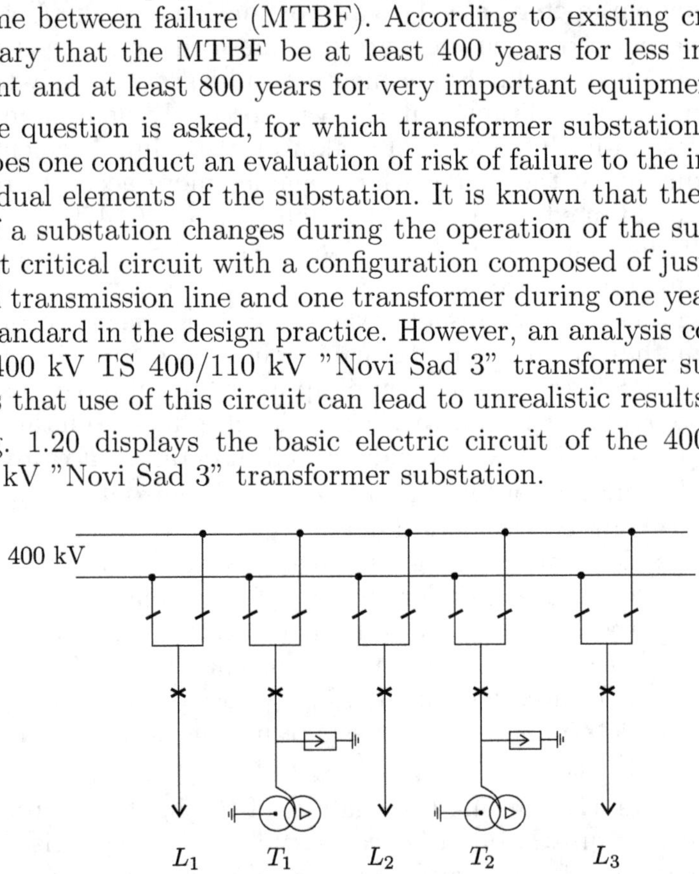


Fig. 1.20 - Basic electric circuit of the 400 kV TS 400/110 kV "Novi Sad 3" transformer substation

In order to evaluate the risk of failure to the insulation equipment, it is very important to determine the probability of individual configurations of transformer substations [55, 57, 58]. This conclusion is derived from the substation data which indicates instability in the configurations. Based on the statistics of the substation data, the frequency of individual configurations was reconstructed. By statistically processing the data for the considered substation, the conclusion was reached that the substation with a complete configuration (three transmission lines and two transformers operational) was most common in the substation during the year. The probability of a certain configuration was determined as the relation between the duration of that configuration and the total duration (1 year or 1 month).

Table 1.6 displays the definitions of individual configurations. Table 1.7 provides the probability of individual configurations determined on the basis of substation events for 1985, as well as for a period of 5 years (1986-1990). The mean probabilities, provided for a period of five years, are provided in parentheses. The year 1985 was selected as the year with the largest number of days with lightning (49 days). This is at the same time an extreme case in the past 50 years in the region of the considered transformer substation.

Table 1.6 - Analyzed configurations of the 400 kV TS 400/110 kV "Novi Sad 3" transformer substation

| Configuration | Status of Circuit Element | | | | |
|---|---|---|---|---|---|
| | L1 | L2 | L3 | T1 | T2 |
| I | ON | ON | ON | ON | ON |
| II | ON | OFF | ON | ON | ON |
| III | OFF | ON | ON | ON | ON |
| IV | ON | ON | OFF | ON | ON |
| V | ON | ON | ON | ON | OFF |
| VI | ON | OFF | OFF | ON | ON |

It can be seen from Tables 1.6 and 1.7 that configuration I, with all of its transmission lines and all transformers operational, is represented the most in the functioning of the considered substation. For this reason, calculations on the basis of the most critical configuration (only one transmission line and one transformer operational) leads to unrealistic results.

Table 1.8 provides a comparative display of the mean time between failure of at least one transformer in the function of the configuration and tower-footing resistance for lightning activity in 1985. The calculation was done for the variant with a built-in metal oxide surge arrester in the transformer bays (Fig. 1.20). The calculation methodology is displayed in [55, 57].

Table 1.7 - Substation configuration monthly probability distributions in 1989 and for a period of 5 years (1986-1990)

| Month | Probability of Configuration in % | | | | | |
|---|---|---|---|---|---|---|
| | $P_I$ | $P_{II}$ | $P_{III}$ | $P_{IV}$ | $P_V$ | $P_{VI}$ |
| January | 98 (91) | 2 (3) | 0 (1) | 0 (2) | 0 (1) | 0 (2) |
| February | 99 (93) | 1 (2) | 0 (2) | 0 (1) | 0 (1) | 0 (1) |
| March | 100 (90) | 0 (4) | 0 (2) | 0 (3) | 0 (1) | 0 (0) |
| April | 93 (88) | 0 (2) | 7 (5) | 0 (2) | 0 (1) | 0 (2) |
| May | 93 (87) | 0 (1) | 0(2) | 7 (6) | 0 (2) | 0 (2) |
| June | 100 (94) | 0 (2) | 0 (1) | 0 (1) | 0 (1) | 0 (1) |
| July | 89 (84) | 0 (4) | 0 (3) | 11 (4) | 0 (2) | 0 (3) |
| August | 86 (87) | 3 (2) | 0 (5) | 0 (2) | 11 (2) | 0 (2) |
| September | 47 (65) | 43 (13) | 0 (12) | 0 (4) | 10 (2) | 0 (4) |
| October | 96 (93) | 4 (3) | 0 (1) | 0 (0) | 0 (1) | 0 (2) |
| November | 97 (93) | 0 (1) | 0 (1) | 1 (2) | 0 (1) | 2 (2) |
| December | 99 (95) | 1 (1) | 0 (2) | 0 (0) | 0 (1) | 0 (1) |

Table 1.8 - Comparative display of the mean time between failure of at least one transformer in the function of the configuration and tower-footing resistance for lightning activity in 1985

| | $R_g$ (Ω) | | | | |
|---|---|---|---|---|---|
| | 10 | 20 | 30 | 40 | 50 |
| $\mathrm{MTBF_k}$ (year) | 14286 | 1042 | 289 | 156 | 91 |
| MTBF (year) | 26316 | 2703 | 1163 | 769 | 488 |

The symbols in Table 1.8 have the following meanings: $R_g$ – tower-footing resistance, $MTBF_k$ – mean time between failure of at least one transformer for the most critical configuration, $MTBF$ – mean time between failure of at least one transformer determined on the basis of the configurations provided in Table 1.6.

Based on the results displayed in Table 1.8, it can be concluded that the consideration of the variability of the configurations during operation provides a realistic picture of the risk of failure to the insulation of the substation equipment in relation to the calculation based on the critical configuration. This example also illustrates the significance of introducing a statistical approach in the methodology of calculation within the field of power systems.

**Example 1.8.** The influence of the numerical process for solving systems of regular differential equations and the size of the integration step on the accuracy of calculation is illustrated in the example of a grounding grid in an impulse regime [59].

The system of regular differential equations given by relation (1.13) describes the behavior of the grounding grid in the domain of time. This system can be solved by applying various numerical methods [7-9, 12-13]. Along with the selection of a numerical model, the accuracy of calculation is also significantly affected by the integration step. This step can be adaptive or fixed.

Every professional software tool for solving regular differential equations should have a built-in variable integration step which is adaptively adjusted during the process of numerically solving the equations. The reason for this approach lies in the fact that on smooth sections of the functions, the process of solving can be accelerated by selecting a larger integration step, while on sections with sudden variations it must be decreased in order to achieve the desired accuracy. Adjusting the integration step is done depending on the local discretization error, so that the evaluation of local errors appears as a basic problem in each integration step. When this error is evaluated, the integration step can be adjusted in several various ways, but always in accordance with whether the desired accuracy (tolerance) is achieved (increasing or maintaining the same step) or not (decreasing the step). The procedure for adaptive adjustment of the integration step is based on a twofold increase/decrease of the step and on the size of the truncation error [9].

In **Example 1.6**, the Runge-Kutta 4th order method is used where the error is proportional to the fifth order of the integration step. This subprogram is titled ode45 (Fig. 1.17) and is based on an adaptive integration step. The calculation results of the grounding grid impulse characteristics are displayed in Fig. 1.18.

For the purpose of illustrating the influence of a fixed integration step on the accuracy of calculation, the same problem was solved using the Simulink® module. This module allows the user to select from a larger number of methods in order to numerically solve systems of regular differential equations and to select the size of the fixed integration step. The methods for solving regular differential equations which are used by MATLAB® technical computing software are described in [13].

In contrast to **Example 1.6**, in which the grounding grid model is formed in the domain of time, by using Simulink® software, the grounding grid model is formed in state space.

Relation (1.13), following the application of the Laplace transform and consideration of the zero initial conditions, becomes:

$$s \cdot y(s) = -A^{-1}B \cdot y(s) + A^{-1} \cdot ii(s) \tag{1.44}$$

where: $s$ – Laplace operator.

After arranging expression (1.44), the final formula is obtained which is necessary for developing the model in state space:

$$y(s) = \frac{1}{s} \cdot (-A^{-1}B \cdot y(s) + A^{-1} \cdot ii(s)) \tag{1.45}$$

Based on formula (1.45), the Simulink® model titled *example8.mdl* is formed (Fig 1.21). The calculation and error analysis are conducted for the injection of an impulse current in the form of a Heaviside function. The parameters of calculation are defined in the m-file titled *parameters.m* (Fig 1.22).

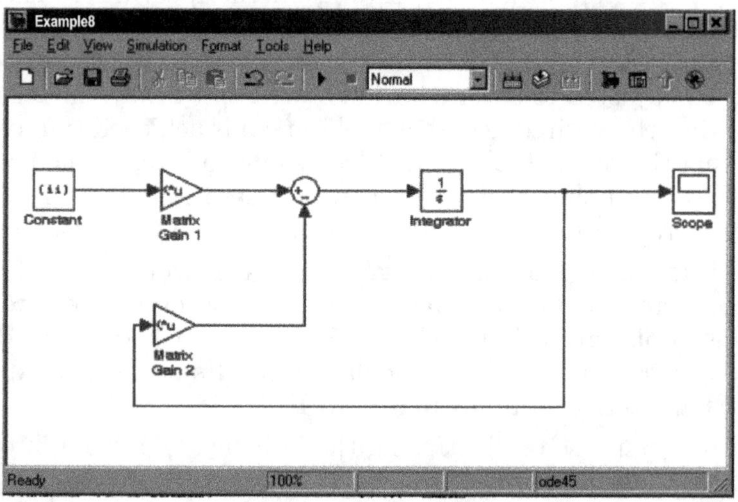


Fig. 1.21 - Appearance of the Simulink® grounding grid model in state space

In this m-file the parameters of the blocks displayed in Fig. 1.21 are defined. The Constant block takes the value of vector ii (Fig 1.23). The contents of the Matrix Gain 1 block represents the matrix $P = A^{-1}$ (Fig. 1.24), while the Matrix Gain 2 block takes the value of the matrix $P * B$, displayed in Fig. 1.25.

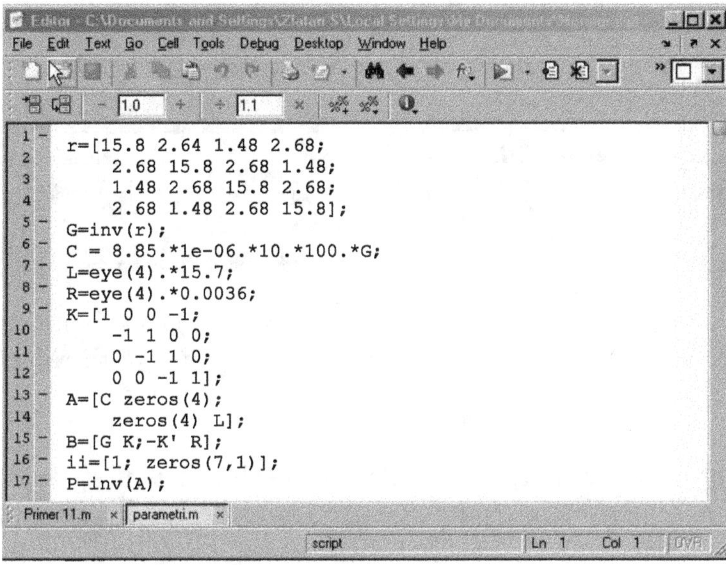


Fig. 1.22 – Appearance of *m*-file *parameters.m*

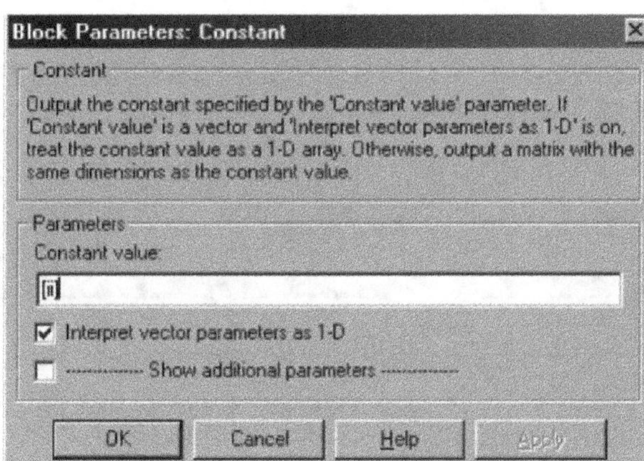


Fig. 1.23 – Window for adjusting the parameters of the Constant block

It is then necessary to select one of the available methods for numerically solving regular differential equations and define the simulation parameters. The option Simulation Parameters is selected for this purpose (Fig. 1.26). A fixed integration step is selected instead of a variable integration step.

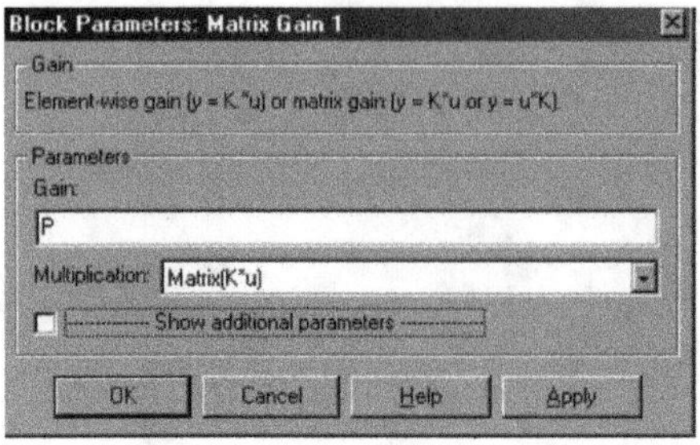


Fig. 1.24 - Window with the contents of the Matrix Gain 1 block

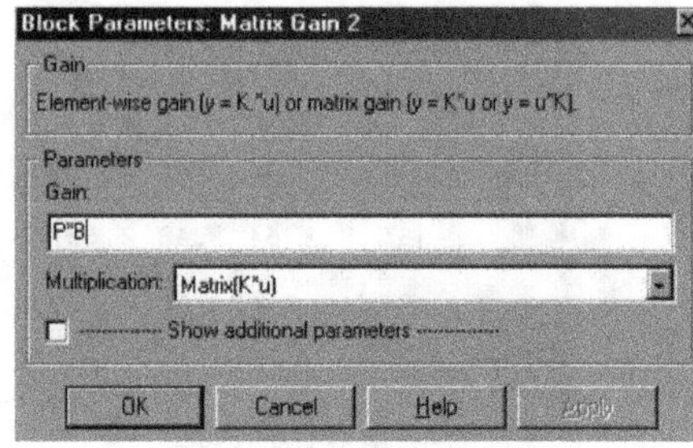


Fig. 1.25 - Window with the contents of the Matrix Gain 2 block

In order to obtain the calculation results, it is first necessary to reference the m-file titled *parameters* within the MATLAB® technical computing environment and then initiate the Simulink® model titled *example8*. Fig. 1.27 displays the dependencies of the relative value of maximum voltage at the location of current injection into the grounding

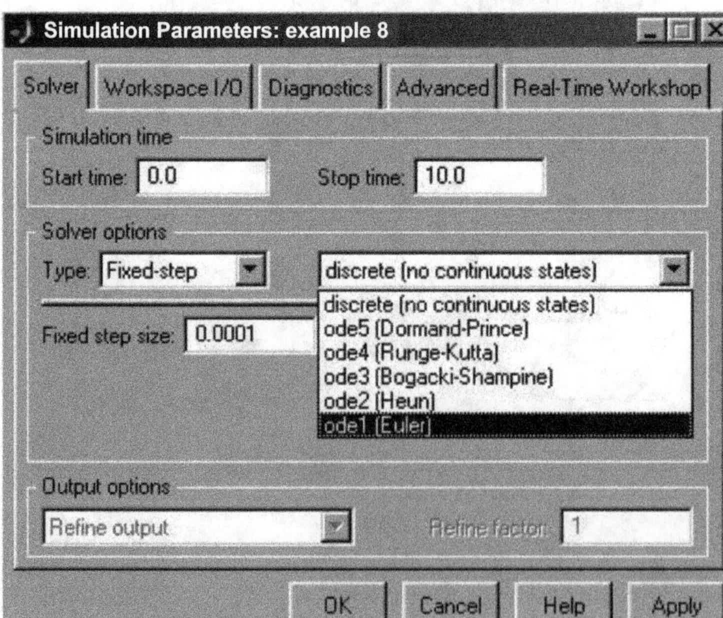


Fig. 1.26 – Appearance of Simulation Parameters options

grid in the function of an integration step for various numerical methods. The value of the fixed step of calculation varies within a range of 0.002 $\mu$s to 0.0275 $\mu$s.

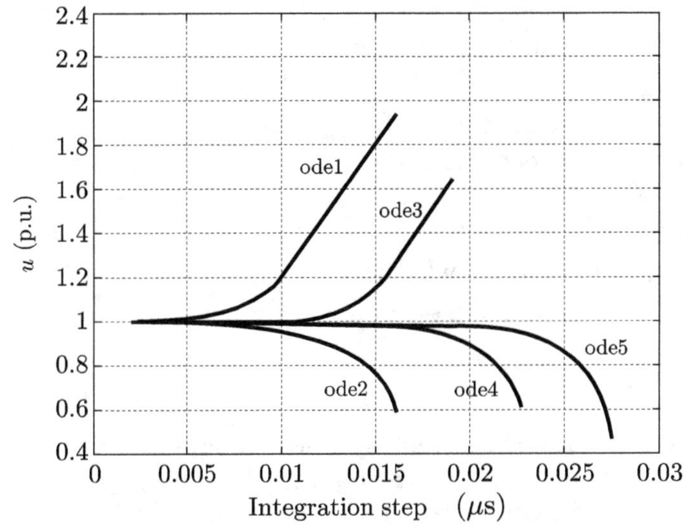


Fig. 1.27 - Dependency of the relative value of maximum voltage at the location of current injection into the grounding grid in the function of an integration step for various numerical methods.

The relative value of the maximum voltage at the location of current injection into the grounding grid is determined by using the following relation:

$$u = \frac{U_{\max\, i}}{U_{\max}} \tag{1.46}$$

where: $U_{\max\, i}$ – maximum value of voltage at the location of current injection into the grounding grid for an $i$ value of the integration step, $U\mathrm{max}$ – reference value of the maximum voltage at the location of current injection into the grounding grid determined for the smallest value of the integration step.

Based on the graphs displayed in Fig. 1.27, it can be concluded that the integration step has an important influence on the accuracy of the calculation results. With certain integration step values, so-called local instability can occur. Those values depend on the selected numerical method.

### 1.6.6. Simulation

#### 1.6.6.1 Aspects of performing simulation

Simulation is performed for the purpose of resolving the tasks related to the following aspects:

- safety,
- technical-economic analysis,
- analysis of sensitivities,
- approximation of data,
- optimization,
- acquiring new knowledge.

The aforementioned aspects are considered briefly below.

*a) Safety*

Safety is an important factor in designing, planning and exploitation of power systems. The carrying out of simulation leads to results which are analyzed in order reach a safe solution which in the social and economic sense. The social aspect relates to the protection of humans and animals, and is based on legal regulations and norms. The economic aspect implies the protection of material resources and bringing possible harm to within the smallest possible measure.

The significance of analyzing safety factors is illustrated in the examples for designing a general and special purpose lightning rod protection structure [60-64]. These examples are shown in Section 2.4.

*b) Technical-economic analysis*

A project, in its essence, is every process through which a goal or group of goals is reached. An investment project is a complex technical-technological, organizational, financial and legal undertaking with the goal of constructing and equipping a structure or structures which have been shown and proven, through previously performed investment studies, to be necessary for the investor or as such provide the investor with a profit.

Performing simulation enables the investor to, within the most delicate phase, the phase of forming the conception, obtain the proper project documentation and studies based on which a final decision can be made on whether or not invest in the project.

*c) Sensitivity analysis*

During simulation it is very important to attempt to obtain answers to the following questions:

- What will happen if the working conditions change (e.g. temperature, pressure, humidity)?
- How will uncontrollable factors affect the calculation results?
- How sensitive is one factor to a change in another factor?

The answers to these questions can be obtained by performing a sensitivity analysis. This analysis is conducted by carrying out calculation multiple times in such a way that in each group of calculations only the value of one parameter is changed while the values all other parameters remain unchanged. One of the alternative possibilities comes in the form of conducting experiments instead of calculations. However, it is most often difficult or impossible to isolate a change in just one of the parameters of interest. Furthermore, experiments are a valuable, but expensive source of information and results, especially when performed under actual conditions.

Conducting an sensitivity analysis makes it possible to realize two goals:

- The first goal being to show the effect of individual parameters on the calculation results. For the case when it can be concluded that a certain parameter has absolutely no effect on the results, the model can then be simplified, which is illustrated in **Example 1.2**.
- The second goal of this analysis relates to the testing of the model. If it is known, based on the physical model and physical laws, that variable $A$ affects variable $B$, and the results of the analysis show no indication of such, then that is a sign that a modification to the model must be made.

The procedure for conducting a sensitivity analysis is displayed in the following example.

**Example 1.9.** This example shows the methodology of calculating the line flashover rate due to lightning discharges in grounding sections on a 400 kV transmission line. The line flashover rate has been determined for six typical configurations of a transmission line tower grounding grid. The configuration of the grounding grid, location where the discharge impulse current enters the grounding grid and the characteristics of uniform soil are varied with the goal of quantifying their effect on the line flashover rate.

Defining the parameters of the equivalent circuit of the line represents the first step in calculating the line flashover rate. Fig. 1.28 shows the corresponding equivalent circuit of a 400 kV line with parameters determined according to the procedure shown in [26]. For the purpose of analyzing the grounding grid model for the line flashover rate, the grounding grid impulse model was used with its equivalent parameters $R_1$, $R_2$ and $L$ being determined according to the procedure described in **Example 1.5**.

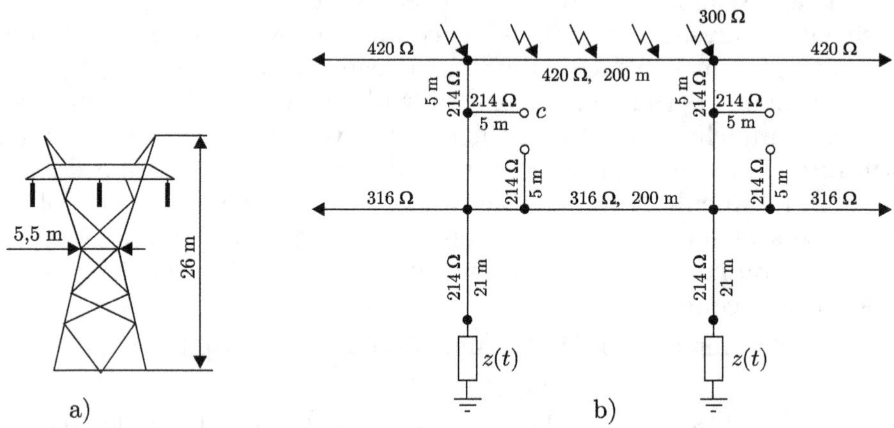


Fig. 1.28 – Overhead 400 kV line,
a) Tower configuration $Y$, b) Equivalent circuit for lightning performance estimation

The calculation of the line flashover rate is determined with the assumption that the line has identical characteristics for its entire length, meaning the same tower geometry and grounding grid impulse impedance. For this reason it is sufficient to consider lightning discharges along just one span. Direct strikes to the ground wire or tower cause flashovers on the line. In order to simulate the random effect of striking location on grounded sections during simulation of a flashover, it is necessary to vary the lighting strike location along the span. Instead of a continuous location change, it is sufficient to consider just a few points on an identical span.

The annual number of flashes along the length part $\Delta d_i$ (km) causing flashover to the line is calculated from the expression:

$$N_{1\,\mathrm{year},\Delta \mathrm{d_i}} = \frac{n_{1\,\mathrm{year},100\,\mathrm{km}}}{100} \cdot \Delta d_i \cdot \frac{P_i + P_{i+1}}{2} \tag{1.47}$$

where: $P_i(P_{i+1})$ – flashover probability to the line if the lightning hits point $i$ $(i+1)$, $n_{1\,\text{year},100\,\text{km}}$ – annual number of lightning flashes to the line per 100 km. This parameter is determined using the following expression:

$$n_{1\,\text{year},100\,\text{km}} = 0,1 \cdot N_g \cdot (b + 28 \cdot h_{ev}^{0,6}) \tag{1.48}$$

where: $N_g$ – annual ground discharge density per km$^2$; $b$ – effective line width (m), $h_{ev}$ – mean height of the ground wire (m).

The total annual number of the line flashover due to flashes along the whole span length $\Delta d$ (km) is:

$$N_{1\,\text{year},\Delta\text{d}} = \sum_{i=1}^{i=m-1} N_{1\,\text{year},\Delta d_i} \tag{1.49}$$

where $m$ – number of equidistant discharge points. The total annual line flashover rate per 100 km is defined as:

$$N_{1\,\text{year},100\,\text{km}} = \frac{100}{\Delta d} N_{1\,\text{year},\Delta d} \tag{1.50}$$

Fig. 1.29 shows six typical configurations of the transmission line tower grounding grid. In configurations II, IV, V and VI the discharge impulse current is injected through point 1 or point 2.

Grounding grids are formed from copper conductors with a 50 mm$^2$ round cross section. As can be concluded on the basis of Fig.1.29, grounding grid configurations II, III and IV have identical lengths. The same conclusion can be made for grounding grid configurations V and VI. The uniform soil resistivity equals $\rho = 50$ Ωm, 100 Ωm and 300 Ωm, respectively. The assumed value of the relative dielectric constant of soil amounts to $\varepsilon_r = 10$.

A sensitivity analysis for the total annual line flashover rate per 100 km on the grounding grid configuration, discharge impulse current injection location and uniform soil resistivity was conducted in the following manner [18, 65]:

- Grounding grid configuration is varied (I to VI) for one discharge impulse current injection location and one value of uniform soil resistivity,
- Location of impulse current injection is varied for one configuration and one value of uniform soil resistivity,

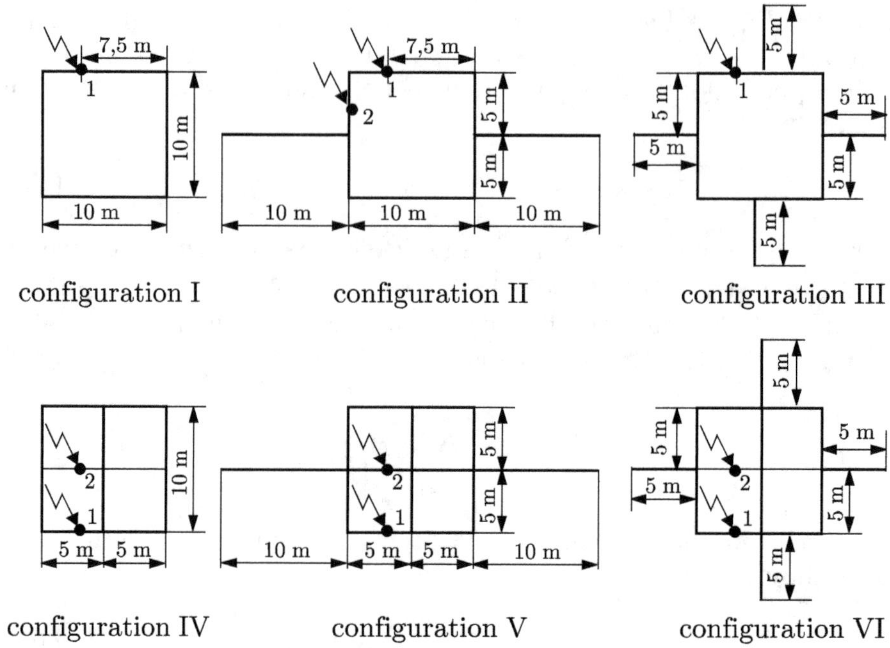


Fig. 1.29 - Typical configurations of the transmission line tower grounding grid

– Value of uniform soil resistivity is varied for one configuration and one location of impulse current injection into the grounding grid.

Calculations were carried out for the following data: $b = 11.6$ m; $h_{ev} = 26$ m, $N_g = 2.81$ 1/km$^2$, basic insulation level 1425 kV. The results of the simulation are displayed in Fig. 1.30.

The symbol for the configuration $ij$ on Fig. 1.30 represents the $i$ grounding grid configuration ($i = \text{I} - \text{VI}$) for current injection location $j$ ($j = 1, 2$).

Based on the calculation results displayed in Fig. 1.30, the following conclusions can be made:

– The total annual line flashover rate per 100 km varies from a maximum value for grounding grid configuration I1 to a minimum value for grounding grid configuration V1;
– Injection of discharge impulse current through point 1 of the grounding grid when compared to point 2 of the grounding grid leads to an unknown increase in the total annual line

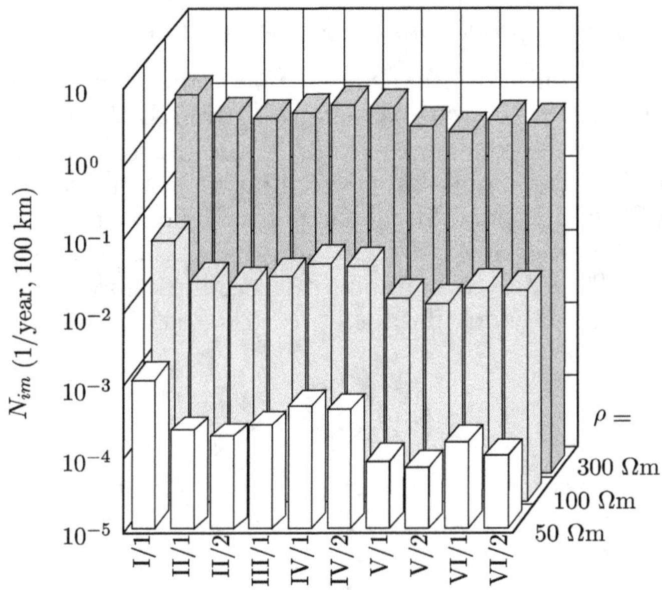


Fig. 1.30 - Dependence of the line flashover rate per 100 km in the function of the grounding grid configuration, discharge impulse current injection location and uniform soil resistivity

flashover rate per 100 km;

- Increasing the value of the uniform soil resistivity leads to an increase in the total annual line flashover rate per 100 km.

### *d) Approximation of data*

Based on the collection of data obtained through measurements or calculations, it is possible to test the basic case and obtain information about the borderline cases for which the developed model is valid. For this reason, an approximation of the line data is made (regression analysis). This is the procedure of determining the function which best approximates a certain collection of data. The function which is used for approximating the collection of data may be linear, polynomial, degree, exponential, logarithmic, reciprocal, etc.

There are several programs which can be used to conduct the approximation of data. Illustrated below is the approximation of data using MATLAB® technical computing software.

*d.1 Approximation of data using a polynomial line*

Approximation of data using a polynomial line in MATLAB® technical computing software is achieved through the polyfit function which uses the least squares method. According to that method, the coefficients of the polynomial are determined through reducing the sum of squares of the differences in all the points. The residual in each point is defined as the difference between the value of the polynomial and the value of data. The polyfit command determines the coefficient of the polynomial through which data is approximated and has the following syntax [13, 41]:

$$>> \text{polyfit}(x, y, n)$$

where $x$ is the vector with the $x$ coordinates of the point, $y$ is the vector with the $y$ coordinates of the point and $n$ is the degree of the polynomial.

**Example 1.10.** Table 1.9 displays the values of a conventional impulse impedance of a square shaped grounding grid $Z_k$ in the function of the relation of the soil resistivity of the upper and lower layers $\rho_1/\rho_2$ [66]. The thickness of the upper layer is $H = 1$ m and the amplitude of the discharge impulse current is 5 kA, while the nominal rise time of the impulse current front and nominal time to half-value of the impulse current wave is 1 $\mu$s and 20 $\mu$s, respectively. The soil resistivity of the lower layer is $\rho_2 = 100$ Ωm.

Table 1.9 - Values of conventional impedance of a square shaped grounding grid $Z_k$ in the function of the relation of the soil resistivity of the upper and lower layers $\rho_1/\rho_2$, for $H = 1$ m, $I = 5$ kA, $T_1 = 1$ $\mu$s, $T_2 = 20$ $\mu$s and $\rho_2 = 100$ Ωm

| $\rho_1$ (Ωm) | $\rho_2$ (Ωm) | $\rho_1/\rho_2$ | $Z_k$ (Ω) |
|---|---|---|---|
| 10 | 100 | 0,1 | 3,16 |
| 50 | 100 | 0,5 | 6,64 |
| 200 | 100 | 2 | 11,41 |
| 1000 | 100 | 10 | 25,64 |

Approximate the provided collection of second-order polynomial points. Draw the given points and line which is obtained through approximation using the polynomial.

Approximate the given collection of points with the second-degree polynomial and obtain a drawn graph by entering the following commands into the MATLAB® command line:

```
>> x = [0.1  0.5  2  10];
>> y = [3.16  6.64  11.41  25.64];
>> p = polyfit(x, y, 2)
p =
-0.2359  4.5783  3.4396
>> xp = 0.1 : 0.01 : 10;
>> yp = polyval(p, xp);
>> plot(x, y, 'o', xp, yp)
>> axis([0  12  0  30])
```

For the purpose of forming a line which is obtained through approximation using the polynomial, a new vector $xp$ was formed with elements that increase by a small step. The vector $xp$ is references the polyval function in order to form vector $yp$ which represents the line obtained through approximation using the polynomial.

The approximation formula for calculating the conventional impulse impedance has the following form:

$$Z_k = -0.2359 \cdot \left(\frac{\rho_1}{\rho_2}\right)^2 + 4.5783 \cdot \left(\frac{\rho_1}{\rho_2}\right) + 3.4396 \tag{1.51}$$

Fig. 1.31 displays the line obtained through approximation of the given collection of points using the polynomial of the defined relation (1.51).

*d.2 Approximation of data using non-polynomial functions*

The approximation of data can be achieved using functions which are non-polynomial. The most commonly used are degree, exponential, logarithmic and reciprocal functions. The procedure for approximating data using these functions is performed through the polyfit command and is reduced to where the function is written in the form of a linear polynomial:

$$y = mx + b \tag{1.52}$$

and within the polyfit command $x$ and $y$ are replaced by the arguments displayed in Table 1.10.

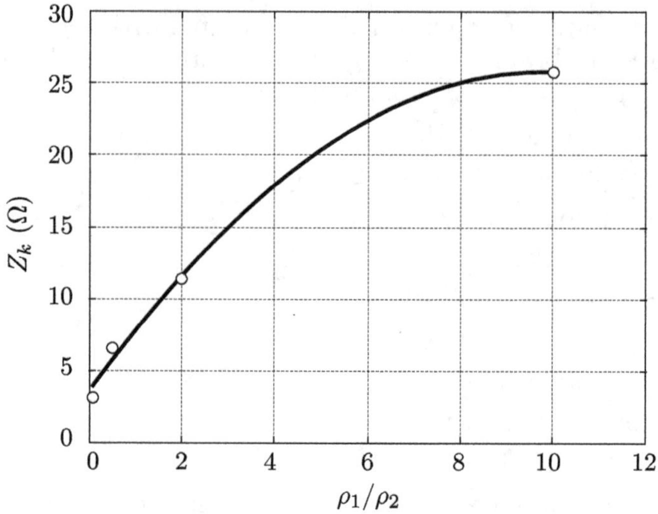


Fig. 1.31 - Line obtained through approximation of the given collection of points using the second-degree polynomial of the defined relation (1.51)

For the given collection of points, the function which provides the best approximation can be envisaged by graphing the data for several combinations of linear and logarithmic divisions of dependent and independent variables. If the data on a certain graph corresponds to the actual line, the corresponding function will provide the best approximation. Table 1.11 presents the functions which provide the best approximation of data when the data corresponds to the actual line for the determined combination of linear and logarithmic divisions of variables $x$ and $y$.

*d.3 Interpolation/extrapolation*

When a line by which data is approximated crosses through all points from the collection of points, the problem is reduced to determining the values between the points. This process is called interpolation. Expanding the application of the model outside of the measured or calculated values is conducted using the process of extrapolation. In this manner, on the basis of existing data, a trend of occurrences can be defined without further use of the developed model.

If a polynomial function is used for interpolation, it is necessary for the degree of the polynomial to be one less than the number of

Table 1.10 - Form of polyfit command for approximation of data using non-polynomial functions

| Function | | Linear Polynomial | Form of Polyfit Command |
|---|---|---|---|
| degree | $y = bx^m$ | $\ln(y) = m\ln(x) + \ln(b)$ | $\text{polyfit}(\log(x), \log(y), 1)$ |
| exponen-tial | $y = be^{mx}$<br>$y = 10e^{mx}$ | $\ln(y) = mx + \ln(b)$<br>$\log(y) = mx + \log(b)$ | $\text{polyfit}(x, \log(y), 1)$<br>$\text{polyfit}(x, log10(y), 1)$ |
| logari-thmic | $y = m\ln(x) + b$<br>$y = m\log(x) + b$ | $y = m\ln(x) + b$<br>$y = m\log(x) + b$ | $\text{polyfit}(log(x), y, 1)$<br>$\text{polyfit}(log10(x), y, 1)$ |
| reciprocal | $y = \dfrac{1}{mx + b}$ | $\dfrac{1}{y} = mx + b$ | $\text{polyfit}(x, 1./y, 1)$ |

Table 1.11 - Functions which provide the best approximation of data when the data corresponds to the actual line for the determined combination of linear and logarithmic divisions of variables $x$ and $y$

| x | y | function |
|---|---|---|
| linear | linear | linear |
| logarithmic | logarithmic | degree |
| linear | logarithmic | exponential |
| logarithmic | linear | logarithmic |
| linear | linear (graph $1/y$) | reciprocal |

points in the collection of data. However, such a polynomial will not always provide a good approximation of the data between the known points. Better interpolation can be achieved if instead of all points from the collection of data, one considers just a few points in close proximity to the place where interpolation is required. This method, which is used by a larger number of lower-degree polynomials, and which is only valid within a limited domain, is known as *spline* interpolation.

One-dimensional interpolation can be performed within the MATLAB® technical computing environment using the interp1 function which has the following form:

$$yi = \text{interp1}(x, y, xi, \text{'method'}) \tag{1.53}$$

where: $y_i$ – interpolated value, $x$ – monotone vector with the horizontal coordinates of the provided points, $y$ – vector with vertical coordinates of the provided points, $x_i$ – value (scalar or vector) which is being interpolated, and 'method' - method of interpolation written as text (not mandatory). Some of the methods of interpolation are:

- 'nearest' – interpolation on the basis of the closest point
- 'linear' – linear interpolation
- 'spline' – partial cubic spline interpolation
- 'pchip' – partial Hermite interpolation

The following example illustrates the process of interpolation using MATLAB® technical computing software.

**Example 1.11.** For the data provided in Table 1.9, **Example 1.10,** interpolate the values between the given points using the nearest, linear, spline and pchip methods. Draw a graph for each interpolation method. Display the given points and line which is obtained through interpolation on each graph.

Fig.1.32 displays the m-file Example11.m which when executed carries out the process of interpolation using the aforementioned methods and graphical displays.

Initiating the aforementioned m-file prompts the results displayed in Fig. 1.33 to be obtained. Each of the four graphs relates to one of the methods of interpolation (*nearest*, *linear*, *spline* and *pchip*). Each of the graphs displays the line obtained through interpolation and the collection of points which represent the given data.

*e) Optimization*

One of the most important tasks of the engineering analysis is to figure out the best way in which to complete the given task. This process contains a certain form of optimization. Optimization methods are methods for finding the optimum solution and can be reduced to the process of minimization and maximization of the target functions. These methods are studied in detail within the field of mathematics called operations research.

MATLAB® technical computing software contains the following commands for optimization [13, 41]:

- fminunc (method of unconstrained optimization),
- lsqnonlin, lsqcurvefit (optimization methods based on the least squares method),

```
x=[0.1 0.5 2 10]
y=[3.16 6.64 11.41 25.64]
xi=[0.1:0.01:10]
yin=interp1(x,y,xi,'nearest')
yil=interp1(x,y,xi,'linear')
yis=interp1(x,y,xi,'spline')
yip=interp1(x,y,xi,'pchip')
subplot(221)
plot(x,y,'o',xi,yin)
xlabel('x')
ylabel('y')
title('nearest')
subplot(222)
plot(x,y,'o',xi,yil)
xlabel('x')
ylabel('y')
title('linear')
subplot(223)
plot(x,y,'o',xi,yis)
xlabel('x')
ylabel('y')
title('spline')
subplot(224)
plot(x,y,'o',xi,yip)
xlabel('x')
ylabel('y')
title('pchip')
```

Fig. 1.32 - *m*-file Example11.m which when executed carries out the process of interpolation using the aforementioned methods and graphical displays

– fsolve (method for solving a nonlinear system of equations),
– fmincon, fminimax, fgoalattain and fseminf (optimization method with constraints).

*f) Acquiring new knowledge*

New knowledge about the system is acquired through the process of simulation. For example, the results of the sensitivity analysis can show corresponding trends which are not normal or indicate that the assumptions on which the development of the model were based are improper.

The capability of a simple modification has been built into the simulation model. Calculation of lightning overvoltages in lines and transformer substations of varying voltage levels represents a typical example. The equivalent circuit formed for a 400 kV line (Fig. 1.28) can be modified in a simple way for analyzing the overvoltage protection for lines of a lower voltage level (110 kV, 35 kV, 10 kV). Modification of the parameters of the equivalent circuit implies the consideration of

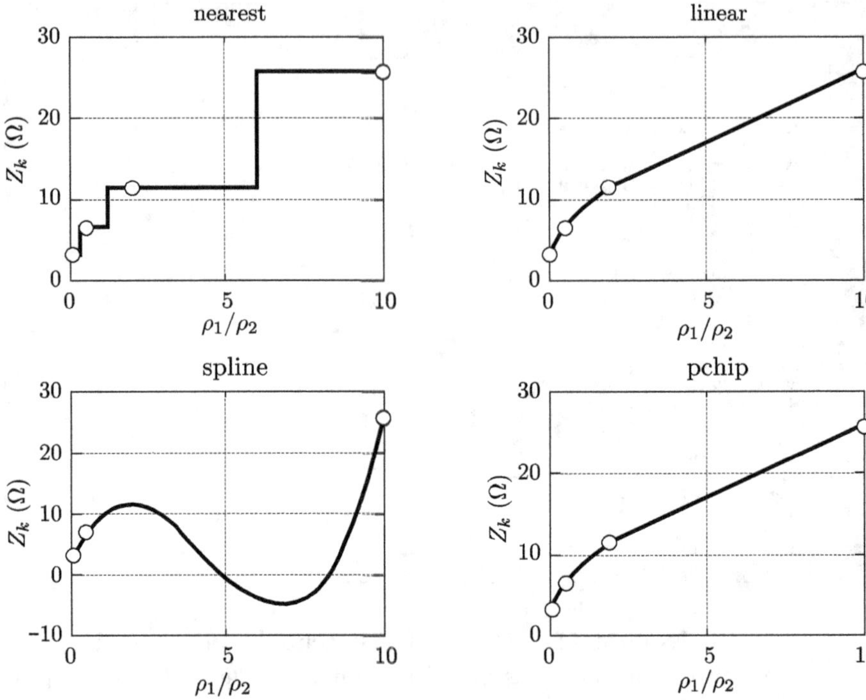


Fig. 1.33 - Interpolation of the values of conventional impulse impedance for a square shaped grounding grid $Z_k$ in the function of the relation of soil resistivity of the upper and lower layers $\rho_1/\rho_2$ using the *nearest*, *linear*, *spline* and *pchip* methods

the geometry and type of tower, characteristics of insulation equipment, characteristics of surge arresters, etc. Along with a simple modification of the equivalent circuit, through the application of one of the existing programs, it is possible to design overvoltage protection for lines and transformer substation of various voltage levels.

### 1.6.6.2 Advantages of simulation

The advantage of simulation is that it is a more economic way of obtaining answers when compared to conducting experiments. Experiments are a valuable, but expensive source of information and results, especially when performed under actual conditions. Based on the results of simulation, a "what if" analysis can be formed. Simulation helps in correcting or confirming a hunch when solving a certain problem. This is especially important when two team members have contradictory hunches about the behavior of a system under a given set of conditions.

The advantages achieved through conducting simulation are:

- increased flexibility,
- increased precision,
- formation of new results,
- improvement of existing results,
- increased understanding of the essence of the model.

## 1.7. Supplement: GIC – software tool for calculating the grounding grid impulse characteristics

### 1.7.1. Purpose of the program

The knowledge of the grounding impulse characteristics (GIC) is of a great significance for a proper lightning performance estimation of transmission lines and high-voltage substations.

The development of a simple and accurate grounding grid model in the impulse regime is a complex task due to the large number of various parameters. The constructive parameters of the grounding grid, electrical characteristics of the soil, as well as the shape, amplitude and location of injection of the impulse current have the most important influence on these characteristics.

Based on the developed mathematical model described in Section 1.6.3 the software tool **GIC** (**G**rounding **I**mpulse **C**haracteristics) was created for calculating impulse characteristics. Formed using the modern MATLAB® technical computing software and Visual C++, this user-oriented software tool is designed to analyze grounding grid impulse characteristics in a way which is suitable for the design practice.

This software tool enables the user to:

- Calculate the distribution of voltage and current along a grounding grid when an impulse current in injected;
- Calculate the impulse impedance of a grounding grid for cases without and with soil ionization;
- Calculate the parameters of a model of an equivalent grounding grid parameters, which when incorporated into any of the

existing programs for calculating lightning overvoltages can more accurately estimate the reliability of transmission lines of various voltage levels. Models can be formed for cases without and with soil ionization, which is exceptionally important for the configuration of a grounding grid in soil with a large resistivity in a zone with significant lightning activity;
- Calculate the effective length of a grounding grid. This parameter is important from the aspect of the design practice and is represented by the length of the grounding grid from which the entire discharge impulse current is released into the earth. The consideration of this effect enables an economic solution in designing linear and complex grounding grids;
- Elements of the expert system contained within the program enable the optimal process of modeling a grounding grid from the aspect of precision and duration of calculation. The concept of the software tool also enables less experienced users to reach an economical solution in designing a grounding grid and overvoltage protection of power and electronic equipment. The capabilities of this software tool are illustrated in the calculation examples.

### 1.7.2. Mathematical model of a grounding grid in an impulse regime

The GIC software tool is based on the corresponding mathematical models of grounding grids in an impulse regime which are described in detail in [17], as well as in the examples provided in Section 1.6.2. The sequence of calculation is provided below.

### 1.7.3. Sequence of calculation

The application of the model of a grounding grid in an impulse regime, encompassing the effect of ionization, consists of the following steps:

1. In the first step, the impulse model of a grounding grid for a case without soil ionization is formed in accordance with the process defined in **Example 1.1**, Section 1.6.3.
2. The arrester currents from individual elements of the grounding grid are established for the defined impulse current which is injected into the grounding grid.

3. The critical strength of the electrical field $E_k$ is adopted.
4. A verification of the condition for the occurrence of soil ionization ($E_i > E_k$) is performed. If this condition is fulfilled, the effective cross sections of all grounding grid elements are determined according to expression (1.29).
5. A matrix of self and mutual resistances is formed for the effective cross sections of grounding grid elements determined in the previous step. On the basis of this matrix, a model is defined for the grounding grid in an impulse regime.
6. A calculation of impulse impedance of the grounding grid is performed while taking into consideration the soil ionization.
7. In accordance with the method of equivalent lines, the elements of equivalent grounding grids are replaced by corresponding lines.
8. An equivalent circuit of the line is formed for analyzing lightning overvoltages.
9. Through the application of any of the existing programs, an estimation is made of the voltage level at the tower due to a lightning overvoltage.

### 1.7.4. Description of the software

The GIC software tool was created using Visual C++ and MATLAB® technical computing software. By double-clicking on the GIC icon, a command menu is opened which enables communication between the user and the computer. The command menu consists of two menu groups.

The first group contains the standard commands for Windows-based software. This group consists of the Save and Open commands which have the same function as in any other software created for the Windows operating system. The second group is the system menu through which one of the following options can be selected:

1. Define parameters for calculating grounding grid impulse characteristics;
2. Execute calculation;
3. Display calculation results.

A detailed explanation of the options from the first and second group is provided below.

#### 1.7.4.1 Defining the parameters for calculating grounding grid impulse characteristics

This option represents an important part of the software and includes the following operations:

– selecting the mathematical model,
– defining the configuration of the grounding grid,
– entering the constructive parameters of the grounding grid,
– defining the soil characteristics,
– selecting the impulse current parameters,
– defining the duration of calculation,
– selecting the display of output variables.

Fig. 1.34 shows the command window for defining the aforementioned parameters.

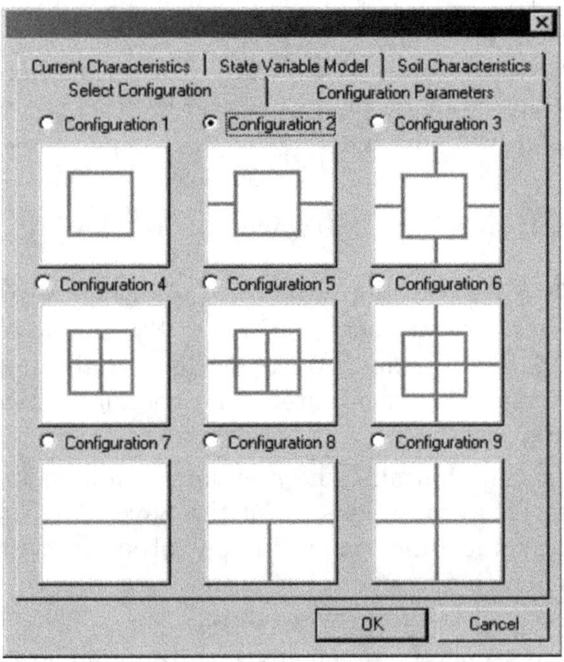


Fig. 1.34 - Command window for defining the parameters for calculation of impulse characteristics

Various methodological approaches have been used in the papers published on the problems related to impulse characteristics of grounding grids. Therefore, the first step is to select the mathematical model. It is possible to select one of the following models:

- a linear grounding grid model with frequency independent parameters,
- a frequency dependent grounding grid model,
- a nonlinear grounding grid model.

The formation of a linear grounding grid model with frequency independent parameters and the methodology of calculation are discussed in detail in [17]. The formation of a frequency dependent model is based on the modeling of grounding grid elements using the corresponding equivalent circuit with lumped parameters. A detailed display of the analytical expressions of individual parameters is provided in [16].

A nonlinear grounding grid model refers to the encompassing of occurrences of soil ionization around the grounding grid. This nonlinear effect occurs during the injection of a large-amplitude impulse current and its release from a grounding grid placed in poorly conductive soil. The procedure for forming a nonlinear grounding grid model is displayed in detail in [22, 23].

Defining the configuration of the grounding grid is represented by the selection of the prototype configuration. The term prototype configuration of the grounding grid implies a typical configuration, while taking into consideration the local effect of the grounding grid impulse characteristics [21], which represents a sufficiently accurate equivalent to the actual grounding system. Some prototype configurations are displayed in Fig. 1.34.

After selecting the corresponding configuration, the constructive parameters of the grounding grid are entered. For the selected configuration 2 on Fig. 1.34, the command window in Fig. 1.35 appears. In this dialog box the following grounding grid parameters are defined:

- Lengths of sides $a$, $b$ and $c$. Lengths of sides $d$, $e$ and $f$ are identical to $a$, $b$ and $c$ respectively;
- Type of material. The available options are copper (Cu) or galvanized steel tape (Fe-Zn);
- Surface of cross-section. A round cross-section of the grounding grid is assumed. Grounding grids with a rectangular cross-section are approximated using a round cross-section of a corresponding surface;
- Depth of grounding grid placement.

Within this dialog box it is also necessary to define the node into which the impulse current is injected. The potential nodes into which

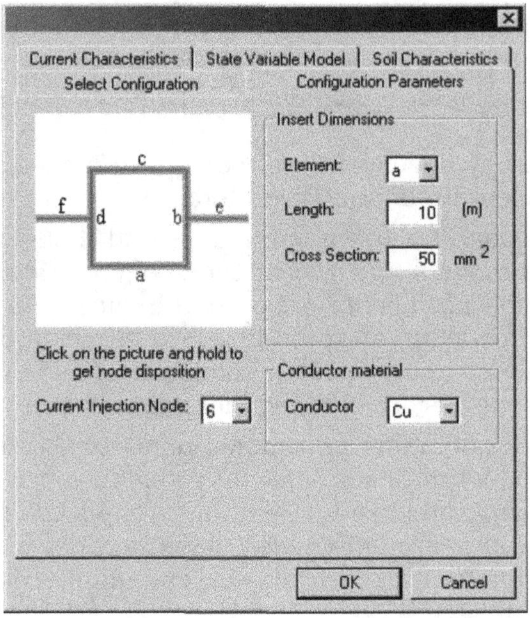


Fig. 1.35 - Command window for entering constructive parameters of the grounding grid for the selected prototype configuration 2

the impulse current may be injected are displayed on the prototype configuration (Fig. 1.36).

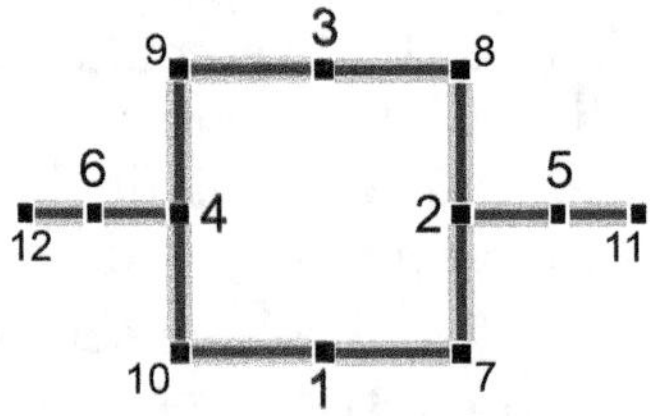


Fig. 1.36 – Display of nodes for prototype configuration 2

Before calculating the impulse characteristics of the grounding grid, it is necessary to define the following soil characteristics:

- soil resistivity,
- relative dielectric constant of soil.

The value of the relative dielectric constant of soil depends on the type of soil and falls within the range of 1-100. For a normal soil moisture content, this constant has a value of 10.

The form and size of the impulse current have an exceptional influence on the impulse characteristics of the grounding grid. This software enables the calculation of impulse characteristics during the injection of the following impulse currents:

– impulse current in the form of a Heaviside function,
– impulse current in the form of a double-exponential function.

The calculation of impulse characteristics of the grounding grid during injection of an impulse current in the form of a Heaviside function is performed with the goal of determining the parameters of an equivalent grounding grid [17, 18].

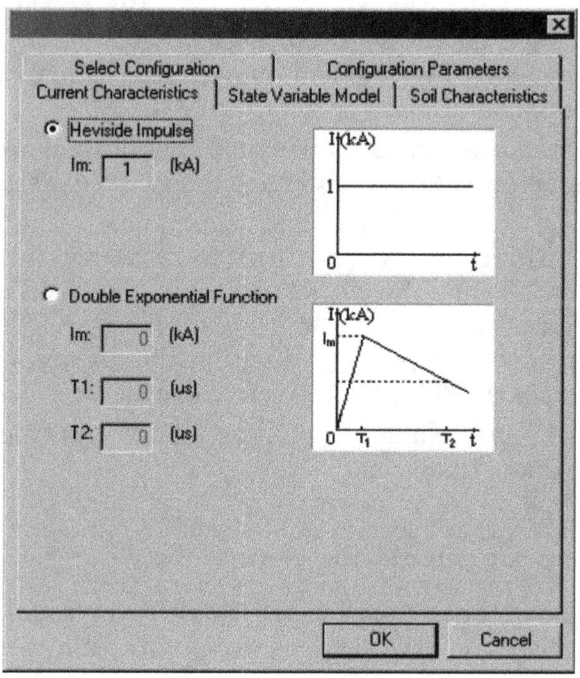


Fig. 1.37 – Selection of impulse current parameters

In order to define the other form of the impulse current, it is necessary to enter data about the amplitude of the impulse current ($I_m$) and the nominal rise time of the impulse current front $T_1$ and nominal time to half-value of the impulse current wave $T_2$ (Fig. 1.37). Due to the simplified graphical display, this impulse current has been linearized on the nominal rise time of the impulse current front and nominal time to half-value of the impulse current wave.

One of the parameters of calculating the impulse characteristics of a grounding grid is also the duration of calculation. This parameter is defined by the instants of time which mark the beginning and end of calculation. The duration of calculation depends on the form of the impulse current and, for normal values $T_1$ and $T_2$, does not exceed ten $\mu$s.

#### 1.7.4.2 Executing the calculation

The calculation of grounding grid impulse characteristics is achieved using MATLAB® technical computing software. The sequence of calculation depends on the selected mathematical model. For a linear model with frequency independent parameters, a matrix of self and mutual resistances is formed.

Based on the resistances matrix, matrices of conductances, capacitances and inductances of grounding elements are determined.

In the case of a frequency dependent model, the parameters of the grounding grid are determined for the dominant frequency of the transient process, defined by the expression $1/(4 \cdot T_1)$. For parameters which are determined in this way, the calculation of grounding grid impulse characteristics is performed within the domain of time, in the same manner as in the case of the linear frequency independent model.

A nonlinear grounding grid model entails the calculation steps described in Section 1.6.3.

#### 1.7.4.3 Display of calculation results

The results of the calculation are shown graphically and textually. The graphical display shows changes in the following values:

- voltage in the nodes of the grounding grid,
- injected impulse current and currents in the grounding grid branches,
- impulse impedance of the grounding grid and impulse coefficient of the grounding grid for cases with and without the occurrence of soil ionization around the grounding grid.

Fig. 1.38 shows the command window for selecting the display of output values. For the purpose of reviewing the display, a limitation of up to five lines per graph with a maximum of two graphs has been emplaced. The maximum output variables are displayed simultaneously on the graphs.

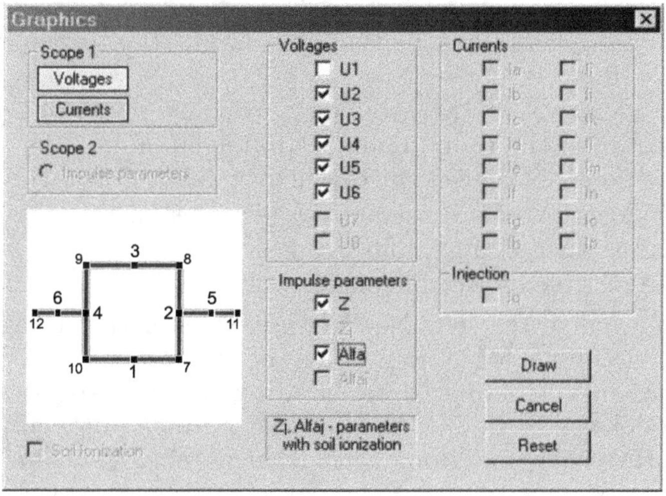


Fig. 1.38 – Command window for selecting the display of output values

The results in the textual presentation are arranged in two columns with the possibility of an additional dialog box for displaying the parameters of an equivalent grounding grid, obtained from the response at the point of injection of the impulse current in the form of a Heaviside function. The first column contains data about the conventional impedance of the grounding grid and the conventional coefficient of the grounding grid in the case without soil ionization. The second column displays the identical output variables for the case with soil ionization. If ionization is not registered this column contains no results.

Results in the additional column exist only in the case of calculation of grounding grid impulse characteristics during injection of an impulse current in the form of a Heaviside function. Parameters which are obtained in such a manner are incorporated simply into some of the existing programs for calculating lightning overvoltages. The graphical and textual displays of calculation results are illustrated in the next Section.

### 1.7.5. Calculation examples

Through the application of the described software tool, a calculation of the impulse characteristics of grounding grid configuration 2 has been performed. In all calculations the following parameters are assumed:

- the length of each of the elements of the grounding grid is $l = 10$ m,
- all elements of the grounding grid are made of copper, the surface of the round cross section is 50 $\text{mm}^2$,
- the depth of placement of the grounding grid is $h = 0.5$ m,
- the location of injection of impulse current into the grounding grid is node 6 (Fig. 1.36),
- the relative dielectric constant of the soil is $\varepsilon_r = 10$.

An analysis of the impulse characteristics of grounding grid configuration 2 was done for various values of uniform soil resistivity, form and amplitude of the impulse current. A frequency independent grounding grid model was used. During injection of the impulse current in the form of a Heaviside function, time changes of voltage in individual nodes were determined for $\rho = 100$ Ωm (Fig. 1.39a), as well as for the impulse impedance of the grounding grid and impulse coefficient of the grounding grid (Fig. 1.39b).

Based on the graphs shown on Figs. 1.39a and 1.39b, it can be concluded that the injection of steep impulse currents leads to a significant increase in impulse impedance of the grounding grid when compared to power frequency resistance. This kind of trend is the consequence of a developed influence of inductance of elements on the impulse characteristics of the grounding grid in soil with a small soil resistivity. According to Fig. 1.39b, the maximum value of the impulse coefficient of the grounding grid for the stated calculation parameters is 3.56.

The parameters of an equivalent grounding grid are contained within the textual presentation of results (Fig. 1.40). The considered grounding grid, in the aforementioned case, can be likened to an $RL$ circuit with parameters amounting to $R_1 = 15.247$ Ω, $R_2 = 5.95$ Ω and $L = 19.89$ μH.

Fig. 1.41 shows the distribution of amplitude of the voltage in individual nodes of grounding grid configuration 2 depending on the form of the impulse current for $\rho = 100$ Ωm. Along with an impulse current in the form of a Heaviside function, an impulse current in the form of a double-exponential function, amplitude 5 kA, was also used.

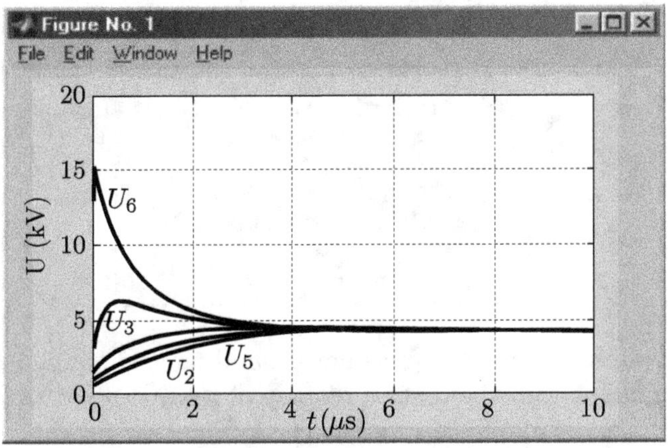


a)

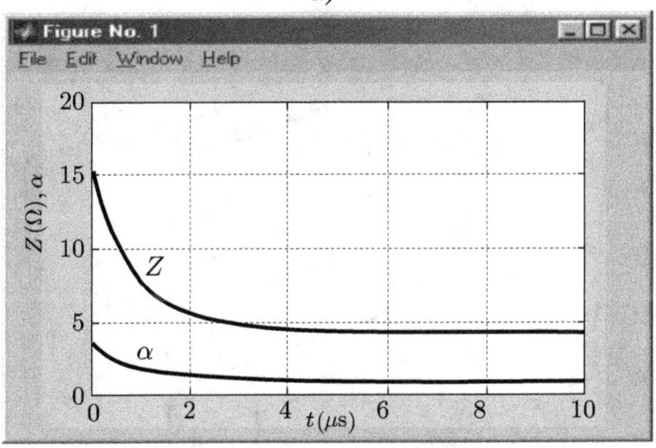


b)

Fig. 1.39 - a) Voltage waveshapes in individual nodes of grounding grid configuration 2, b) Waveshapes of impulse impedance of the grounding grid and impulse coefficient of the grounding grid; $\rho$ = 100 Ωm, impulse current in the form of a Heaviside function

The nominal rise time of the impulse current front is 1 $\mu$s or 5 $\mu$s, while the nominal time to half-value of the impulse current wave is 30 $\mu$s.

The graphs shown in Fig. 1.41 indicate the following conclusions:

– The change of voltage is most apparent in the first several meters and depends on the form of the applied impulse current. The voltage in node 4 located 5 m from the injection point

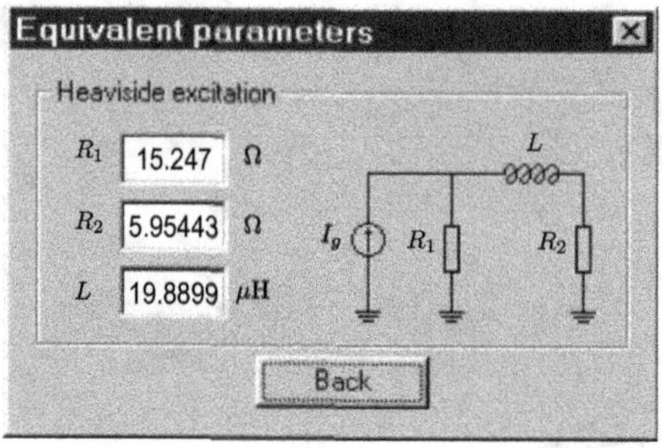


Fig. 1.40 - Parameters of the equivalent grounding grid configuration 2 contained within the textual display of results; $\rho = 100\ \Omega\text{m}$, impulse current in the form of a Heaviside function

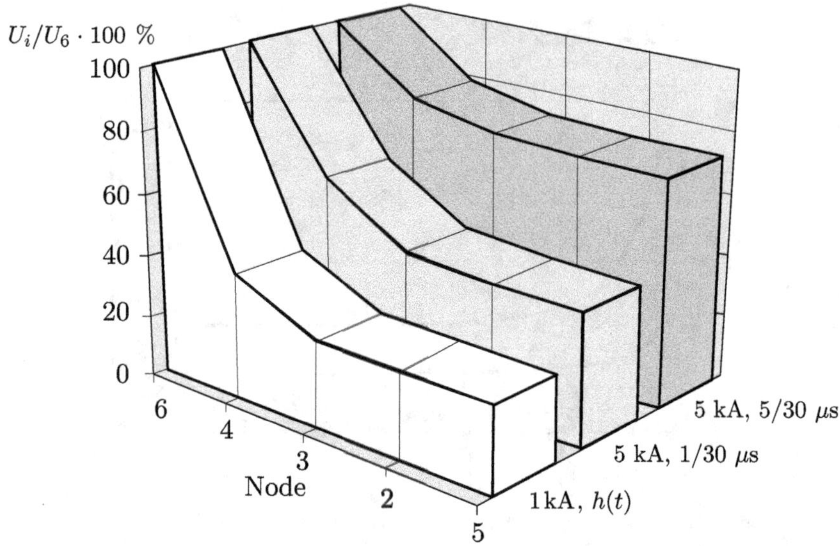


Fig. 1.41 - Distribution of amplitude of voltage in individual nodes of grounding grid configuration 2 depending on the form of the impulse current for $\rho = 100\ \Omega\text{m}$

in the aforementioned case amounts to 41% for an impulse current in the form of a Heaviside function, 80.3% for an impulse current in the form 5/30 $\mu$s. This kind of trend of change in voltage is the consequence of the inductances of grounding grid elements which have a greater influence on the flow of the transition process as the injected impulse current gets steeper.

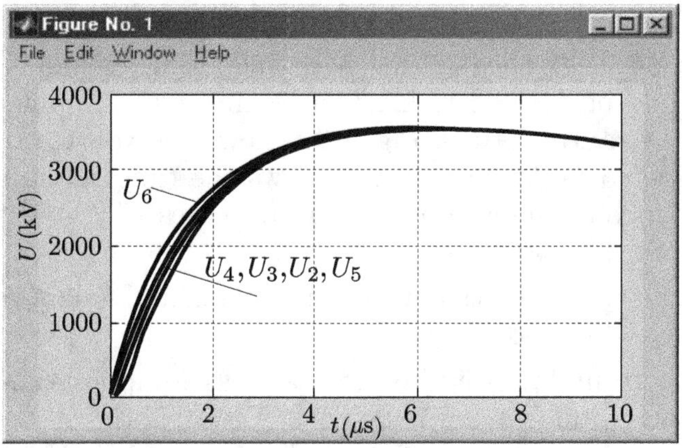


a)

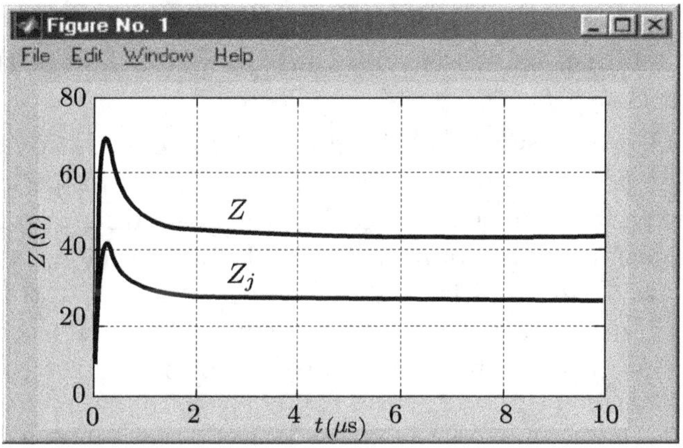


b)

Fig. 1.42 - a) Voltage time changes in individual nodes of grounding grid configuration 2, b) Time changes of impulse impedance of the grounding grid for cases without ($z$) and with the occurrence of soil ionization ($z_j$); $\rho = 1000$ Ωm, $I_m = 100$ kA, $T_1/T_2 = 5/30$ μs

– After a certain distance the voltages become practically constant, regardless of the form of the applied impulse current. In the considered case, this distance amounts to 10 m, measured from the injection point.

In this calculation there was no noticed occurrence of the effect of soil ionization, which is the consequence of small values of uniform soil resistivity and amplitude of the impulse current. For that reason, the calculation was repeated for $\rho = 1000\ \Omega$m and an impulse current amplitude of $I_m = 100$ kA. The form of the impulse current is defined by the ratio 5/30 $\mu$s. Fig. 1.42 shows the voltage time changes in the considered nodes. Fig. 1.42b shows the time changes of the impulse impedance of the grounding grid for cases with and without the occurrence of soil ionization.

Based on the graphs shown in Figs. 1.42a and 1.42b, the following conclusions can be made:

- The grounding grid becomes almost equipotential in soil with a large uniform soil resistivity (Fig. 1.42a).
- The occurrence of soil ionization around the grounding grid causes a decreased impulse impedance of the grounding grid. This decrease depends, at first, on the values of the product of the uniform soil resistivity and the amplitude of the impulse current ($\rho \cdot I_m$).

Fig. 1.43 provides a textual display of the calculation results. By comparing the corresponding values of the impulse impedance of the grounding grid and impulse coefficient of the grounding grid for the case with soil ionization (right column on Fig. 1.43) to the case without soil ionization (left column on Fig. 1.43), the positive effect of this nonlinear phenomenon can be seen clearly. In the considered case, the reduction of values of the impulse impedance of the grounding grid amounts to 27%.

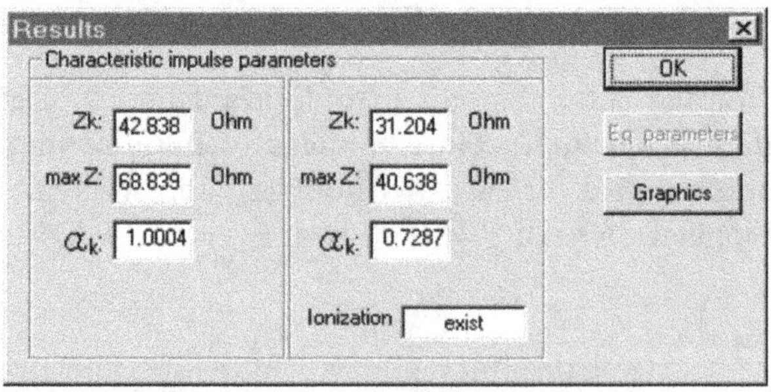


Fig. 1.43 - Textual display of the calculation results for the case with soil ionization around grounding grid configuration 2; $\rho = 1000\ \Omega$m, $I_m = 100$ kA, $T_1/T_2 = 5/30\ \mu$s

### 1.7.6. Input-output values and possible errors

This section provides input-output values and their units. Input-output values are provided in the line which appears before the description of the mathematical model. The values marked with $*$ are those which the user must provide within the individual dialog boxes within the software. Additionally, the possible errors are described along with their corresponding messages.

#### 1.7.6.1 Input-output variables

$\sigma_i$ – specific electrical conductance of the conductor material ($1/\Omega$m)

$S^*$ – surface of the cross section (mm$^2$), this value is internally calculated in the software in m$^2$

$l^*$ – length of the conductor (m)

$R_i$ – series resistance of the conductor ($\Omega$)

$r_{ij}$ – mutual resistance of conductors $i$ and $j$ ($\Omega$)

$r_{ii}$ – self-resistance of conductor $i$ ($\Omega$)

$n$ – total number of grounding grid elements

$R_s$ – power frequency grounding resistance of the grounding grid ($\Omega$)

$I$ – $n$-dimensional vector of units

$G$ – matrix of self and mutual conductances ($1/\Omega$)

$C$ – matrix of self and mutual capacitances of elements of the grounding grid ($\mu$F)

$\varepsilon$ – dielectric constant of soil ($\mu$F/m)

$\varepsilon_r^*$ – relative dielectric constant of soil

$\rho^*$ – uniform soil resistivity ($\Omega$m)

$\mu$ – magnetic permeability of soil ($\mu$H/m)

$L_{ij}$ – mutual inductance of conductors $i$ and $j$ ($\mu$H)

$L_s$ – self-inductance of linear conductor ($\mu$H)

$L_i$ – internal inductance of conductor ($\mu$H)

$L_e$ – external inductance of conductor ($\mu$H)

$\mu_m$ – magnetic permeability of conductor material ($\mu$H/m)

$[K]$ – incidence matrix

$[K]^t$ – transposited incidence matrix

| | |
|---|---|
| $[0]$ | – zero matrix |
| $[u_1]$ | – vector of the voltage of the grounding grid nodes (kV) |
| $[i_2]$ | – vector of the current of the grounding grid inductive branches (kA) |
| $[I_C]$ | – vector of the current of equivalent current sources connected to the grounding grid nodes (kA) |
| $[e_L]$ | – vector of electromotive force of equivalent voltage sources in the grounding grid branches (kV) |
| $T_1^*$ | – nominal rise time of the impulse current front ($\mu$s) |
| $T_2^*$ | – nominal time to half-value of impulse current wave ($\mu$s) |
| $I_m^*$ | – amplitude of discharge impulse current injected into the grounding grid (kA) |
| $h(t)^*$ | – impulse current in the form of a Heaviside function (kA) |
| $J_g(t)$ | – waveshape of discharge impulse current injected into the ground grid (kA) |
| $t$ | – time ($\mu$s) |
| $t_0^*$ | – time of beginning of calculation ($\mu$s) |
| $t_k^*$ | – time of end of calculation ($\mu$s) |
| $z(t)$ | – grounding grid impulse impedance in time t ($\Omega$) |
| $Z_k$ | – conventional impedance ($\Omega$) |
| $\alpha(t)$ | – grounding grid impulse coefficient in time $t$ |
| $\alpha_k$ | – grounding grid conventional coefficient |
| $E_k$ | – critical strength of the electric field (kV/m) |
| $a_{ei}$ | – effective radius of the $i$ element (m) |
| $h^*$ | – depth of grounding grid placement (m) |
| $R_1$, $R_2$ | – equivalent grounding resistances ($\Omega$) |
| $L$ | – equivalent grounding inductance ($\mu$H) |

#### 1.7.6.2 Possible Errors

*Reading the cross section of the conductor in $m^2$.* Providing the cross section of the grounding grid conductor in this way will lead to incorrect values of the grounding grid parameters being provided and, in accordance with that, also incorrect results. The cross section of the conductor must be read in $mm^2$.

*Reading the duration of calculation times in s.* The times for the beginning of calculation $t_0$ and end of calculation $t_k$ must be provided in $\mu$s.

**Example:**

$$t_0 = 0; \quad t_k = 10, \quad \text{and not} \quad t_k = 10e - 06 \text{ s}.$$

*Calculation of the parameters of an equivalent grounding grid based on an impulse current in the form of a double-exponential function.* The parameters of an equivalent grounding grid $R_1$, $R_2$ and $L$ can be determined only on the basis of the calculation of the voltage response of the grounding grid to an impulse current in the form of a Heaviside function. For this reason, the software has been programmed to prevent the possibility of calculating the parameters of an equivalent grounding grid on the basis of an impulse current in the form of a double-exponential function.

*Calculation of grounding grid impulse characteristics with consideration of the effect of ionization in an impulse current in the form of a Heaviside function.* Considering that an impulse current in the form of a Heaviside function assumes a step function of amplitude 1 kA, the occurrence of soil ionization around the grounding grid conductor is not possible. For this reason, the software has been programmed to prevent the possibility of entering an impulse current in the form of a Heaviside function for a previously selected nonlinear grounding grid model.

*Improper selection of grounding grid model.* The analysis of the effect of soil ionization on grounding grid impulse characteristics often leads to an improper grounding grid model. The user most often makes a change to the amplitude of the discharge impulse current and/or the value of the uniform soil resistivity, and as such the first model – a linear model, remains active, which assumes the lack of nonlinear phenomena such as the effect of soil ionization. As a consequence of an improper model, following the calculation, a message appears indicating a lack of soil ionization, despite the fact that under the given conditions it most often exists.

Because of this it is necessary to select the proper grounding grid model in the first step of analysis of the effect of soil ionization on grounding grid impulse characteristics. From the available options of *State Variable Model*, it is necessary to select a nonlinear grounding grid model.

### 1.7.6.3 Messages when working with the software tool GIC

*Reading the depth of grounding grid placement h.* Considering the values of depth for grounding grid placement which are used in practice, these values are limited to a range of 0.01-10 m. When entering a value $h$ which is outside of this range, a message will appear indicating the recommended values for this parameter.

*Reading the relative dielectric constant of soil $\varepsilon_r$.* Its value depends on the type of soil and falls in the range of 1-100. For a normal soil moisture content, this constant has a value of 10. When entering a value $\varepsilon_r$ which is outside of this range, a message will appear indicating the recommended values for this parameter.

*Reading the uniform soil resistivity $\rho$.* The program has envisaged the smallest value for the uniform soil resistivity to be one tenth of Ωm, referring to the most common configuration of soil. Due to practical reasons, the upper value of uniform soil resistivity is limited to 2000 Ωm. For this reason, when entering a value $\rho$ which is larger than 2000 Ωm, a message appears indicating the aforementioned limitation.

*Message indicating the occurrence of soil ionization.* Within the textual display of calculation results, the following two messages may appear:

**1. Ionization exists**. This message indicates the occurrence of soil ionization around the grounding grid conductor.

**2. Ionization doesn't exist**. This message indicates a lack of soil ionization around the grounding grid conductor, which regularly occurs during the injection of a discharge impulse current with a small amplitude into a grounding grid placed in soil with a small uniform soil resistivity.

However, as mentioned in Section 1.7.6.2, this message may be the consequence of improper selection of a grounding grid model.

*Reading the value of cross section S.* Due to practical reasons, the value of the cross section has been limited to a range of 5–1000 mm$^2$.

*Reading the value of length of element l.*

Due to practical reasons, the value of length has been limited to a range 0.5–200 m.

*Reading parameters of current in the form of a double-exponential function.* The maximum value of current may be selected from a range of 0.01–10000 kA. Time $T_1$ is limited to 0.1–30 $\mu$s, while the limitation for $T_2$ is 1–100 $\mu$s.

*Reading the duration of calculation.* The beginning of calculation is fixed at $t_0 = 0\ \mu s$, while the time for the end is calculated together with the length of calculation, and may be selected from a range of 0.01—50 $\mu$s.

## 1.7.7. Using windows commands

### 1.7.7.1 Opening the program

The program is opened by placing the mouse pointer over the GIC icon and double-clicking the left mouse button.

### 1.7.7.2 Appearance of screen with windows commands

By activating the program, a screen appears with the following menus:

- File
- View
- Run
- Window
- Help.

The following icons also appear on the screen:

- Open an existing project
- Save
- Set parameters of the configuration
- Computation parameters
- Run calculation
- Graphic parameters
- Helper
- Results
- Help
- About.

A detailed description of the aforementioned options is provided below by using the Help menu.

### 1.7.7.3 Description of individual commands using the help menu

 **File -> Open**

The keyboard shortcut for this command is Ctrl+O.

This option activates a dialog box for opening an existing file from previously recorded parameters for calculating a configuration – GIC Config File (*.gcf).

**File -> Save**

The keyboard shortcut for this command is Ctrl+S.

This option opens a dialog box in which the user can choose to record data into a configuration set, results display and graphical display (Fig. 1.44).

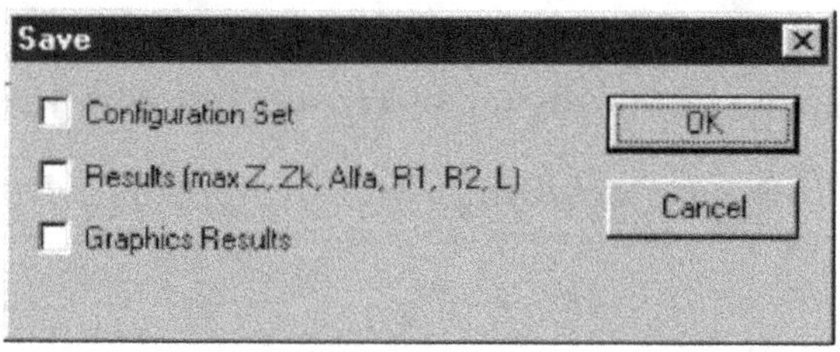


Fig. 1.44 - Dialog box for recording data into a configuration set, results display and graphical display

Three options are available:

- Configuration Set.
- Results (maxZ, Zk, Alfa, R1, R2, L).
- Graphics Results.

The selection of the option Configuration Set prompts the opening of a dialog box for recording a file in the format GIC Config File (*.gcf). Previously set parameters of a configuration can be recorded in this way. These files can be opened later, at which time only the configuration parameters will be read. The calculation parameters and results are not saved in this file, so when opening a *.gcf file, it is necessary to repeat the setting of time parameters of calculation as well as execution of the calculation itself.

The selection of the option Results (maxZ, Zk, Alfa, R1, R2, L) prompts the opening of a dialog box for recording a text file (*.txt). In this way the calculation results are saved in a regular text file which can be viewed using any text editing software.

Selection of the option Graphics Results prompts the opening of a dialog box for recording a bitmap file (*.bmp). This file contains the graphically represented results of the calculation.

**File -> Print**

The keyboard shortcut for this command is Ctrl+P.

This option opens the standard dialog box for printing documents.

**File -> Print Setup**

This option opens the standard dialog box for adjusting printing parameters.

**File -> Exit**

This option closes the application.

**View -> Toolbar**

Clicking on this item will cause the toolbar, meaning the icons which provide quick access to individual commands, to either appear or disappear.

**Toolbar**

The toolbar provides quick access to the menu options (Fig. 1.45).

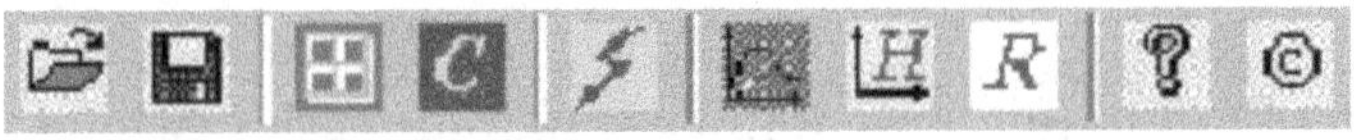

Fig. 1.45 – Toolbar

Fig. 1.46 shows the Toolbar with corresponding text.

**View -> Statusbar**

This option shows or hides the Statusbar.

 **Run -> Parameters**

This option opens a dialog box which enables the user to do the following:

- define the grounding grid configuration (Select Configuration),
- enter in the constructive parameters of the grounding grid (Configuration Parameters),
- select the impulse current parameters (Current Characteristics),
- select the mathematical model (State Variable Model),
- define the characteristics of the soil (Soil Characteristics).

The option Select Configuration enables the user to select a prototype configuration. The term prototype configuration of the grounding grid implies a typical configuration, while taking into consideration the

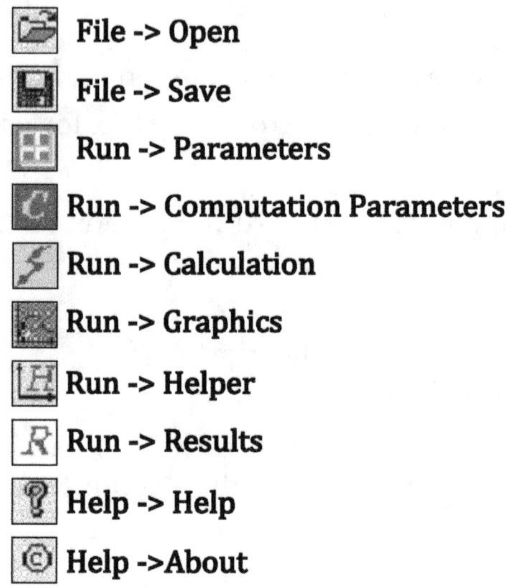


Fig. 1.46 – Toolbar with corresponding text

local effect of grounding grid impulse characteristics, which represents a sufficiently accurate equivalent to the actual grounding system. A prototype configuration is shown on Fig. 1.34.

Following the selection of the corresponding configuration, it is necessary to enter in the constructive parameters of the grounding grid using the option Configuration Parameters. Fig. 1.47 shows the dialog box for entering the constructive parameters for grounding grid configuration 2.

This option enables the user to define the following grounding grid parameters:

- Lengths of sides $a$, $b$ and $c$. The sides for which lengths can be defined are highlighted in red on the grounding grid configuration. The lengths of the remaining sides ($d$, $e$ and $f$) are identical to sides $a$, $b$ and $c$, respectively. First, select the desired side in the Element drop-down list, then enter its length in meters into the Length field.
- Surface of the cross section - Cross Section. A round cross section for the grounding grid is assumed. Grounding grids with a rectangular cross-section are approximated using a round cross-section of a corresponding surface;
- Type of material. The Conductor drop-down list provides the

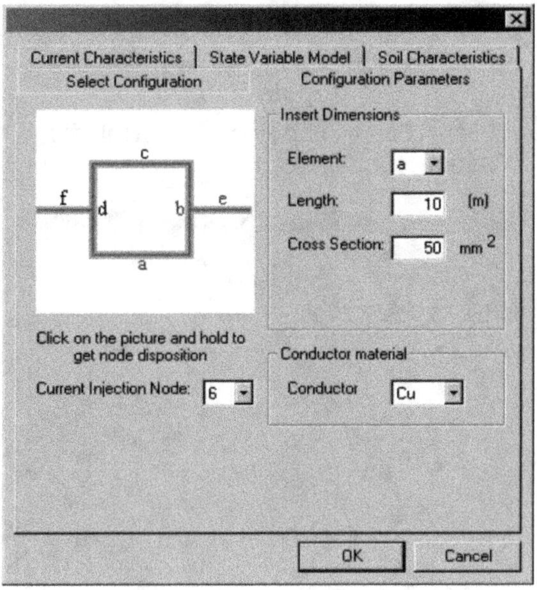


Fig. 1.47 - Dialog box for entering constructive parameters of grounding gird configuration 2.

following options: copper (Cu) or galvanized steel tape (Fe-Zn);

– The node into which the impulse current is injected – Current Injection Node. A display of the potential nodes into which the impulse current can be injected appears when the user clicks and holds down the left mouse button on the grounding grid configuration (Fig. 1.48).

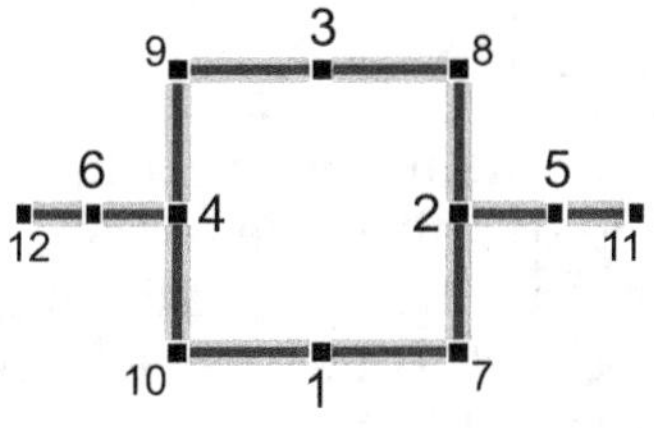


Fig. 1.48 – Display of potential nodes for injecting the impulse current

The shape and size of the impulse current have an exceptional influence on the impulse characteristics of the grounding grid. These values are defined using the Current Characteristics option (Fig. 1.49). This software enables the calculation of impulse characteristics during the injection of the following impulse currents:

- an impulse current in the form of a Heaviside function,
- impulse current in the form of a double-exponential function.

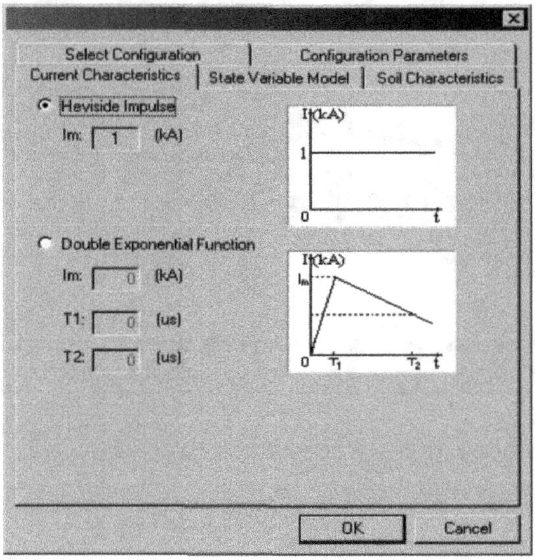


Fig. 1.49 – Dialog box for defining impulse current characteristics

The recommended value for the amplitude of an impulse current in the form of a Heaviside function $I_m$ is 1 kA. The calculation of grounding grid impulse characteristics during injection of this current is performed for the purpose of determining the parameters of an equivalent grounding grid.

In order to define the other form of the impulse current, it is necessary to enter data about the amplitude of the impulse current ($I_m$), as well as about the nominal rise time of the impulse current front $T_1$ and nominal time to half-value of the impulse current wave $T_2$. Due to the simplified graphical display, this impulse current has been linearized on the nominal rise time of the impulse current front and nominal time to half-value of the impulse current wave.

Using the State Variable Model option it is possible to select one of the following models (Fig. 1.50):

- a linear grounding grid model with frequency independent parameters (Frequency Independent Model),

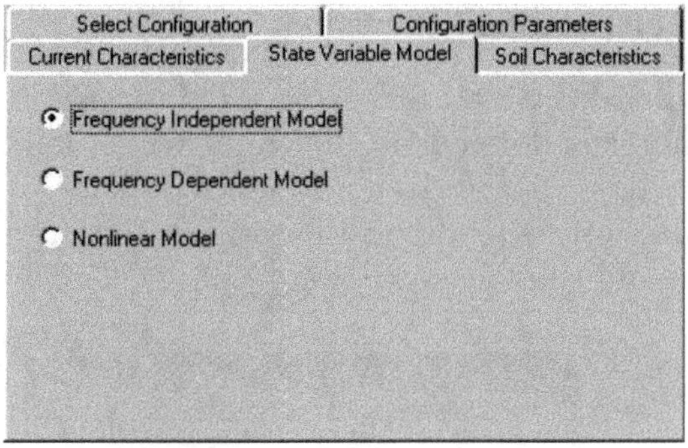


Fig. 1.50 – Dialog box for defining the mathematical model

- a frequency dependent grounding grid model (Frequency Dependent Model),
- a nonlinear grounding grid model (Nonlinear Model).

A nonlinear grounding grid model refers to the encompassing of occurrences of soil ionization around the grounding grid. This nonlinear effect occurs during the injection of a large-amplitude impulse current and its release from a grounding grid placed in poorly conductive soil.

Before calculating the grounding grid impulse characteristics it is necessary to select the Soil Characteristics tab in order to define the following soil characteristics (Fig 1.51):

- uniform soil resistivity (Specific Resistivity),
- relative dielectric constant of soil (Relative Dielectric Constant),
- depth of grounding grid placement (Depth of Burial).

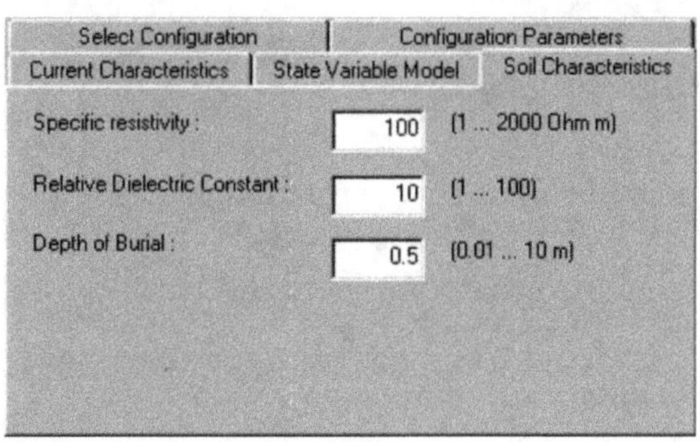


Fig. 1.51 – Dialog box for defining soil characteristics

**Run -> Computation Parameters**

This option opens a dialog box for defining the calculation duration. This parameter is defined by the times which mark the beginning $t_0$ (predefined as $t_0 = 0$), and end of calculation $t_k$ (Fig. 1.52). The duration of calculation depends on the form of the impulse current and for normal values does not exceed twenty $\mu$s.

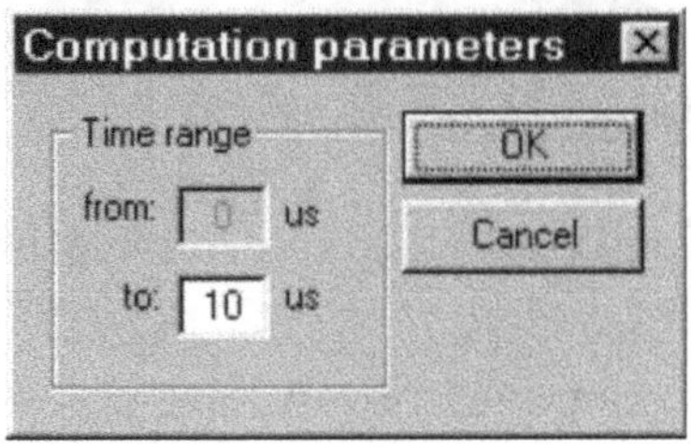


Fig. 1.52 – Dialog box for defining duration of calculation

**Run -> Calculation**

By selecting this option the calculation of grounding grid impulse characteristics is initiated (Fig. 1.53). The calculation, in accordance with the provided parameters and capabilities of the computer, lasts anywhere form fifteen seconds to a few minutes.

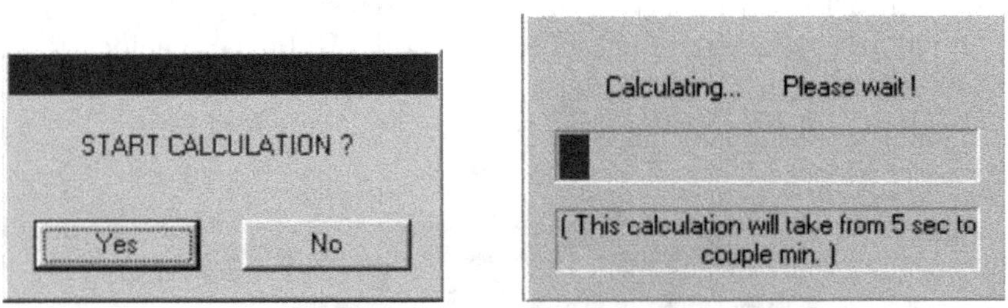


Fig. 1.53 - Beginning of calculation of grounding grid impulse characteristics

After the calculation is complete, the Results dialog box will appear.

**Run -> Results**

Results in the textual display are arranged into two columns in the first dialog box. An additional dialog box appears for calculations of the parameters of an equivalent grounding grid. The first column contains data about the conventional impedance of the grounding grid and the conventional coefficient of the grounding grid in the case of no soil ionization. The second column displays the identical output variables for the case with soil ionization. If ionization is not registered this column contains no results (Fig. 1.54).

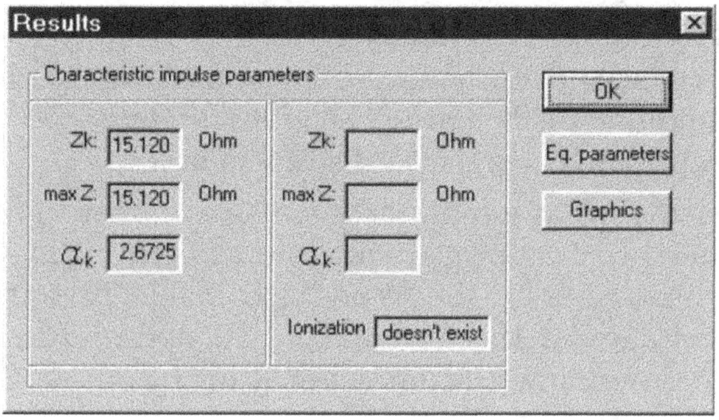


Fig. 1.54 – Textual display of calculation results

For the calculation of grounding grid characteristics during injection of an impulse current in the form of a Heaviside function, the *Eq.* parameters button on Fig. 1.54 becomes active and when clicked on opens a window with the calculated parameters of an equivalent grounding grid for the selected prototype configuration: $R_1$, $R_2$ and $L$ (Fig. 1.55).

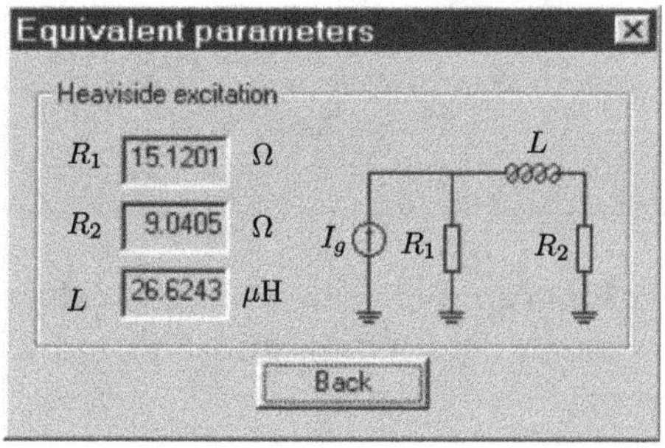


Fig. 1.55 - Display of parameters of an equivalent grounding grid for the selected prototype configuration

Additionally, the Results dialog box can also be used to open a separate Graphics dialog box for creating graphical displays.

###  Run -> Graphics

After executing the calculation, the user can define a graphical display of output values. The graphical display can contain the following output values:

- voltage time changes in the grounding grid nodes,
- time changes of the injected impulse current and currents in the grounding grid branches,
- time changes of the impulse impedance of the grounding grid and impulse coefficient of the grounding grid for cases without and with the occurrence of soil ionization.

Along with that, a simultaneous display of two graphs (Scope1 and Scope2) has been envisaged with a maximum of five lines for the first graph and two lines for the second graph. The first graph (Scope1) displays the voltages in the grounding grid nodes (Fig. 1.56) or the currents in the grounding grid branches (Fig. 1.57).

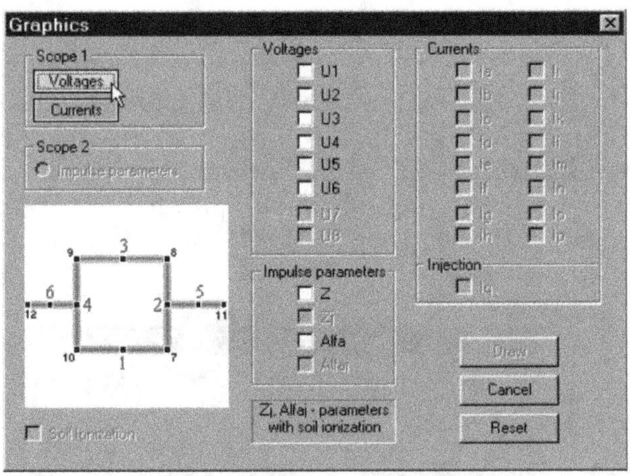


Fig. 1.56 - Dialog box for graphing voltages in grounding grid nodes, impulse impedance of the grounding grid and impulse coefficient of the grounding grid

The second graph (Scope2) displays the time changes of the impulse impedance of the grounding grid and impulse coefficient of the grounding grid for cases without and with the occurrence of soil ionization. The drawing of these graphs is initiated by clicking on the Draw button which remains inactive until the user selects the values which should be displayed. The Reset button clears all previously marked values for display.

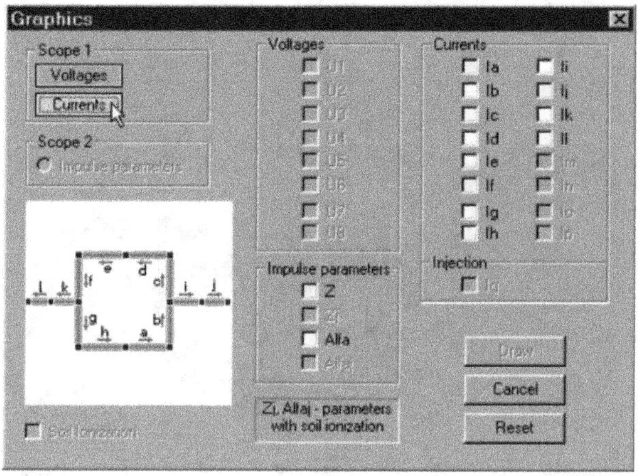


Fig. 1.57 - Dialog box for graphing currents in grounding grid branches, impulse impedance of the grounding grid and impulse coefficient of the grounding grid

**Run -> Helper**

This option opens a dialog box which will make it easier for the user to link the values displayed on the graph to their actual arrangement on the grounding grid prototype configuration. Fig. 1.58 shows the Helper dialog box for configuration 2 and an impulse current injected into node 6.

**Window -> Refresh**

The keyboard shortcut for this command is Ctrl+R. Using this option will refresh the display in the window.

**Help -> Help**

The keyboard shortcut for this command is F1. This command opens the Help application. The option Help/Contents contains the following sections:

- introduction,
- menu commands,
- step by step.

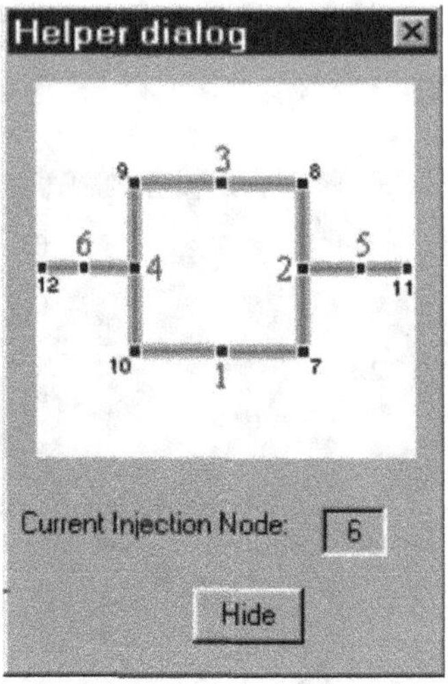


Fig. 1.58 - Helper dialog box for configuration 2 with impulse current injected into node 6

The first two sections have been explained previously and the third provides a detailed description of the calculation algorithm. This algorithm describes the sequence of command execution:

1. Under the Run menu select Run− > Parameters or from the toolbar opening the dialog box for entering the parameters of the grounding grid and impulse current;
2. Click on the Select Configuration tab and select the desired grounding grid prototype configuration;
3. Click on the Configuration Parameters tab and set the constructive parameters of the grounding grid: length of sides, cross section, type of material and node into which the impulse current is injected;
4. Click on the Current Characteristics tab and select the form of the impulse current and enter in the necessary parameters;
5. Click on the State Variable Model tab and select the mathematical model for the grounding grid;
6. Click on the Soil Characteristics tab and set the following soil parameters: uniform soil resistivity, dielectric constant and grounding grid placement depth;
7. Close the dialog box by clicking on the OK button;
8. Under the Run menu select the option Run− >Computation

Parameters or click on the icon from the toolbar and enter the end time of the calculation into the dialog box;

9. Select the option Run– >Calculation or click on the icon in order to initiate the calculation;
10. Click on the Graphics button in the Results dialog box (which is equivalent to the option Run– >Graphics or clicking on the icon on the toolbar) which opens the Graphics dialog box;
11. Click on the button Voltages or Currents in the Graphics dialog box to select whether the graph will display voltages or currents, which then opens the corresponding collection of check boxes;
12. Mark the appropriate check boxes to select the values which will be displayed on the graphs;
13. Click on the Draw button which then closes the Graphics dialog box;
14. Select the option File– >Save, click on the icon or use the keyboard shortcut Ctrl+S to open the Save dialog box and select what should be archived: a text file with results, graphs or prototype configuration parameters.

**Help ->About**

This option provides information about the version of the program and its authors [67].

# LITERATURE


[1] D.M. Ingels: What Every Engineer Should Know About Computer Modeling and Simulation, Marcel Dekker, Inc., New York and Basel, 1985.

[2] J.K. Krouse: What Every Engineer Should Know About Computer-Aided Design and Computer-Aided Manufacturing, Marcel Dekker, Inc., New York and Basel, 1982.

[3] C. Reisbeck: CAD/CAM, Einfuhrung, Praxis, Auswahl, Hoppenstedt Technik Tabellen Verlag, Darmstadt, 1990.

[4] J. Stark: What Every Engineer Should Know About Practical CAD/CAM Applications, Marcel Dekker, Inc., New York and Basel, 1986.

[5] H. Yoshikawa, F. Arbaba, T. Tomiyama: Intelligent CAD, III, Selected and Reviewed Papers and Reports From the IFIP TC5/WG5.2 Third International Workshop on Computer-Aided Design, Osaka, Japan, 26-29 September, 1989, North-Holland, 1991.

[6] U. Butz, D. Heinze, I. Neumann: Softwareentwicklung für die Projektirung von Elektroenergieanlagen, Verlag Technik GmbH Berlin, 1990.

[7] D. Tošić: Introduction to Numerical Analysis, Belgrade, 1997.

[8] D. Radunović: Numerical Methods, Academic Mind, Belgrade, 2004.

[9] V.A. Levi, D.D. Bekut: Application of Computerized Methods in Power Systems, Stylos, Novi Sad, 1997.

[10] Z. Petrović, S. Stupar: Computer-Aided Design, Finite Differences Method, University of Belgrade, 1996.

[11] S. Moaveni: Finite Element Analysis, Theory and Application with ANSYS, Second edition, Pearson Education, Inc., 2003.

[12] I. Mendaš, P. Milutinović, D. Ignjatijević: 100 Most Useful FORTRAN Subprograms, IP Dragon, Grupacija Mikro Knjiga, 1991.

[13] J.H. Mathews, K.D. Fink: Numerical Methods Using MATLAB, Fourth Edition, Pearson Education International, 2004.

[14] B. Ivković, Ž. Popović: Project Management in Civil Engineering, Faculty of Civil Engineering, Nauka, Belgrade, 1995.

[15] Z. Stojković: Grounding Grid Impedance Calculation of the High-Voltage Substations in Uniform and Two-Layer Soil, 22nd Conference of JUKO-CIGRE, R 33-01, Vrnjačka Banja, May 1995.

[16] Z. Stojković: Modeling the Impulse Impedance of a Grounding Grid in the Analysis of the Protection of Transformer Substations from Lightning Overvoltages, PhD dissertation, Faculty of Electrical Engineering, Belgrade, 1995.

[17] Z. Stojković: Computer-Aided Design in Power Engineering – Software Tools, Monography, II edition, Faculty of Electrical Engineering, Belgrade, Academic Mind, Belgrade, March 2003.

[18] Z. Stojković, M. S. Savić: Influence of Transmission Line Tower Grounding Impedance to the Line Flashover Rate, European Transactions on Electrical Power - ETEP Vol. 9, No. 4, July/August 1999, pp. 261-270.

[19] Z. Stojković, M. S. Savić, J. M. Nahman, D. Salamon, B. Bukorović.: Experimental Investigation of Grounding Grid Impulse Characteristics, European Transactions on Electrical Power - ETEP Vol. 8, No. 6, Nov/Dec 1998, pp. 417-421.

[20] Z. Stojković, M. S. Savić, J. M. Nahman, D. Salamon, B. Bukorović: Sensitivity Analysis of Experimentally Determined Grounding Grid Impulse Characteristics, IEEE Transactions on Power Delivery, Vol. 13, No. 4, Oct 1998, pp. 1136-1142.

[21] Z. Stojković, M. S. Savić, Lj. Gerić: The Local Effect of Grounding Grid Impulse Characteristics, Elektroprivreda, No. 3, Belgrade, 2000, pp. 20-28.

[22] Z. Stojković: The Soil Ionization Influence on the Lightning Performance of Transmission Lines, Electrical Engineering (Archiv fur elektrotechnik) Vol. 82, No. 1, Sep. 1999, pp. 49-58.

[23] Z. Stojković, M. S. Savić: A Dynamic Model of Transmission Line Tower Grounding Impulse Characteristics, 25th Conference of JUKO-CIGRE, Herceg Novi, R33-02, September 2001.

[24] Z. Stojković, M. S. Savić: Transmission Line Tower Grounding Impedance Influence to the Line Flashover Rate, Selected Chapter XIV in Monography "Power System Operation, Control and Measurement", Publisher Prof. Dr. Nahman, J., Belgrade 1998, pp.145-157.

[25] Z. Stojković, S. Jakovljević, J. Lošić: Improving the Process of Calculating Grounding Grid Impulse Resistance Proposed in TR-9, 3rd Conference of JUKO CIRED, R2.4., Vrnjačka Banja, October 2002.

[26] M. Savić, Z. Stojković: High-Voltage Technique - Lightning Overvoltages, Monography, II corrected and amended edition, Faculty of Electrical Engineering, Belgrade, 2001.

[27] J. Nahman: Digital Calculation of Earthing Systems in Nonuniform Soil, Archiv fur Elektrotechnik, Vol. 62, 1980, pp. 19-24.

[28] J. Nahman, D. Salamon: Earthing System Modelling by Element Aggregation, Proc. IEE, Pt.C, No. 1, Vol. 133, 1986, pp. 54-58.
[29] J. Nahman: Numerical Method for Determining Self-Resistances of Thin Cylindrical Linear Conductors, Elektrotehnika, Vol. 27, No. 1-2, 1984, pp. 27-32.
[30] R. Velasquez, D. Mukhedkar: Analytical Modelling of Grounding Electrodes Transient Behavior, IEEE Transactions on Power Apparatus and Systems, Vol. PAS-103, No. 6, June 1984, pp. 1314-1322.
[31] E. Ja. Rjabkova: Grounding Grids in High-Voltage Substations, Energija, Moscow, 1978, (in Russian).
[32] B. Reljin: The Theory of Electrical Circuits I - Solving Circuits in the Domain of Time, Nauka, Belgrade, 1990.
[33] R. Natarajan: Computer-Aided Power System Analysis, Marcel Dekker, Inc., New York and Basel, 2000.
[34] M. A. Mousa: The Soil Ionization Gradient Associated with Discharge of High Currents into Concentrated Electrodes, IEEE Transactions on Power Delivery, Vol. 9, No. 3, July 1994, pp. 1669-1676.
[35] E. E. Oettle: A New General Estimation Curve for Predicting the Impulse Impedance of Concentrated Earth Electrodes, IEEE Transactions on Power Delivery, Vol. 3, No. 4, Oct 1988, pp. 2020-2029.
[36] C.H. Edwards, D.E. Penney: Differential Equations and Boundary Value Problems, Computing and Modeling, Third edition, Pearson Education, Inc., 2004.
[37] ATP - EMTP Rule Book, Canadian-American EMTP Users Group, 1997.
[38] Electromagnetic Transients Program (EMTP) - Primer, Westinghouse Electric Corporation, 1985, p. 211.
[39] L. Prikler, H. Kr. Hoidalen: ATPDraw™ for Windows 5.6, Users' Manual, Nov 2009, p. 270.
[40] MATLAB® – The Language of Technical Computing, Version 7.10.0 (R2010a), The Math Works Inc., 2010.
[41] Z. Stojković, J. Mikulović, Z. Stojanović: Workshop for Software Tools in Power Engineering, Faculty of Electrical Engineering, Belgrade, Academic Mind, Belgrade, June 2006.
[42] H. Bode: Matlab-Simulink, Analyse und Simulation dynamischer Systeme, 2. Auflage, Teubner Verlag, Wiesbaden, 2006.
[43] L. Kraus: Programming Language C++ with Solved Tasks, Academic Mind, Belgrade, 2007.
[44] L. Kraus: Solved Tasks for the Programming C++, Academic Mind, Belgrade, 2006.

[45] R. C. Leinecker, T. Archer: Visual C++ 6 Bible, Mikro Knjiga, Belgrade, 1999.

[46] A. M. Miri, Z. Stojković: Transient Electromagnetic Phenomena in the Secondary Circuits of Voltage- and Current Transformers in GIS (Measurements and Calculations), IEEE Transactions on Power Delivery, Vol. 16, No. 4, Oct. 2001, pp. 571-575.

[47] A.M. Miri, Z. Stojković: Transient Electromagnetic Phenomena in the Secondary Circuits of Measuring Transformers in the $SF_6$ Substation, Elektroprivreda, No. 4, 1999, pp. 61-66.

[48] Z. Stojković, A.M. Miri, G. Mitrić: Estimation of Metal Clad Gas $SF_6$ Insulated Substation Potential Pickup Caused by Switching Disconnector Operations, Elektroprivreda, No.2, 2003, pp. 54-59.

[49] W.A. Perry: Quality Assurance for Information Systems: Methods, Tools, and Techniques, QED Information Sciences, Inc, Wellesley, MA 1991.

[50] P. Jiantao: Software Testing, Carnegie Mellon University, Spring 1999, (Online), Available, http://www.ece.cmu.edu

[51] S. Boštjančić , M. Stojanović, R. Nedeljković: Methodology of Software Testing, 15th Telecommunications Forum, TELFOR 2007, Belgrade, R9-03, Nov. 2007, pp. 577-580.

[52] IEEE Std 610.12-1990: IEEE Standard Glossary of Software Engineering Terminology, 1990.

[53] E. D. Sunde: Earth Conduction Effects in Transmission Systems, Dower publications, New York, 1968.

[54] M.S. Savić, Z. Stojković: High Voltage Substation Equivalent Circuits in the Lightning Performance Estimation, IEE Proceedings - Generation, Transmission and Distribution, Vol. 141, No. 2, March 1994, pp. 99-105.

[55] Z. Stojković, M.S. Savić: The High-Voltage Substation Configuration Influence on the Estimated Lightning Performance, Electrical Engineering, Archiv fur elektrotechnik, Vol. 80, No. 4, August 1997, pp. 275-283.

[56] Z. Stojković, M.S. Savić: Selection of Type Equivalent Circuits of Substations for the Analysis of the Lightning Overvoltage Protection, 20th Conference of JUKO CIGRE, R33-02, Neum, April 1991.

[57] Z. Stojković: An Improved Method for HV Substation Lightning Performance Estimation, IEEE Transactions on Power Delivery, Vol. 14, No. 3, July 1999, pp. 917-922.

[58] Z. Stojković, M.S. Savić: Involvement of High-Voltage Substation Configurations in the Assessment of Stresses of Isolation From Lightning

Overvoltages, 21st Conference of JUKO CIGRE, R33-03, Vrnjačka Banja, May 1993.

[59] B. Dugošija, Z. Stojković: Application of Numerical Methods and MATLAB Technical Computing Software in Calculation of Grounding Systems, Faculty of Electrical Engineering, University of Belgrade, 2009.

[60] Z. Stojković, Ž. Stankić: AutoCAD-Based Concept for Estimating Lightning Protection Zone of Transmission Lines and Structures; International Journal of Electrical Engineering Education (IJEEE), Vol.43, No.4, Oct 2006, pp. 299-317.

[61] Z. Stojković: Evaluation of Lightning Protection Zone Using AutoCAD-Based Software Tool, Institute of Power Transmission and High Voltage Technology, University of Stuttgart, Annual Report 2006, pp. 64-67.

[62] Z. Stojković, Ž. Stankić: Lightning Protection Design of General and Special Buildings, Elektroprivreda, No. 3, 2005, pp. 84-91.

[63] Z. Stojković, Ž. Stankić: Lightning Protection Design of General and Special Buildings, 27th Conference of JUKO-CIGRE, RC4-01, Zlatibor, 29 May - 3 June 2005.

[64] Z. Stojković, A.Grujić, S. Tenbohlen: Lightning Protection Design of Substations and Transmission Lines, 28th Conference of JUKO-CIGRE, RC4-01, Vrnjačka Banja, 30 September - 5 October 2007.

[65] Z. Stojković: Transmission Line Tower Grounding Grid Influence to the Line Flashover Rate, 24th Conference of JUKO CIGRE, R33-03, Vrnjačka Banja, October 1999.

[66] M. Marković, Z. Stojković: Analysis of Power Frequency and Impulse Characteristics of a Grounding Grid in Equivalent Uniform Soil, Faculty of Electrical Engineering, University of Belgrade, 2009.

[67] Z. Stojković, V. Cvejić, S. Jokić: GIC - Software Tool for Calculation of Grounding Grid Impulse Characteristics, Faculty of Electrical Engineering, Belgrade, 2009.

[68] G. Brechmann, C.W. Dzieia, R.E. Hornemann, H.H. Hubscher, L.D. Jagla, N.J. Klaue: Westerman's Electrical Engineering Handbook, Građevinska Knjiga, Belgrade, 2000.

# 2. APPLICATION OF SOFTWARE TOOLS IN POWER ENGINEERING CALCULATIONS

## 2.1. MATLAB® /Simulink®

### 2.1.1. Introduction

MATLAB® technical computing software is a software tool for solving mathematical problems, analyzing data and visualization. This tool integrates numerical analysis, matrix calculation, data processing and graphical display. It is characterized by its ability to solve all mathematical problems. The advantage of this software tool is in its simple expression of mathematical problems and solutions as they are written in mathematics, by which traditional programming is avoided [1-4].

MATLAB® technical computing software is also represented by a programming language which makes many mathematical problems easier to solve making earlier programming languages such as FORTRAN, BASIC or C obsolete. As a result of the simple approach in programming, MATLAB® technical computing software makes it possible to solve mathematical and engineering problems in a significantly shorter amount of time, which is exceptionally important in the field of engineering and science [5, 6]. For this reason MATLAB® technical computing software has become a dominant software tool in universities and scientific research institutions across the entire world. Additionally, it is also being used more and more by design firms in order to solve several engineering problems.

The unbelievable success of this software tool is also a consequence of its conceptual organization. MATLAB® technical computing software has the capability of being upgraded with modules. These additional modules are called Toolboxes. The family of additional software tools contains functions which are useful for several mathematical and engineering disciplines. Along with the aforementioned modules, MATLAB® technical computing software is also constantly being updated with other new modules. The names of these modules which solve

Z. Stojković, *Computer-Aided Design in Power Engineering*,
DOI: 10.1007/978-3-642-30206-0_2, © Springer-Verlag Berlin Heidelberg 2012

Table 2.1 - Additional modules for MATLAB® technical computing software

| Module | Module |
|---|---|
| Parallel Computing Toolbox | SimPowerSystems |
| Simulink | Filter Design Toolbox |
| Aerospace Toolbox | Fixed-Income Toolbox |
| Bioinformatics Toolbox | Fixed-Point Toolbox |
| Communications Toolbox | Global Optimization Toolbox |
| RF Toolbox | Image Acquisition Toolbox |
| Control System Toolbox | Image Processing Toolbox |
| Curve Fitting Toolbox | Instrument Control Toolbox |
| Signal Processing Toolbox | Statistics Toolbox |
| Data Acquisition Toolbox | Mapping Toolbox |
| SimBiology | Symbolic Math Toolbox |
| Database Toolbox | System Identification Toolbox |
| SimElectronics | Vehicle Network Toolbox |
| Datafeed Toolbox | Model-Based Calibration Toolbox |
| SimHydraulics | Wavelet Toolbox |
| Econometrics Toolbox | Optimization Toolbox |
| SimMechanics | |

a wide spectrum of engineering and scientific problems are displayed in Table 2.1.

MATLAB® technical computing software also enables the user to form additional functions which is yet another advantage of open-source systems such as this software tool [7]. MATLAB® technical computing software was formed for use on various computers, of which the most common is the personal computer.

Simulink® software is a module specialized for simulation of dynamic systems in a graphical environment. Using Simulink® software enables the analysis of linear, nonlinear, time-continuous or discrete multivariable systems with concentrated parameters [8, 9].

Realization of simulation is achieved by forming a Simulink® model and using the Simulink® function which solves systems of first-order ordinary differential equations. The fundamental advantage of Simulink® software is the simplicity of operation which is reflected in the display of the Simulink® model in the shape of a block diagram.

The illustrative nature and simplicity of presenting the problem which is required to be solved are the fundamental advantages of this form of display.

A block diagram is one form of a mathematical model of a system which illustrates the dynamic characteristics of the system, the main variables of the system and the link between those variables. The functional relation between the parts of the system can be viewed in the displayed block diagram. By displaying every element of the system, the principle of the "black box" approach is represented. This approach is based on every element being displayed by a specific block which contains the mathematical relation between the input and output variables of that element. Several blocks are mutually connected by orientational lines which indicate the flow of the signal from one block to another. Additionally, the signals indicate the input and output variables.

Each Simulink® model corresponds to an m-file which is a series of MATLAB® and Simulink® commands and functions which create Simulink® models.

This chapter shows the basic groups of blocks within Simulink® software and the additional module which is used in power system calculations – SimPowerSystems™. The application of these tools is illustrated in three examples. The first example shows the application of the additional module in Simulink® software in order to calculate the characteristic values of the fault current which is necessary for the selection of high-voltage equipment. The application of this module is displayed within a portion of an equivalent circuit of a power system. The obtained results are discussed and compared with results determined using the classic approach to calculation.

The second example shows the personally developed software tool SPLCAD (Software Power Line CAD) for designing medium-voltage overhead lines. The tool was realized using MATLAB® technical computing software as the development platform for creating the user interface (Graphical User Interfaces Toolbox - GUI). Creating and working with databases is realized through use of the software MS Access and MS Excel.

The third example relates to the display of a Simulink® model of a turbine regulator within the "Kokin Brod" hydroelectric power plant. A comparative analysis is shown of the numerically and experimentally determined results for the considered turbine regulator.

## 2.1.2. Basic groups of blocks in Simulink® software*

The blocks within Simulink® software are classified into eight basic groups according to the type of system which they portray. Opening the window with the basic icons can be done in two ways:

– through the command

>>simulink

which is entered into the command line of MATLAB® technical computing software, or

– by selecting the option File > New > Model.

Fig. 2.1 shows the window which contains the basic blocks of Simulink® software. The icon is activated by placing the cursor over the desired icon and then double-clicking. This opens a new window which contains the blocks which belong to that group. Fig 2.2 shows the window with blocks of Simulink® sources.

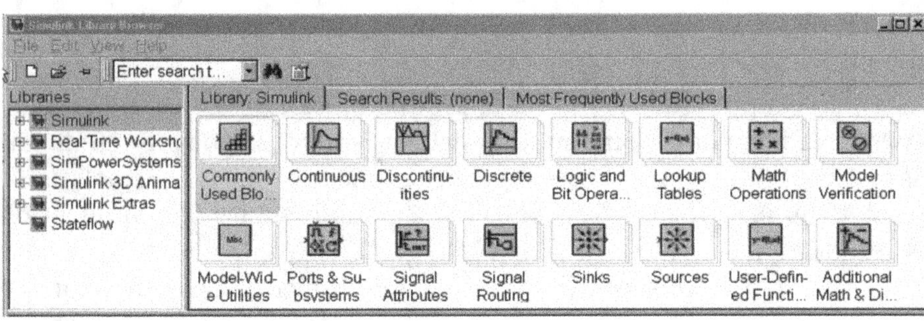


Fig. 2.1 – Window with basic blocks

Some of the most commonly used blocks from this group have the following functions:

– Constant – constant value generator,
– Signal Generator – generator of various types of signals,
– Step – step function generator,
– Ramp – ramp function generator,
– Sine Wave – sine function generator,
– From File – generates functions defined by data in a mat-file,
– From Workspace – generates functions defined by data from the workspace,

*Reprinted with permission from MathWorks. Simulink and SimPowerSystems are registered trademarks of the MathWorks, Inc.

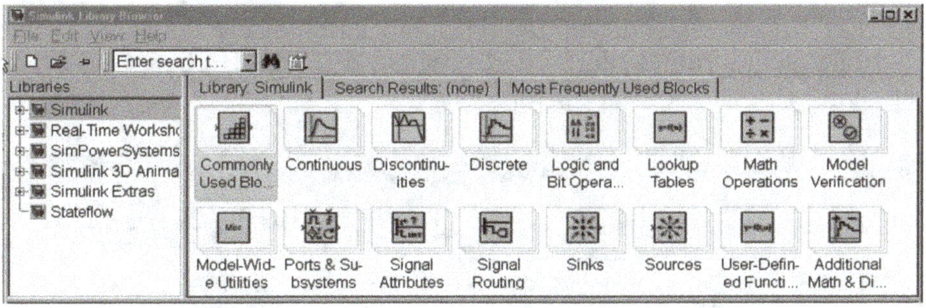


Fig. 2.2 – Window with blocks of sources

– Random Number – generates a series of random numbers.

Fig 2.3 shows the window with blocks of sinks.

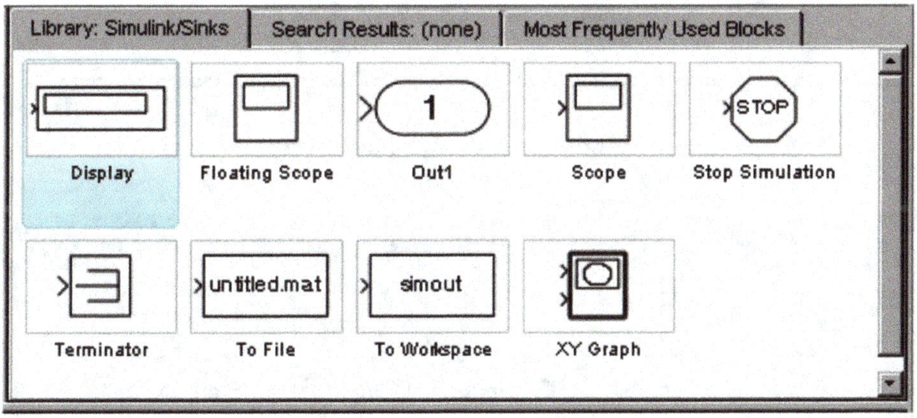


Fig. 2.3 – Window with blocks of sinks

Some of the most commonly used blocks from this group have the following functions:

- Scope – opens a graphical window which displays the output results in a diagram which has no visible markings on the axes,
- XY Graph – opens a graphical window which displays the results in the phase plane,
- To File – places the output results in a mat-file,
- To Workspace – places the output results in a matrix which can be referenced from the workspace.

Fig 2.4 shows the window with blocks of discrete systems. The functions of individual blocks can be viewed in the Help system or by displaying the Block Description option.

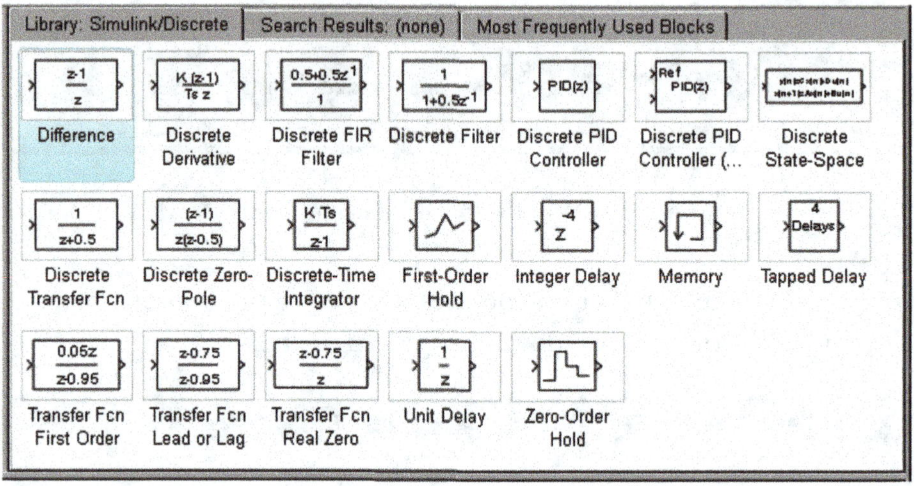


Fig. 2.4 – Window with blocks of discrete systems

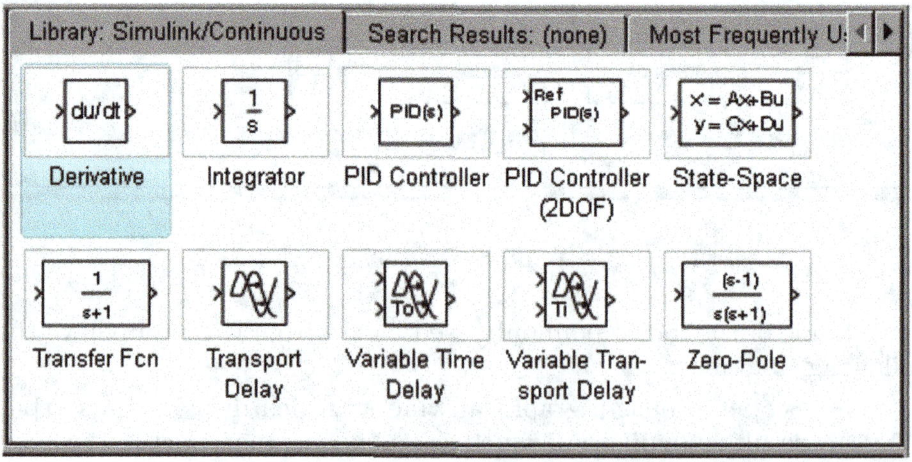


Fig. 2.5 – Window with blocks of continuous systems

Fig 2.5 shows the window with blocks of continuous systems.

Fig 2.6 shows the window with blocks of discontinuous systems.

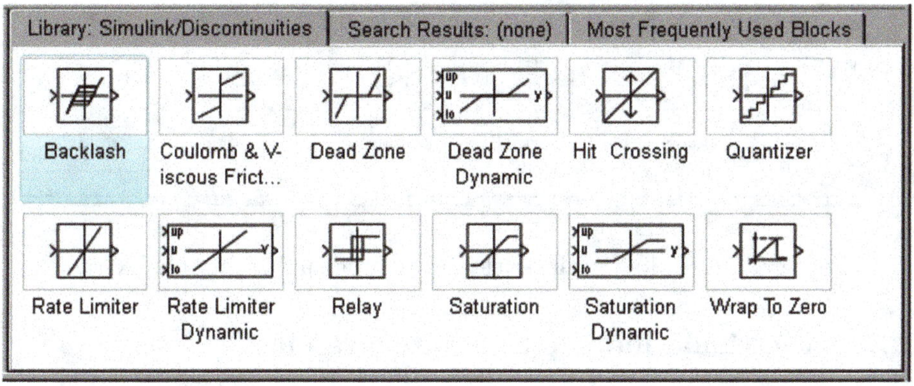


Fig. 2.6 – Window with blocks of discontinuous systems

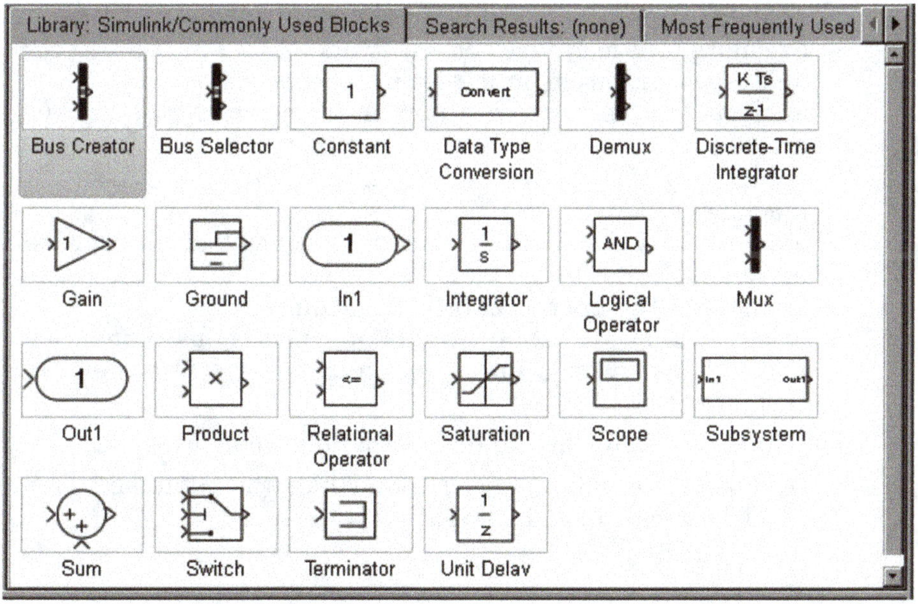


Fig. 2.7 - Window with the most commonly used blocks of the Simulink® module

Fig. 2.7 shows the window with the most commonly used blocks of the Simulink® module.

The window with blocks from the subgroup Simulink Extras is shown in Fig. 2.8.

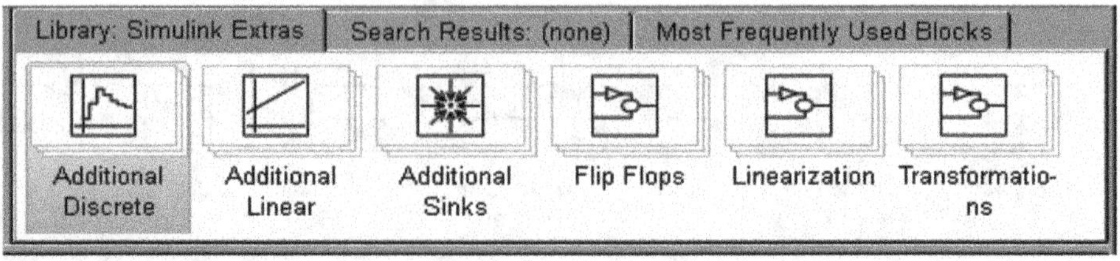


Fig. 2.8 – Window with blocks from subgroup Simulink Extras

### 2.1.3. Additional module SimPowerSystems™*

SimPowerSystems™ is a component of Simulink® software. This module can be used for calculation in several areas within power systems. The applicable fields for this module are:

- analysis of power system networks,
- calculation of load flows of a network,
- analysis of transient processes in a network,
- analysis of a network with non-linear elements, such as surge arresters,
- analysis of statistical and dynamic stability of synchronous machines,
- analysis of a circuit for vector management of asynchronous drives,
- analysis of the operation of DC machines,
- calculation of excitation circuits of synchronous generators,
- analysis and calculation of parameters for steam and hydro turbines.

This module is launched in one of the following two ways:

- by typing the following command into the command line of MATLAB® technical computing software:
  >>powerlib

or

- by launching the Simulink® software and clicking on the SimPowerSystems™ icon.

Sfter launching the additional module SimPowerSystems™, a window appears as in Fig. 2.9.

The additional module itself consists of several blocks:

*Reprinted with permission from MathWorks. Simulink and SimPowerSystems are registered trademarks of the MathWorks, Inc.

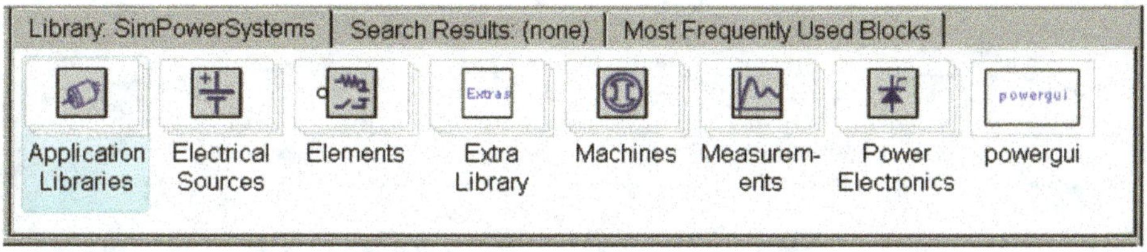


Fig. 2.9 - Window with blocks of the additional module SimPowerSystems™

- Electrical Sources,
- Elements,
- Power Electronics,
- Machines,
- Measurements,
- Extra Library,
- Powergui,
- Application Libraries.

The most commonly used elements of the aforementioned blocks are described below.

*a) Elements of the Electrical Sources block*

The elements of the Electrical Sources block are shown in Fig. 2.10.

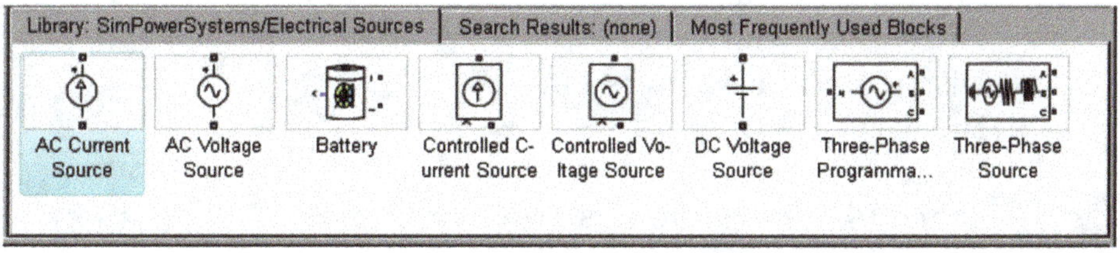


Fig. 2.10 – Window with elements of Electrical Sources block

*b) Elements of the Elements block*

The elements of the Elements block are shown in Fig 2.11.

The most commonly used elements of this block are:

- Series RLC Branch,

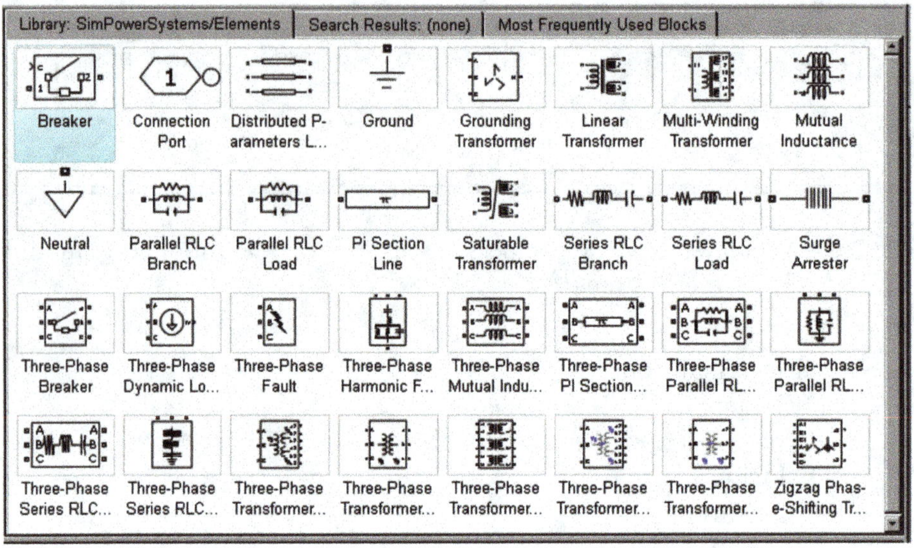


Fig. 2.11 – Elements of the Elements block

- Series RLC Load,
- Parallel RLC Branch,
- Parallel RLC Load,
- Linear transformer,
- Saturable transformer,
- Mutual inductance,
- PI Section Line,
- Distributed Parameters Line,
- Breaker,
- Surge Arrester,
- Three Phase Transformer (two windings),
- Three Phase Transformer (three windings),

Some of the elements of the Elements block are described below.

*b.1 $\pi$ section line model*

The dialog box for entry of parameters of a $\pi$ section line model is shown in Fig. 2.12.

The following parameters of the $\pi$ section line model can be set:

- Frequency used for RLC specification (Hz),

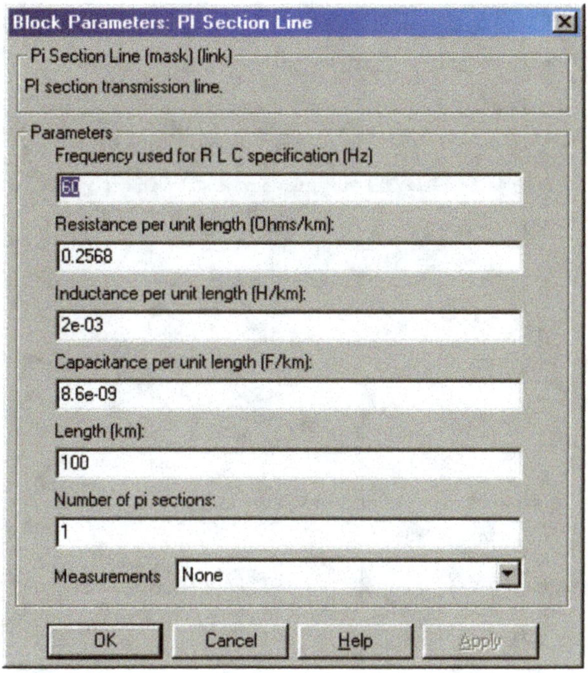


Fig. 2.12 – Dialog box for entry of parameters of $\pi$ section line model

- Resistance per unit length (Ohm/km),
- Inductance per unit length (H/km),
- Capacitance per unit length (F/km),
- Length (km),
- Number of pi sections,
- Measurements (can measure input current and voltage, output current and voltage, all current and voltage).

*b.2 Distributed parameters line model*

The dialog box for entry of parameters of a distributed parameters line is shown in Fig. 2.13.

The following elements are entered into the dialog box for a distributed parameters line:

- Number of phases,
- Frequency used for RLC specification (Hz),
- Resistance per unit length (Ohm/km),

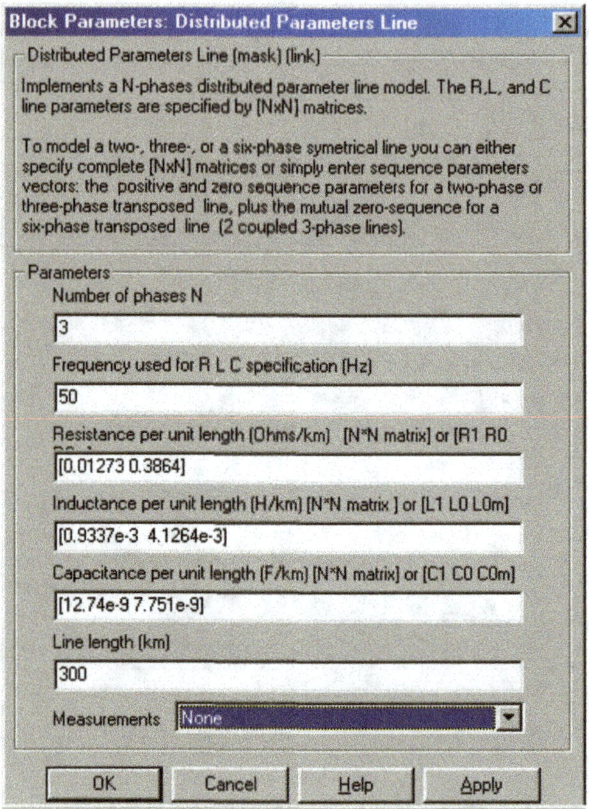


Fig. 2.13 - Dialog box for entry of parameters of a distributed parameters line

- Inductance per unit length (H/km),
- Capacitance per unit length (F/km),
- Line length (km),
- Measurements (can measure phase voltages).

*b.3 Breaker model*

The dialog box for entry of breaker parameters is shown in Fig. 2.14. It is possible to set the following parameters:

- Breaker Resistance Ron (Ohm) – value is not allowed to equal zero,
- Initial state (0 for 'open', 1 for 'closed'),

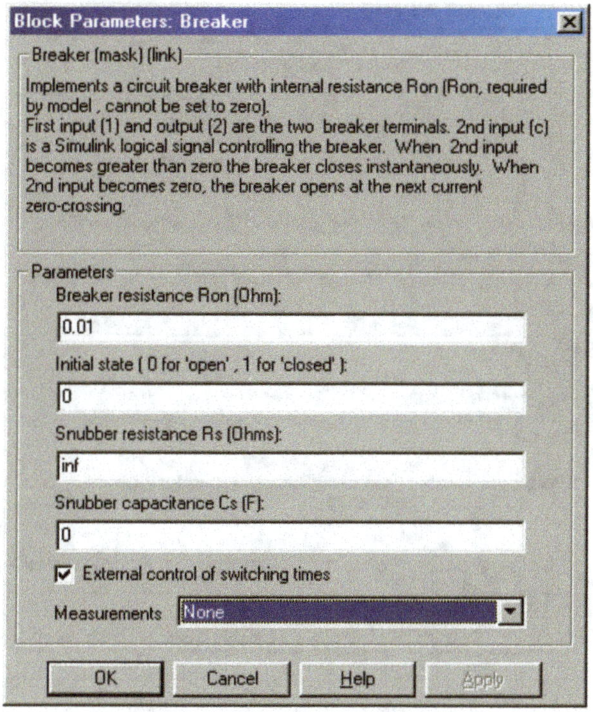


Fig. 2.14 – Dialog box for entry of breaker parameters

- Snubber resistance Rs (Ohm) – disregard by entering inf,
- Snubber capacitance Cs (F) – disregard by entering 0 or inf,
- Switching times (switching time which is defined by the vector $[t_1\ t_2]$ in which the first instant relates the opening/closing depending on the initial state and the second instant relates to the inverse process),
- External control of switching times (if this field is checked it enables the external control of the breaker via a Simulink® signal),
- Measurements (can measure branch voltage, branch current or both values simultaneously).

*b.4 Power transformer model*

Three-phase transformers represent an important element of a power system. Fig. 2.15 shows the dialog box for entering the parameters of a three-phase transformer (three windings).

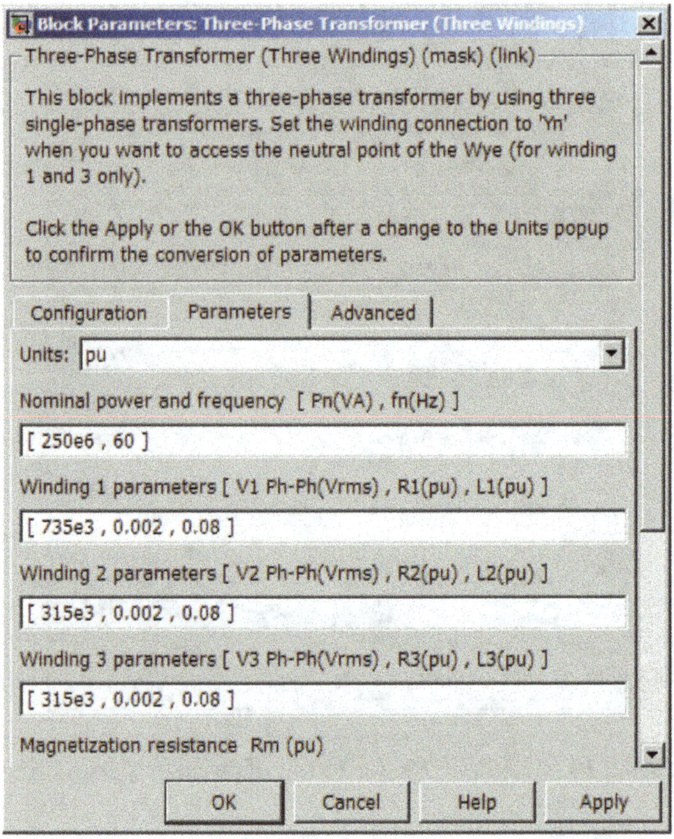


Fig. 2.15 - Dialog box for entering parameters of a three-phase transformer (three windings)

It is possible to set the following parameters for this transformer:

- Units,
- Nominal power and frequency [Pn(VA), fn(Hz)],
- Winding 1 parameters [V1 Ph-Ph (V), R1(pu), L1 (pu)],
- Winding 2 parameters [V2 Ph-Ph (V), R2(pu), L2 (pu)],
- Winding 3 parameters [V3 Ph-Ph (V), R3(pu), L3 (pu)]
- Magnetization resistance Rm (pu).

The dialog box for entering the configuration of a three-phase transformer (three windings) relates to the:

- Winding connections,
- Saturable core,

– Simulate hysteresis,
– Specify initial fluxes,
– Measurements (can measure winding voltages, winding currents, fluxes and excitation currents, fluxes and magnetization currents and all measurements).

*b.5 Model of 3-Phase Fault block*

In engineering calculations it is necessary to model a corresponding fault. A model of a 3-Phase Fault enables the simulation of interphase faults and ground faults, as well as their combination. A view "beneath the mask" of this block is provided in Fig. 2.16.

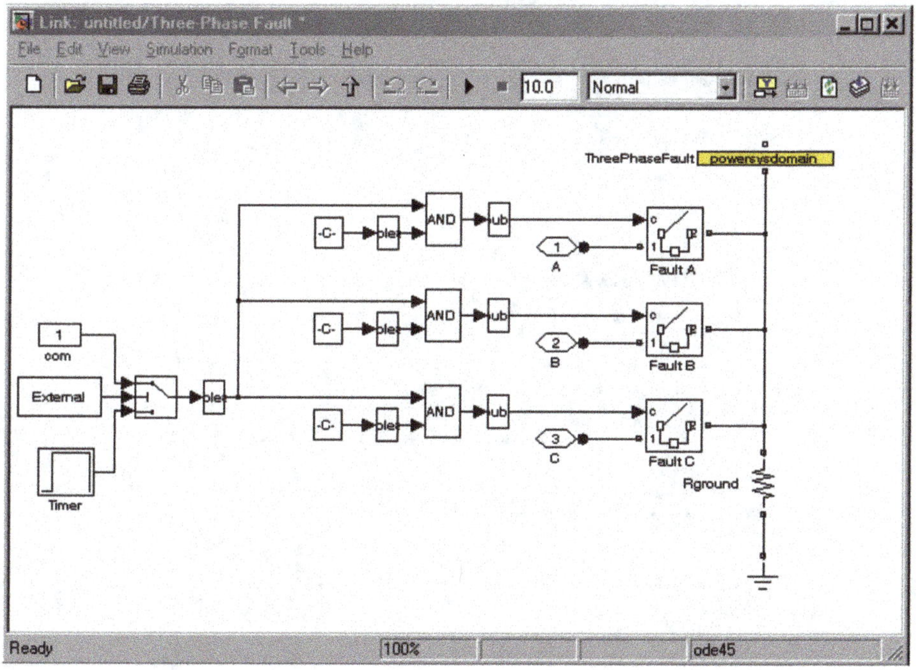


Fig. 2.16 – A view "beneath the mask" of a 3-Phase Fault block

It can be concluded that this block is composed of three breakers, through which faults are simulated, as well as breaker management circuits. The control process can be achieved in two ways:

– by using the internal counter which defines the fault timing
– by using an external Simulink® signal.

Fig. 2.17 shows the dialog box for entering the parameters of a 3-Phase Fault block.

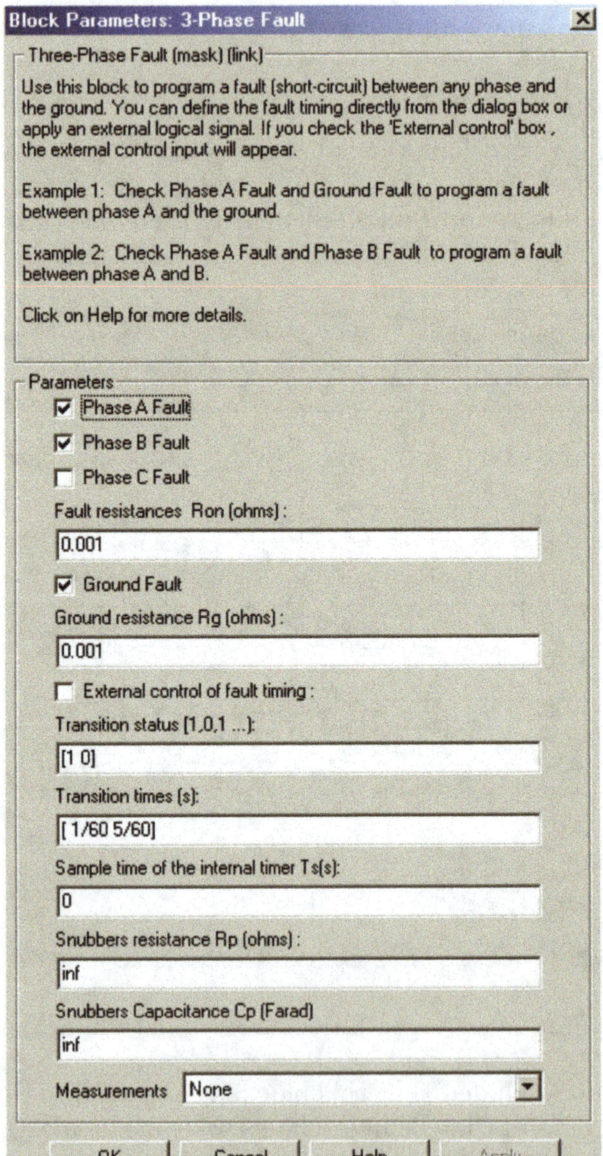


Fig. 2.17 – Dialog box for entering parameters of a 3-Phase Fault block

It is possible to set the following parameters for this block:

- by placing a check in the corresponding fields, the phases in which the faults occur are defined:
- Phase A Fault,
- Phase B Fault,
- Phase C Fault,
- Fault resistance Ron (Ohm),
- Ground Fault (if this field is checked the fault coincides with the presence of an ground fault),
- Ground resistance Rg (Ohm),
- External control of fault timing,
- Transition status [1, 0, 1, ... ],
- Transition times (s),
- Sample time of the internal timer T (s),
- Snubbers resistance Rp (Ohm),
- Snubber Capacitance Cp (Farad),
- Measurements (can measure voltage, current or both values simultaneously).

*c) Elements of Power Electronics block*

A display of the elements of the Power Electronics library is provided in Fig. 2.18. The details of modeling of these elements can be viewed in the selected literature [1, 8].

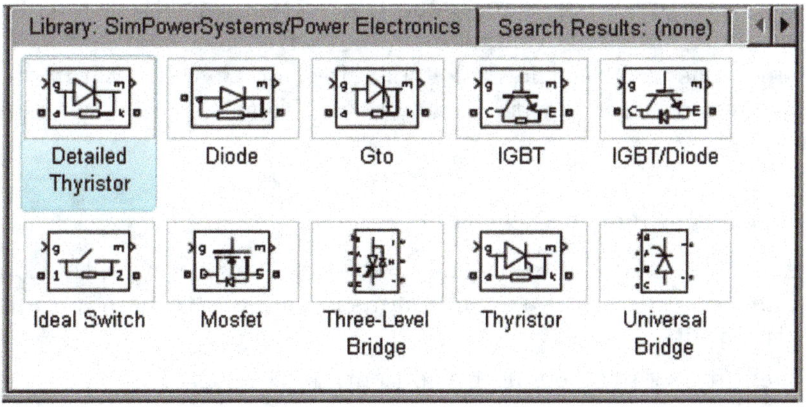


Fig. 2.18 – Window with elements of Power Electronics block

*d) Elements of Machines block*

This block contains models of various electrical machines.

The following elements are located in this library:

- Asynchronous Machine SI Units,
- Asynchronous Machine pu Units,
- DC Machine,
- Discrete DC Machine,
- Excitation System,
- Generic Power System Stabilizer,
- Hydraulic Turbine and Governor,
- Machines Measurement Demux,
- Multi-Band Power System Stabilizer,
- Permanent Magnet Synchronous Machine,
- Simplified Synchronous Machine SI Units,
- Simplified Synchronous Machine pu Units,
- Single Phase Asynchronous Machine,
- Steam Turbine and Governor,
- Stepper Motor,
- Switched Reluctance Motor,
- Synchronous Machine SI Fundamental,
- Synchronous Machine pu Fundamental,
- Synchronous Machine pu Standard.

The details of modeling of these elements can be viewed in the selected literature [1, 8]. The dialog box for entering the configuration and parameters of synchronous machines with basic parameters in relative units, as well as the dialog box for entering the parameters of asynchronous machines with parameters in relative units are displayed below.

*d.1 Synchronous machine model*

Figs. 2.19 and 2.20 show the dialog boxes for entering the configuration and parameters of a synchronous machine with basic parameters in relative units, respectively.

Definition of the configuration entails (Fig. 2.19):

- model selection (Preset model),
- mechanical input, which refers to defining data on mechanical power and speed,

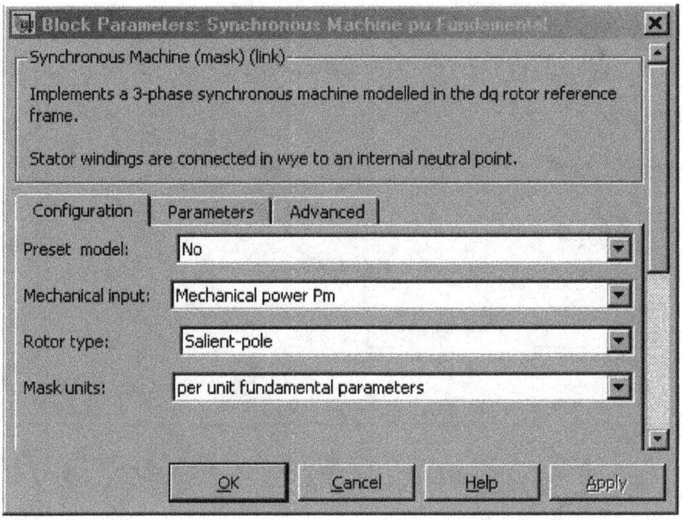


Fig. 2.19 - Dialog box for entering data on the configuration of a synchronous machine with basic parameters in relative units

- selection of rotor type (salient-pole or round),
- mask units.

In accordance with Fig. 2.20, the following options are provided for setting the parameters of a synchronous machine with basic parameters in relative units:

- Nom Power, Line-to-line voltage and frequency [Pn(VA) Vn(Vrms) fn(Hz)],
- Stator [Rs Ll Lmd Lmq] (pu),
- Field [Rf Llfd] (pu),
- Dampers [Rkd Llkd Rkq1 Llkq1] (pu),
- Inertia coeficient, friction factor and pole pairs [H(s) F(pu) p()],
- Initial conditions [dw (%) th(deg), ia, ib, ic (pu), pha, phb, phc (deg) Vf (pu)],
- Simulate saturation (if this option is checked saturation parameters are involved).

*d.2 Asynchronous machine model*

The dialog boxes for entering the configuration and parameters of an asynchronous machine with parameters in relative units are displayed in Fig. 2.21 and Fig. 2.22, respectively.

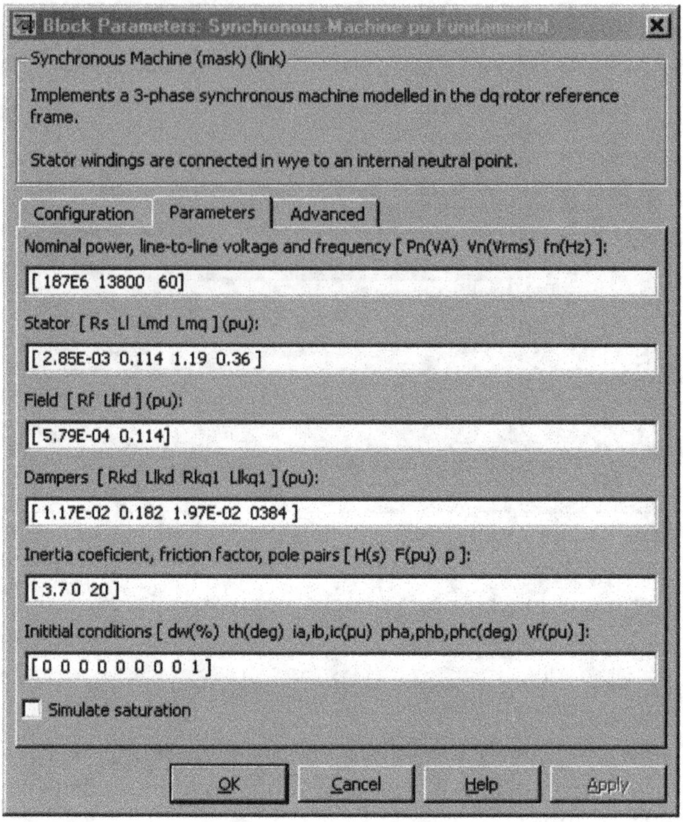


Fig. 2.20 - Dialog box for entering parameters of a synchronous machine with basic parameters in relative units

Definition of the configuration entails (Fig. 2.21):

– model selection (preset model),
– mechanical input, which refers to defining data on torque Tm and speed,
– selection of rotor type (wound or squirrel-cage),
– reference frame (rotor, stationary, synchronous),
– mask units.

The dialog box in Fig. 2.22 contains the following parameters of an asynchronous machine in relative units:

– Nominal power, voltage (line-line) and frequency [Pn(VA) Vn(Vrms) fn(Hz)],
– Stator resistance and inductance [Rs Lls] (pu),

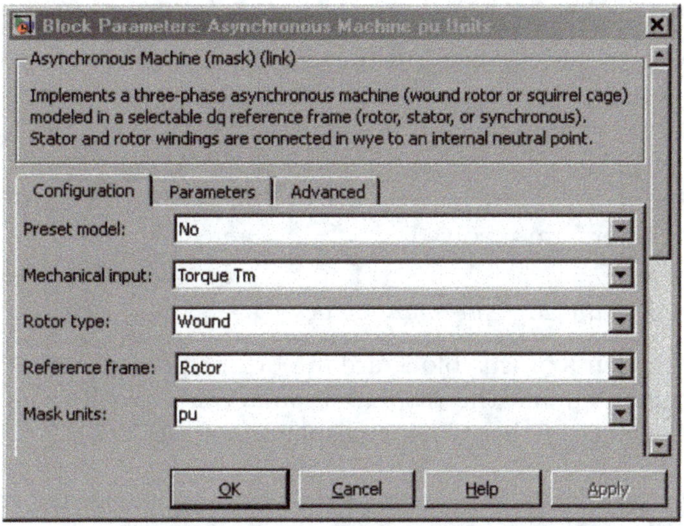


Fig. 2.21 - Dialog box for entering data on the configuration of an asynchronous machine with basic parameters in relative units

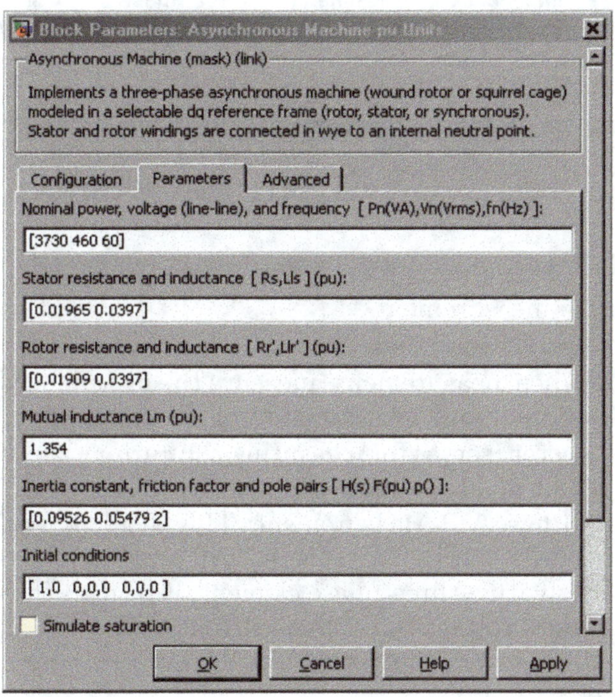


Fig. 2.22 - Dialog box for entering parameters of an asynchronous machine in relative units

- Rotor resistance and inductance [Rr' Llr'] (pu),
- Mutual inductance Lm (pu),
- Inertia constant, friction factor and pole pairs [H(s) F(pu) p()],
- Initial conditions [s() th (deg), isa, isb, isc (pu), pha, phb, phc (deg)],
- Simulate saturation (if this option is checked saturation parameters are involved).

*e) Elements of the Measurements block*

The elements of this block are important in engineering analyses. These elements enable the display of simulation results. The window with the elements of this block is displayed in Fig. 2.23.

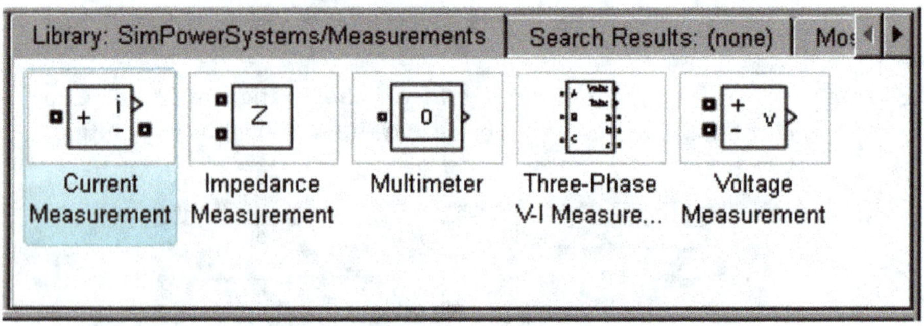


Fig. 2.23 – Window with elements of the Measurements block

This block encompasses the following elements:

- Voltage measurement,
- Current measurement,
- Impedance measurement (measure the impedance between two nodes of a circuit as a function of the frequency),
- Multimeter (serves for measuring the current and voltage which are specified in the model),
- Three-Phase V-I Measurement: This block is used to measure three-phase voltages and currents in a circuit. When connected in series with a three-phase element, it returns the three phase-to-ground voltages and line currents. The block can output the voltages and currents in per unit values or in volts and amperes.

*f) Elements of the Extra Library block*

The Extra Library block contains the following elements:

- Control Blocks,
- Discrete Control Blocks,
- Discrete Measurements,
- Measurements,
- Phasor Library.

Fig. 2.24 shows the elements of the Extra Library/Measurements block. This block enables the registration of some actual values such as:

- 3-Phase Sequence Analyzer: This block outputs the positive-, negative-, zero- or all sequence component(s) (Magnitude and Phase) of a set of three balanced or unbalanced signals which may contain harmonics,
- Fourier: The Fourier block performs a Fourier analysis of the input signal over a running window of one cycle of the fundamental frequency. First and second outputs return respectively the magnitude and phase (degrees) of the harmonic component specified,
- RMS: The RMS block computes the true RMS value (including fundamental, harmonic, and DC components) of input signal. The RMS value is calculated over a running window of one cycle of the specified frequency,
- Total Harmonic Distortion: This block measures the total harmonic distortion (THD ) of a periodic instantaneous voltage or current connected to the input,
- abc_to_dq0 Transformation: This block performs the abc to dq0 transformation on a set of three-phase signals. It computes the direct axis Vd, quadratic axis Vq, and zero sequence V0 quantities in a two axis rotating reference frame,
- dq0_to_abc Transformation: This block transforms three quantities (direct axis, quadature axis and zero-sequence components) expressed in a two axis reference frame back to phase quantities.

The transformations used for the corresponding calculations are defined in literature [1, 8].

*g) Graphical User Interface Powergui*

The graphical user interface enables:

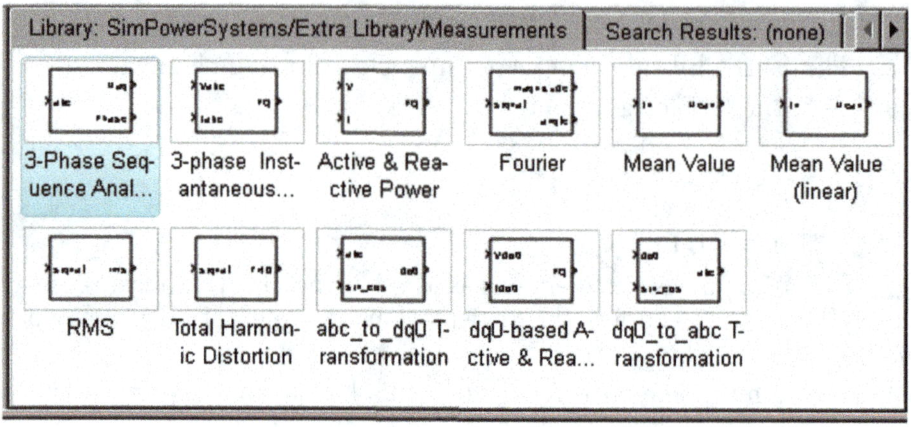


Fig. 2.24 - Window with elements of the Extra Library/Measurements block

- display of measured voltages and currents for all variable states,
- change of initial values of variables,
- calculation of load flows and initial values of three-phase networks,
- display of impedance dependency on frequency in a graphical form,
- generating a model in state space and display of the response of the system in the domain of time and frequency,
- generating the calculation results in state space.

*h) Elements of the Application Libraries block*

This block contains the following modules:

- Distributed Resources Library,
- Electric Drives Library,
- Flexible AC Transmission Systems (FACTS) Library.

Contemporary problems in designing ecological power plants (wind power plants) initiated the forming of the module Distributed Resources Library/Wind Generation, which contains the following elements:

- Wind Turbine,
- Wind Turbine Doubly-Fed Induction Generator (Phasor Type),
- Wind Turbine Induction Generator (Phasor Type).

For example, the following four groups of parameters are provided for Wind Turbine Doubly-Fed Induction Generator (Phasor Type):

– Generator data,
– Converters data,
– Turbine data,
– Control parameters.

Fig. 2.25 shows the window with the generator parameters of the given element of a Wind Turbine Doubly-Fed Induction Generator (Phasor Type).

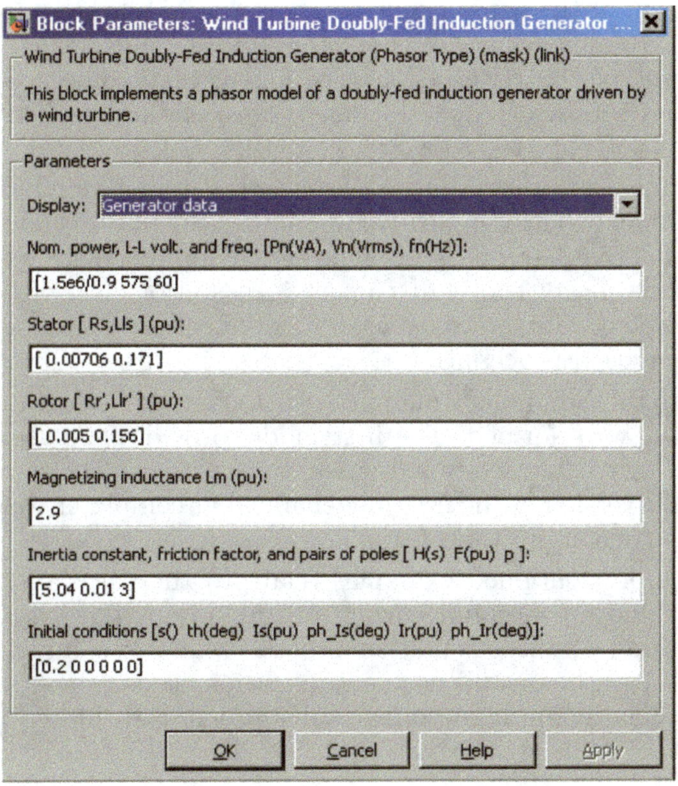


Fig. 2.25 - Window with the generator parameters of the given element of a Wind Turbine Doubly-Fed Induction Generator (Phasor Type)

The parameters of the remaining elements of this block, as well as of the Electric Drives Library and Flexible AC Transmission Systems (FACTS) Library modules, are explained in detail in the corresponding literature [1, 8].

### 2.1.4. Application of MATLAB® technical computing software in calculation of characteristic values of fault current

#### 2.1.4.1 General considerations

The selection of high-voltage equipment is an integral part of designing power system facilities. This selection is conducted on the basis of criteria which include indicative values of the network at the location of installation of the equipment and the rated or allowed values of the equipment [10]. The characteristic values of the fault current represent a component of the criteria for selection of high-voltage equipment. These values can be calculated using various methodological processes [10, 11] and software tools [5, 12-14].

The goal of this example is the show the application of the MATLAB® technical computing software and the Simulink® module in the calculation of the characteristic values of the fault current. The example of the calculation of the fault current in a section of a power system illustrates the advantage of using the aforementioned module over the classic calculation process [15].

Using the Simulink® module, two models for calculation of the fault current were formed. The first model provides the user with the corresponding calculation in individual time periods. The second model is a modified version of the first model aimed at completely determining the fault current in all time periods.

The aforementioned examples relate to the calculation of characteristic values of the fault current outside of any database software. An illustration of the calculation process which is conducted within database software, as well as the application of databases in the automation of designing high-voltage substations, is displayed in Section 2.3 [16].

#### 2.1.4.2 Calculation of characteristic values of the fault current

*a) Classic calculation process*

Fig. 2.26 shows a single-pole diagram of a section of a power system with a three-phase short circuit on busbars C. Breakers 1, 3', 3" 4', 4", 5 and 6 are switched on. The faults on the 220 kV side of the network are switched off in 0.2 s. The considered substation is located

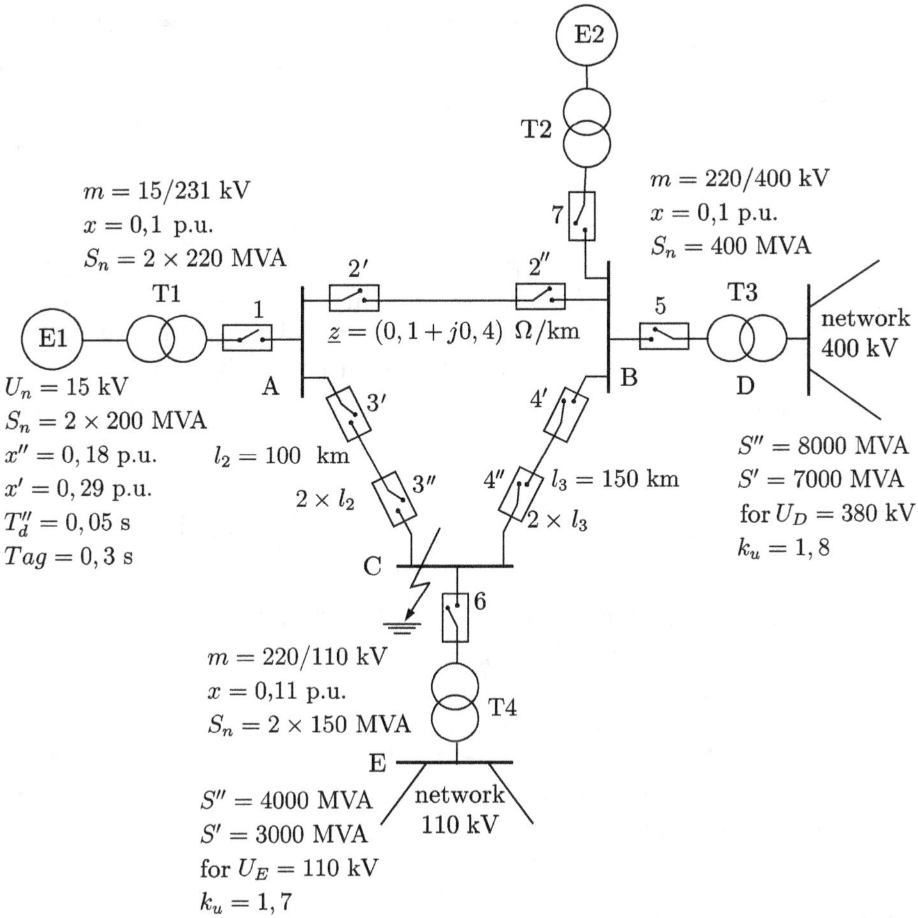


Fig. 2.26 - Single-pole diagram of a section of a power system with a three-phase short circuit on busbars C

within a network with an efficient grounding neutral point. Data on the elements is shown in Fig 2.26.

For the calculation of characteristic values of the current of a three-phase short circuit it is necessary to form a proper equivalent diagram. This process is displayed in detail in the literature [10, 15]. The effective value of the current of a short circuit is determined for a subtransient, transient and steady state period. The active resistances of the elements are used only for determining the time constant of aperiodic components of the fault current [10].

The effective values of the fault current for all three periods are calculated using the expression:

$$I_k = \frac{k \cdot U_{nC}}{\sqrt{3} \cdot X_e} \tag{2.1}$$

The symbols in 2.1 have the following meanings: $k$ - factor with a value of 1.1 for a subtransient period, 1.15 for a transient period, 1.20 for a steady state period; $U_{nC}$ – effective value of phase to phase voltage of the system at the fault location; $X_e$ – equivalent reactance for the corresponding period, observed parallel from the fault location.

The approximate value of the impulse component of the current of a short circuit on busbars C is determined using the expression:

$$i_{imC} = \sqrt{2} \cdot k_{im} \cdot I_k^{''} = \sqrt{2} \cdot \left(1 + e^{-\frac{0.01}{T_{ae}}}\right) \cdot I_k^{''} \tag{2.2}$$

where: $k_{im}$ – impulse coefficient, $T_{ae}$ – time constant of the aperiodic component of the fault current, $I_k^{''}$ – subtransient value of current of a three-phase short circuit.

Through the calculation process, displayed in detail in [10, 15], the characteristic values are determined for the current of the three-phase short circuit. The values of a subtransient fault current and impulse component are shown in Table 2.2.

### 2.1.4.3 Calculation process using the Simulink® module

The single-pole diagram of a section of a power system in Fig. 2.26 has been modified in a way which is appropriate for forming a Simulink® model (Fig. 2.27).

The symbols in Fig. 2.27 have the following meanings: $U_a$, $U_b$, $U_c$ – voltage sources, $x_1$, $x_2$, $x_3$ – reactance of corresponding branches reduced to the value of voltage at the fault location, $I_1$, $I_2$, $I_3$ – current in branches 1, 2 and 3, respectively, $I_k$ – total fault current. Fig. 2.28 shows the appearance of the corresponding Simulink® model. All elements of the model are taken from the Simulink® library within the Simulink® Library Browser.

With the goal of automating the calculation process, an input m-file was formed which contains the data necessary for executing the

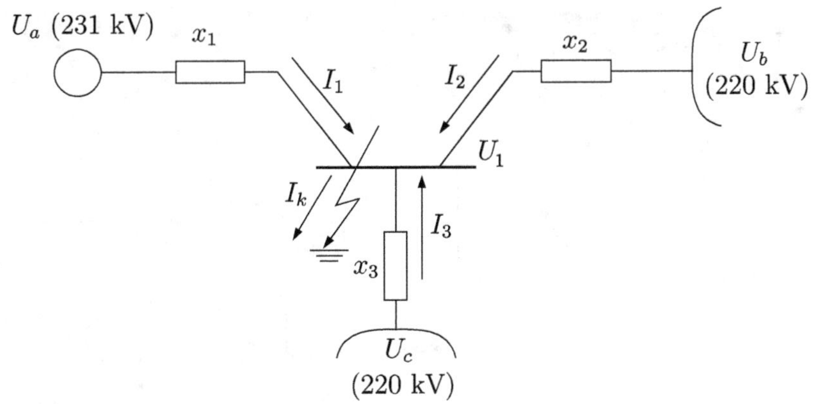


Fig. 2.27 – Equivalent circuit for forming a Simulink® model

simulation in a graphical environment. By using this file and the proposed model, it is possible to individually determine the subtransient, transient and steady state component of the fault current.

The icons marked $U_a$, $U_b$ and $U_c$ in Fig. 2.28 represent the function of sine by which the stated voltage sources are modeled. The existence of four multipliers can be seen in Fig. 2.28. Three multipliers, connected to the voltage sources, contain a coefficient of $1/x$, where $x$ represents the reactance of the corresponding branch within the observed period. The fourth multiplier is connected to the Product2 icon. This multiplier contains a coefficient with a value of $1/R$ where $R$ represents the parallel active resistance at the fault location. In the normal working regime this resistance has an infinitely large value, while during a short circuit its value falls to zero. The fault timing is set with the element marked Step2, which represents the step function generator. The element marked Product2 defines the amount of input data which will be jointly multiplied.

For the purpose of measuring the desired values, it is possible to attach the corresponding measuring instrument (oscilloscope). In the model from Fig. 2.28, four oscilloscopes have been placed through which the indicated variables are registered.

The algorithm for calculation consists of the following steps:

- forming of the input m-file by which the input data is defined within the local memory of MATLAB® technical computing software,
- forming of the Simulink® model (Fig. 2.28), which also in-

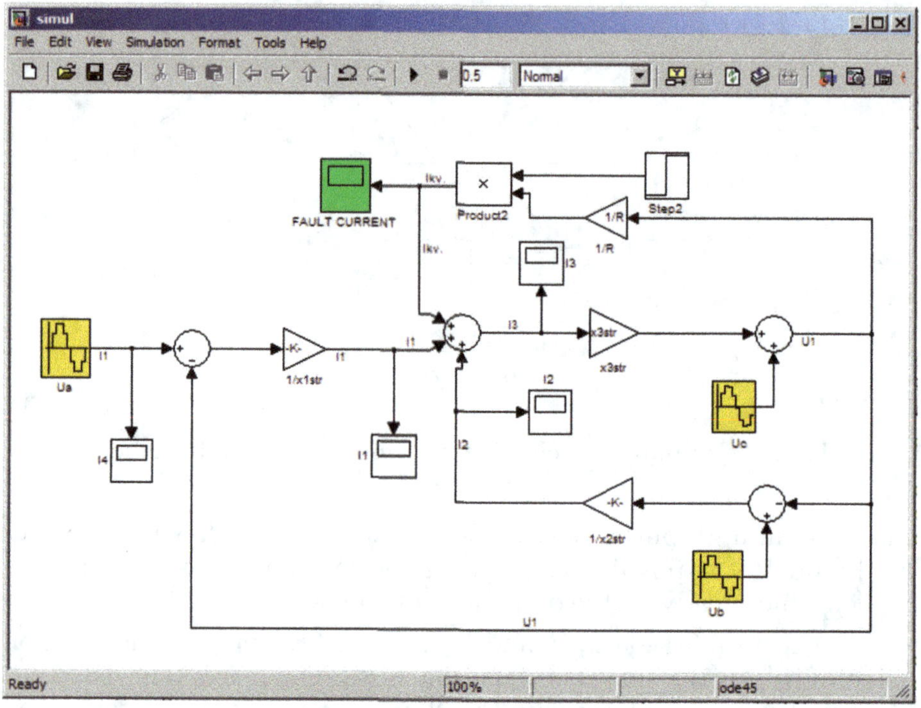


Fig. 2.28 – Appearance of simplified Simulink® model

cludes defining its elements,
- setting the simulation parameters,
- executing the calculation.

All parameters of the equivalent diagram, as well as the input data, are assigned in the form of one *m*-file called *simu.m* which is shown in Fig. 2.29.

Using the described Simulink® model, calculations were executed for the input data from Fig. 2.26. A portion of the calculation results are displayed in Table 2.2. The symbol $I''_{KC}$ in Table 2.2 relates to the effective value of the subtransient component of the current of a three-phase short circuit on busbars C. On the basis of the calculation results displayed in Table 2.2, it can be concluded that both examples provided practically the same calculation results.

Fig 2.30 shows a complete model in Simulink® which enables the entire calculation of the current of a three-phase short circuit for the stated time period. As opposed to the simplified model provided in

C:\Documents and Settings\Zlatan\My Documents\MATLAB\simu.m

```
% Defining the values of Ua, Ub and Uc [V]
Ua=231e+3/sqrt(3)*sqrt(2);
Ub=220e+3/sqrt(3)*sqrt(2);
Uc=220e+3/sqrt(3)*sqrt(2);
w=314;
% Defining the active resistance and reactance values [Ohm] for all periods
xg1str=48.03; xg1tr=77.37; xg1ss=240; Tag1=0.3;
xl2=40; Rl2=10;
xl3=60; Rl3=15;
xt1=26.68;
xt3=12.1;
xt4=35.5;
xn400str=5.46; xn400tr=6.24; xn400ss=xn400tr; Tan110=-0.01/log(1.7-1);
xn110str=12.1; xn110tr=16.13; xn110ss=xm110tr; Tan400=-0.01/log(1.8-1);
Rg1=xg1str/(w*Tag1); Rn110=xn110str/(w*Tan110); Rn400=xn400str/(w*Tan400);
R1=(Rg1+Rl2)/2; R2=Rn400+Rl3/2;
% Defining the impulse coefficient (k=1.1 or 1.15 or 1.2);
k=1.1;
x1str=(xg1str+xt1+xl2)/2; x1tr=(xg1tr+xt1+xl2)/2; x1ss=(xg1ss+xt1+xl2)/2;
x2str=xn400str+xt3+xl3/2; x2tr=xn400tr+xt3+xl3/2; x2ss=xn400ss+xt3+xl3/2;
x3str=xt4/2+xn110str; x3tr=xt4/2+xn110tr; x3ss=xt4/2+xn110ss;
% Calculation of equivalent resistances, reactances and time constant
Re=R1*R2*Rn110/(R1*R2+R1*Rn110+R2*Rn110);
xestr=x1str*x2str*x3str/(x1str*x2str+x1str*x3str+x2str*x3str);
Tae=xestr/(w*Re);
R=10e+10;
Tdstr=0.105;
Tdtr=0.15;
Tkl=0.07;
Tswitch=0.38;
% Calculation of impulse component of the current
Iim=sqrt(2)*(1+exp(-0.01/Tae))*k*220000/sqrt(3)/xestr
```

script | Ln 1 | Col 1 | OVR

Fig. 2.29 – Appearance of *m*-file with assigned values

Fig. 2.28, this model contains several subsystems which enable the automatic transition from one time period to another. The first subsystem relates to the change of factor $k$, and with that, to the change in the value of voltage of the voltage source in various time periods (Fig. 2.31). The second subsystem encompasses the change of reactance of the elements during the fault. This subsystem is displayed in Fig. 2.32.

Based on the displayed results it can be concluded that knowing the characteristic values of the fault current is necessary for the

Table 2.2 - Collective overview of calculation results of the current of three-phase short circuit on busbars C of a single-pole diagram according to Fig. 2.26

| Example of application | $I''_{KC}$ (kA) | $I_{imC}$ (kA) |
|---|---|---|
| Classic process | 10.0 | 25.7 |
| Simulink® | 10.2 | 26.0 |

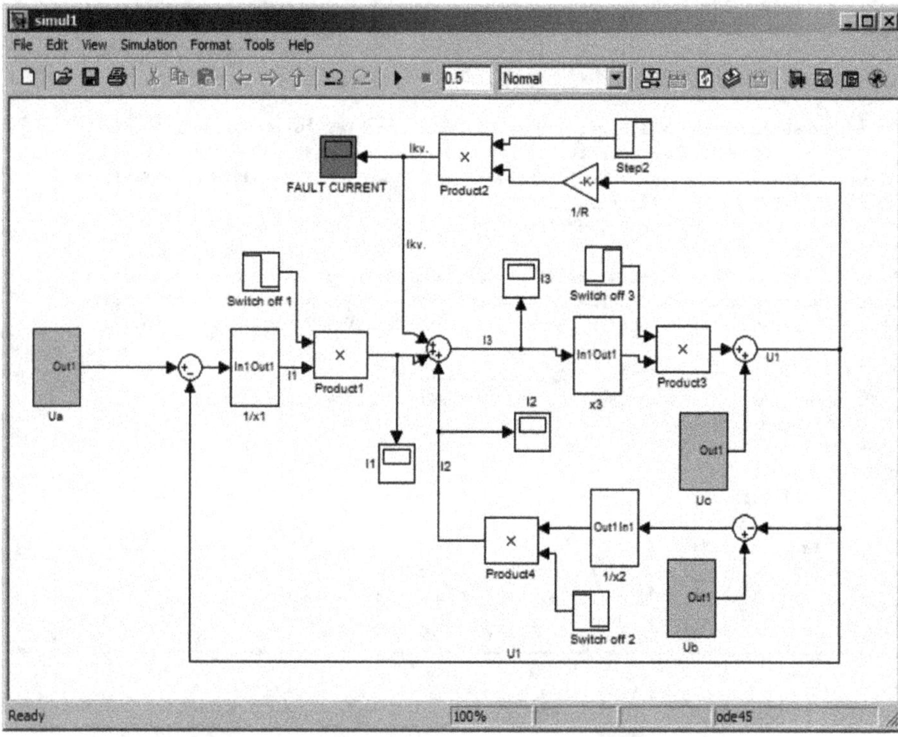


Fig. 2.30 – Appearance of complete Simulink® model

proper selection of high-voltage equipment. These values can be calculated using the existing software tools or a user-developed program. The illustrated example shows the possibility of using a Simulink® model through which automation of the process of calculating the characteristic values of the fault current is achieved.

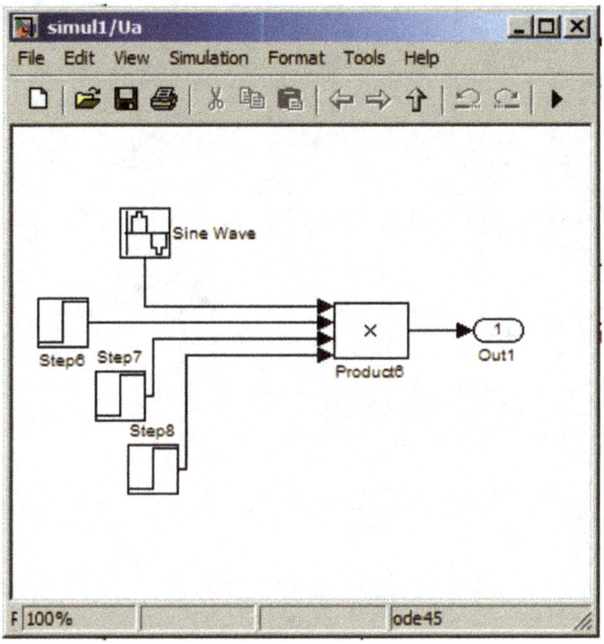


Fig. 2.31 – Appearance of voltage source subsystem

### 2.1.5. SPLCAD software tool for designing medium-voltage overhead lines

#### 2.1.5.1 Introduction

This example shows the software tool SPLCAD (Software Power Line CAD) for designing medium-voltage overhead lines [17]. The tool was realized using MATLAB® technical computing software as the development platform for creating the user interface (Graphical User Interfaces Toolbox – GUI), and the programs MS Access and MS Excel for creating and working with databases. The application of the programs for creating and working with databases is illustrated in section 2.3.

SPLCAD was developed within the framework of technological development project [18] and enables the verification of electric strain of equipment for conductor attachment and complete mechanical calculation of an overhead line.

SPLCAD uses the geo-referenced foundation of the region in which

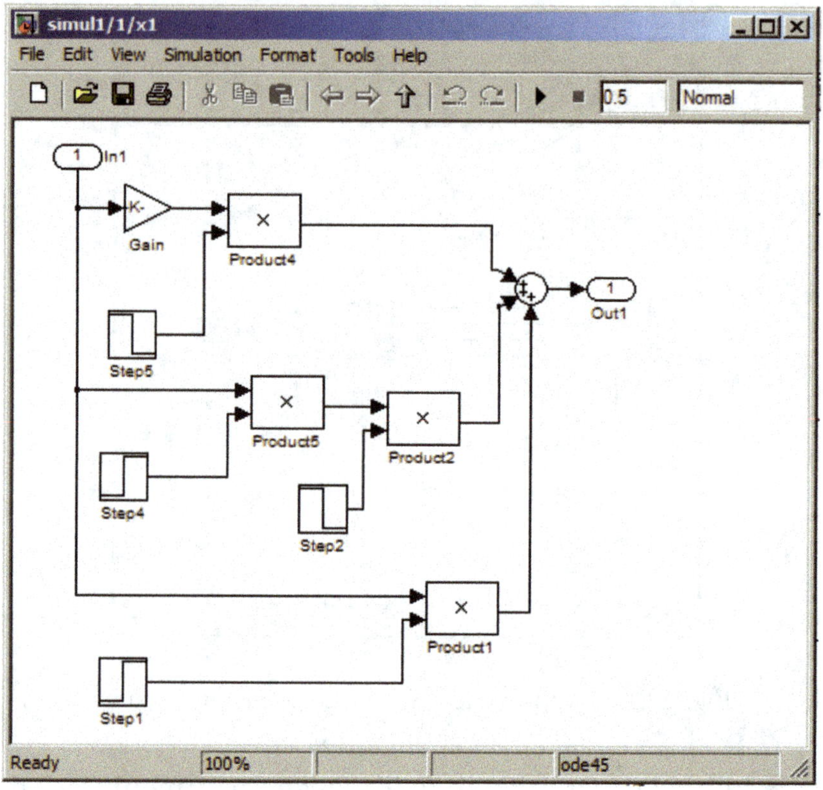


Fig. 2.32 - Appearance of subsystem for change of reactance of elements during fault

the construction of the newly-designed overhead line is envisaged as its basic data. There is also the possibility of reading and working with a 3D model of the terrain for the given area. SPLCAD has also been adapted for reading, displaying and processing results obtained from a GPS (Global Positioning System) device. Using the measuring data from the GPS device with or without an existing 3D model, the tool can create a new 3D model of the area of interest. This enables the complete visualization of the terrain as well as the coordination of work in a 3D space. SPLCAD contains a database with models of standard equipment which is used in electrical energy distribution systems. The same database also enables one to have complete insight into the estimation of works and equipment for a specific section of overhead line.

#### 2.1.5.2 Description of SPLCAD software tool and an example of its application

Working with the tool SPLCAD begins by forming a new project. After creating a new project the program generates a communication window for entering and processing data. The basic input data for the project includes: foundations, 3D models of the terrain and measured GPS device data.

All foundations are divided into two groups. The first group refers to foundations such as cadastre, urbanistic, topographic and other drawn foundations. Pictures of the actual terrain obtained from satellite and airplane photographs represent the second group of foundations. The basic condition for the ability to work with a certain foundation relates to the necessity for it to be georeferenced. Due to the given condition, when attempting to import a certain foundation, SPLCAD automatically verifies whether the foundation is georeferenced, i.e. does the electronic record of the foundation also contain a georeference matrix. The aforementioned matrix determines the resolution, i.e. proportion, orientation and geocoordinates of one point of the foundation.

Because SPLCAD exclusively supports operation in the geo-units UTM (Universal Transverse Mercator) system, it is necessary to determine which UTM zone the given foundation, i.e. area which it represents, belongs to (Fig. 2.33).

Upon selecting the proper UTM zone the next step in the process is creating a georeference matrix. In order to create a valid georeference matrix it is necessary to know the $X$ and $Y$ coordinates for at least two points of the foundation, whereby the given points are not allowed to have the same $X$ or $Y$ coordinate. After graphically marking the points of the foundation for which coordinates are known and entering them into the appropriate row and column of the georeferencing table, it is then necessary to execute the calculation of values of the elements of the georeference matrix. By determining these values the process of georeferencing the foundation is complete.

SPLCAD has been adapted for reading and process data from the standard type of electronic record (ASCII Grid) which is a 3D model of terrain. At the same time the tool itself, on the basis of a certain number of valid points of known $X$, $Y$ coordinates and elevation levels can create a 3D model of a certain area. The created models are saved in MATLAB® files with the extension .mat.

The files of measured data from a GPS device which SPLCAD

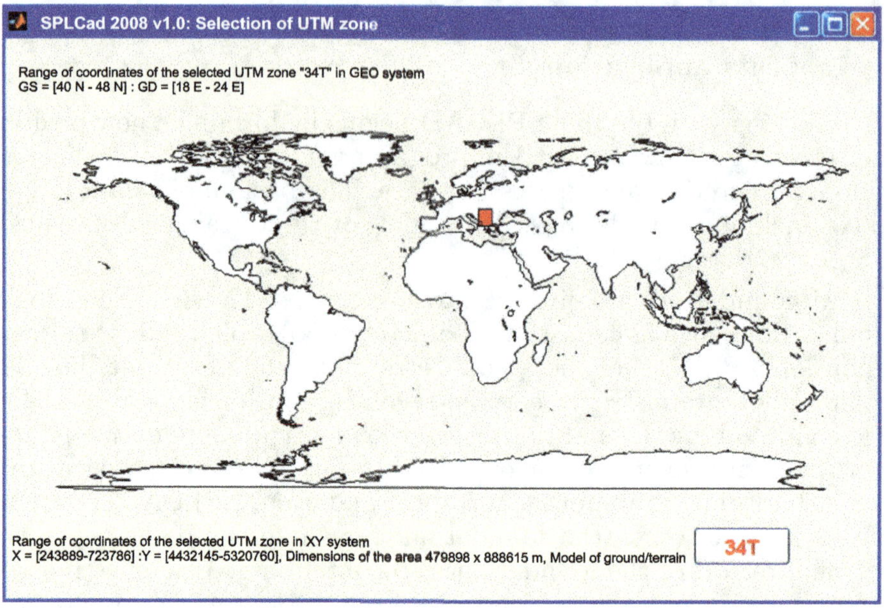


Fig. 2.33 – Selection of UTM zone

can read are in ASCII records (.txt) and records created using Microsoft Office Excel (.xls). SPLCAD offers the capability of importing certain sections which were previously created and elaborated in other projects into the current project. This enables the upgrading of an existing project and forming a real picture of the status of the entire area of interest.

*a) Main working window*

The main working window contains two windows for graphical display, the collection of tools for various purposes and a certain number of function menus (Fig. 2.34). The georeferenced foundation group 1 is entered into the upper window for graphical display while group 2 is entered into the lower window. The same windows serve for entry and display of tower locations. Tower locations can be created, i.e. entered in graphically (by drawing), tabularly (by entering coordinates in the corresponding table), as well as through input from a certain file which contains measurement data of the coordinates of the tower locations. A tower location represents a unique object of the active project. After being created it is assigned an identification number which represent the primary key for the table of created tower locations in the database of the active project.

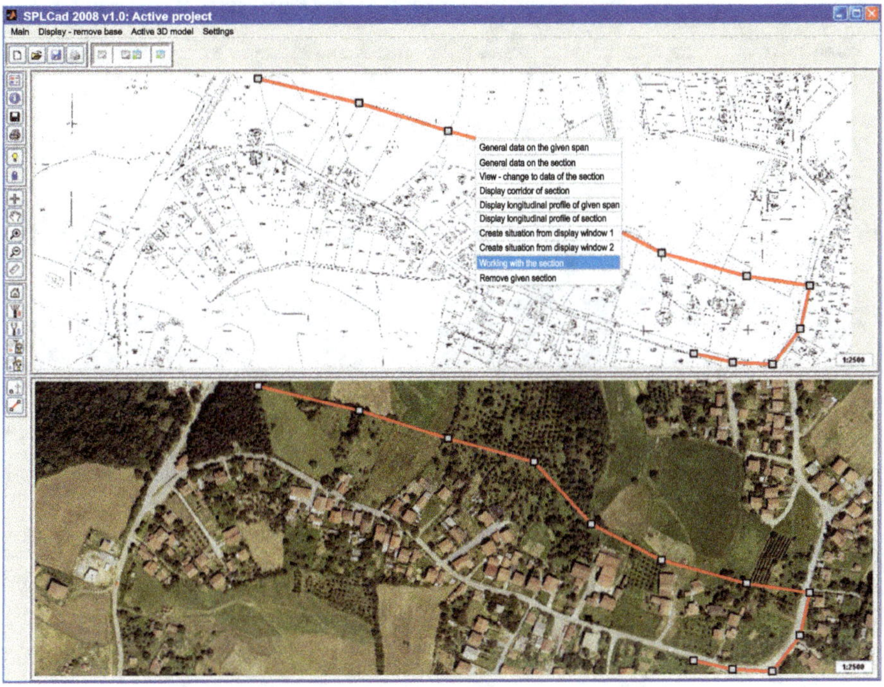


Fig. 2.34 - Main working window with synchronized display of the created section

The next step is to create the section of the overhead line. The section represents a collection of tower locations with physically connected spans. The created section represents a unique object which refers to a certain part or a complete overhead line. In order for the section to be successfully created, it is necessary to adequately fill in the required data on that section, as well as the area to which it belongs. The section is also assigned a corresponding identification number which represents the primary key of the table of created sections. SPLCAD automatically assigns each created section a corresponding UIContextMenu with a collection of functions (Fig. 2.34) which enables further work with the section. By activating the function UIContextMenu with the inscription "work with the section" a new dialog box is opened which enables the design of the overhead line (Fig. 2.35). It is important to note that a majority of the tools in the main working window and working with a section window support a synchronized display of the structure for graphical representation.

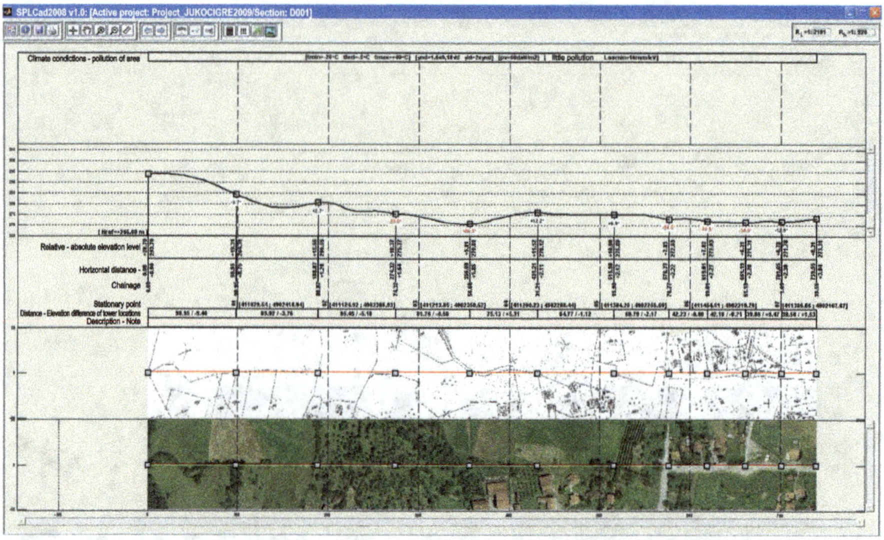

Fig. 2.35 - Longitudinal profile drawing and foundation of a section of overhead line

*b) Working with a section*

SPLCAD enables the entry of a conductor of a new or existing feeder, of a certain type and cross-section. The user assigns the value of the maximum total operational strain of the conductor $\sigma_{F_{\max}}$, after which the program verifies whether the entered value is smaller or equal to the permitted value $\sigma_{nd}$ of the given conductor [19]. The symbol for which feeder the conductor belongs to, the symbols for its beginning and end, as well as the adopted value of $\sigma_{F_{\max}}$, are displayed in the upper object for graphical display. It is possible to change the cross section and type of conductor at the locations where they are mechanically non-tensioned, as well as the number of feeders on the section, which means that working with double circuit lines and mixed lines is possible.

The displayed pictures of the situation along the section are double referenced. The georeference matrices contain the newly created matrix and reference matrix. For the purpose of total coordination within the area, the reference matrix was created in accordance with the distances from the starting point and width of the section corridor.

Each graphical marking of a tower location has a corresponding UIContextMenu with the inscription Branch selection / Foundation / Grounding tower. By selecting the menu, a dialog box is opened which gives the user the option of selecting the branch of the tower depending on material, height, rated tensile strength and manufacturer. After selecting the branch the user also defines the restrained length of tower in foundation.

The user has the option of a roller, prism or block foundation, during which it is necessary to recognize the value of restrained length of tower in foundation and the mechanical characteristics of the soil while fulfilling the condition:

$$M_d \geqslant F_n L \tag{2.3}$$

where: $M_d$ – allowable bending moment of the selected foundation [daN.m], $F_n$ – rated horizontal tensile force at the top of the branch [daN], $L$ – length of branch [m].

When selecting the grounding tower there is the possibility of developing a supplementary ring or radial grounding grid if the foundation grounding in the form of a "reinforced rods" is not implemented or does not fulfill the criteria of protection from a back flashover, or does not fulfill the conditions for safety from touch voltage.

After selecting the aforementioned elements, the existing graphical marking of the tower location is replaced by the new one which recognizes the length of the branch, the value of restrained length of tower in foundation (proper positioning according to elevation level) and the rated horizontal tensile force at the top of the branch. The newly created graphical marking is assigned a new UIContextMenu with a function carrying the inscription "equipping tower, conductor attachment".

SPLCAD offers a separate dialog box for equipping the tower with crossarm and equipment for conductor attachment. During this time it is possible to display a 3D model of the area surrounding the tower location and a model of the installed tower branch. SPLCAD offers the ability to select a crossarm depending on material (concrete, steel and crossarm from aluminum alloy), type (one-leg, two-leg for accepting one, two three or four conductors) and purpose. Another important piece of information which is entered into the database is the crossarm height.

The user selects the crossarm height by entering the value which indicates its distance from the top of the tower. SPLCAD verifies whether the selected crossarm can be installed on the given tower and at the defined height. After installing the crossarm it is necessary to outfit it with equipment for conductor attachment. SPLCAD contains the corresponding modeled types of pin-type and post insulators, as well as insulator string made from glass / porcelain cap and pin insulator units. By selecting the equipment attachment along with defining the crossarm height, the coordinates are clearly defined for the attachment points in the 3D system. Each point receives its own identification number in the corresponding table of the database.

The aforementioned data, which is exceptionally significant for proper and precise mechanical calculation, is entered into the database along with the selected equipment.

*c) Mechanical calculation*

SPLCAD conducts mechanical calculation of the lines in stages for each tension section. The basic conditions for beginning the calculation is that at least one tension section is completely defined which means that the necessary equipment is selected and acceptance of the conductors is carried out at all attachment points of the created tension section.

SPLCAD enables the selection of reference conditions and the following calculations:

- maximum values of horizontal strain of a conductor $\sigma_{\max}$,
- changes of the conditions of a conductor $\sigma(t)$,
- sag and vertical distance of a conductor from the ground,
- span between conductors during asynchronous swinging due to wind,
- mechanical strain of towers, crossarm and equipment for conductor attachment.

*c.1 Calculation of maximum values of horizontal strain of a conductor*

The maximum values of horizontal strain of a conductor are calculated for each span of a tension section, during which time at a suspension point with a larger elevation level the maximum (assigned) value of total strain of the conductor $\sigma_{F_{\max}}$ is reached (Fig. 2.36).

In accordance with the symbols in Fig. 2.36 a system of equations

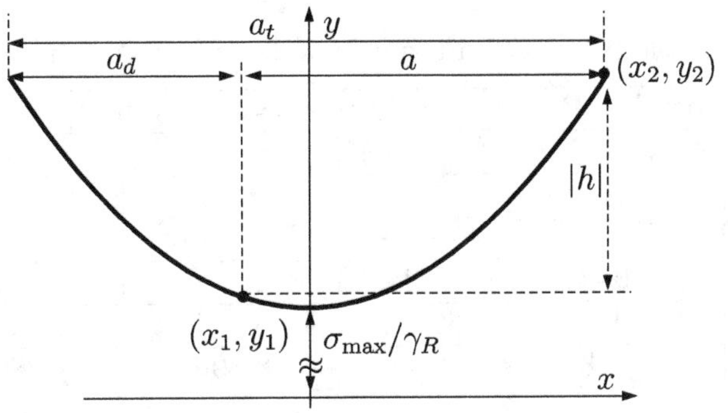


Fig. 2.36 – Self coordinate system span

is formed for all spans of the tension section:

$$F_1(\sigma, x_2) = y_2 - \frac{\sigma}{\gamma_R} \cosh \frac{\gamma_R}{\sigma} x_2 = 0 \tag{2.4}$$

$$F_2(\sigma, x_2) = y_2 - |h| - \frac{\sigma}{\gamma_R} \cosh \frac{\gamma_R}{\sigma} (x_2 - a) = 0 \tag{2.5}$$

Variable $y_2$ is assigned the value $\sigma_{F_{\max}} \gamma_R$ where $\gamma_R = \gamma$ or $\gamma_R = \gamma + \gamma_{nd}$ (various reference conditions). The provided system of equations is solved using the Newton-Raphson method with initial values of unknowns $\sigma(0) = \sigma_{F_{max}}/2$ and $x_2^{(0)} = a/2$.

For all calculated values of the maximum horizontal strain of the conductor, the values of critical span $a_{cr}$ are determined using the following expression:

$$a_{cr} = \frac{\sigma_{\max}}{\cos \psi_{ai}} \sqrt{\frac{24(t_{\min} - t_0)\alpha}{\gamma_R^2 - \gamma^2}} \tag{2.6}$$

where: $t_{min}$ – minimum temperature in the area of the section (°C), $t_0$ – temperature during which there is an additional load on the overhead line (°C), $\alpha$ – coefficient of temperature expansion of the conductor (1/°C), $\cos \psi_{ai}$ – cosine of the slope of the ideal span of the tension section [20].

The examples below of the first tension section of the feeder with the marking "I10A01" of the aluminum conductor steel reinforced ACSR 50/8 mm$^2$ show the way in which SPLCAD executes mechanical calculation. Table 2.3 shows the values $\sigma_{\max}$ and $a_{cr}$ determined through the calculation on the basis of the aforementioned relations.

Table 2.3 – Calculation results of $\sigma_{\max}$ and $a_{cr}$

| Conductor: ACSR 50/8 mm$^2$ /$\sigma_{F_{\max}} = 9$ daN/mm$^2$/$k_{ice} = 1,6$ | | | | |
|---|---|---|---|---|
| Ref. | $t_{\min} = -20°\text{C}$ $(\gamma_R = \gamma)$ | | $t_0 = -5°$ C+ice $(\gamma_R = \gamma + \gamma_{nd})$ | |
| Item No. | $\sigma_{\max}$ (daN/mm$^2$) | $a_{cr}$ (m) | $\sigma_{\max}$ (daN/mm$^2$) | $a_{cr}$ (m) |
| 1 | 8.942 | 39.28 | 8.818 | 38.73 |
| 2 | 8.984 | 39.46 | 8.914 | 39.15 |
| 3 | 8.973 | 39.41 | 8.896 | 39.07 |

From the collection of given values, the span which has a horizontal strain calculated as having the minimum value is singled out (Item No. 1 for $t_0 = -5°$ C + ice $(\gamma_R = \gamma + \gamma_{nd})$). That value is adopted as the unique value of the maximum horizontal strain of the conductor in the tension section. This fulfills the condition for the value of the total strain of the conductor in each attachment point of the tension section to be less or equal to the value $\sigma_{F_{\max}}$.

*c.2 Selection of reference conditions*

The value of critical span $a_{cr}$, which is calculated on the basis of the adopted maximum value of horizontal strain of the conductor, is compared with the value of the ideal span $a_i$ of the given tension section [20]. SPLCAD conducts a selection of reference conditions on the basis of the relation of the values of the aforementioned spans.

*c.3 Calculation of changes of the condition of a conductor*

The calculation of changes of the conditions of a conductor is conducted without using the ideal span during which for each span $(k)$ of the tension section the following system of equations is formed:

$$F_{1k} = L_k - \sqrt{h_k^2 + \frac{4\sigma^2}{\gamma_k^2}\,\text{sh}^2\left(\frac{a_k\gamma_k}{2\sigma}\right)} = 0 \tag{2.7}$$

$$F_{2k} = L_k - L_{0k} - (t - t_0)\alpha L_{0k} - \frac{\sigma_{F_{srk}} - \sigma_{F_{sr0k}}}{E} L_{0k} = 0 \tag{2.8}$$

$$F_{3k} = \sigma_{F_{srk}} - \frac{a_k\sigma}{2L_k} - \frac{L_k^2 + h_k^2}{\dfrac{4L_k}{\gamma_k}\,\text{th}\,\dfrac{a_k\gamma_1}{2\sigma}} = 0 \tag{2.9}$$

where: $L_k$ – length of the conductor in the $k$ span of a tension section, $L_{0k}$ – length of the conductor in the $k$ span of a tension section during zero iteration, $E$ – Young's module of elasticity.

It is necessary to add a supplementary equation to the given system of equations which shows that the sum of spans is equal to the length of the tension section of $n$ spans:

$$F_{3n+1} = a_{01} + \cdots + a_{0k} + \cdots + a_{0n} - a_1 - \cdots - a_k - \cdots a_n = 0 \tag{2.10}$$

The values of variables which represent the initial lengths ($L_{0k}$) and middle strains ($\sigma_{F_{sr0k}}$) are calculated for each span from the expression (2.7) and (2.8) by replacing the values according to the adopted reference conditions. The given system of equations is solved using the Newton-Raphson method with the initial values of unknowns $a_i^{(0)} = a_{0i}$, $\sigma^{(0)} = \sigma_{\max}/2$ and $\sigma_{F_{sri}}(0) = \sigma_{F_{\max}}/2$. Table 2.4 shows the changes of the horizontal strain of the conductor depending on the changes in temperature for the analyzed tension section.

Table 2.4 - Horizontal strain of a conductor depending on temperature changes for the analyzed tension section

| Conductor: ACSR 50/8 mm$^2$ / $\sigma_{\max} = 8.818$ daN/mm$^2$ | | | | |
|---|---|---|---|---|
| $t$ (°C) | −20° C | −5° C | −5° C+ ice | +40° C |
| $\sigma$ (daN/mm$^2$) | 2.588 | 2.196 | 8.818 | 1.585 |

### *c.4 Sags and vertical distances of conductors*

By solving the state equation of the conductor, as well as establishing the elevation levels of the route of the section, the vertical distances of the conductors from the ground are calculated. The values of minimum vertical distances of the conductors from the ground are of significant interest, as are the values of maximum flexion in each span (Table 2.5).

The symbols in Table 2.5 have the following meanings: $f_{\max}$ – maximum value of sag in the span, $D_{\min}$ – minimum space of conductor in the middle of the span [21], $\Delta h$ – minimum vertical distance of the conductor from the ground, $L$ – distance of location with minimum value $\Delta h$ from the initial point of the section.

Table 2.5 - Values of minimum vertical distance of the conductor from the ground and values of maximum flexion in each span

| Conductor: ACSR 50/8 $mm^2$ / $t = +40°$ C $\rightarrow \sigma = 1.585$ daN/$mm^2$ | | | | |
|---|---|---|---|---|
| Item No. | $f_{\max}$ (cm) | $D_{\min}$ (cm) | $\Delta h$ (m) | $L$ (m) |
| 1 | 266 | 117 | 4.07 | 49.47 |
| 2 | 219 | 109 | 10.00 | 98.95 |
| 3 | 198 | 105 | 8.12 | 204.25 |

*c.5 Space between conductors during asynchronous swinging due to wind*

Recognizing the geometry and layout of the crossarms with which the heads of the towers are equipped, as well as the coordinates of the conductor attachment points, the calculation is made for the safety interval between conductors in the middle of the span during asynchronous swinging due to wind [21]. Fig. 2.37 shows the layout of conductors as well as the necessary value of minimum interval between conductors in the middle of the span.

*c.6 Mechanical strain of towers, crossarms and conductor attachment equipment*

First strain of the selected towers is calculated in accordance with the cases of the loads defined in the given literature [21]. If some of the towers sustain greater strain than that which is permitted, the user will be informed which tower that is and what the value of calculated strain is. For the mentioned case the dialog box is shown in which one can see the vector diagram of force of the strain of towers and their resulting values (Fig. 2.38).

When calculating the mechanical strain of the crossarms and conductor attachment equipment the vertical force of strain due to the weight of the conductor is calculated along with the additional load (while recognizing the value of gravitational span $a_{gr}$) and the horizontal component of strain which remains as a consequence of the horizontal strain of the conductor (tensioning). If, for the total added equipment, there is no occurrence of overload from the program, then the display of the longitudinal profile is drawn for the defined value of ambient temperature (Fig. 2.39). This also enables the 3D display of the created section of overhead line.

Based on the illustrated process, it can be concluded that using the software tool SPLCAD enables complete 2D and 3D visualization

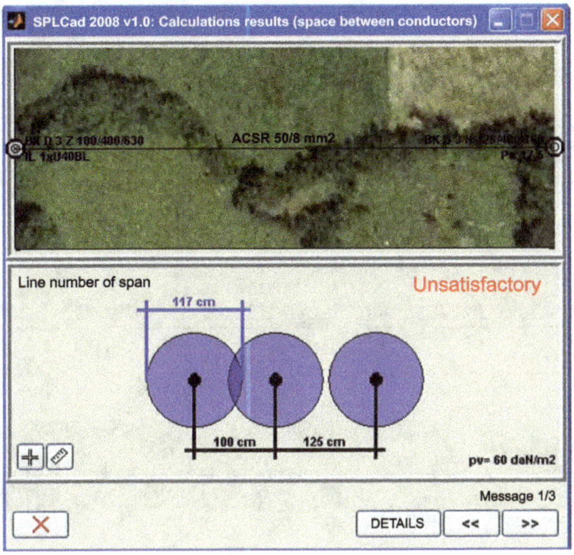


Fig. 2.37 - Layout of conductors and necessary value of minimum interval between conductors in the middle of the span

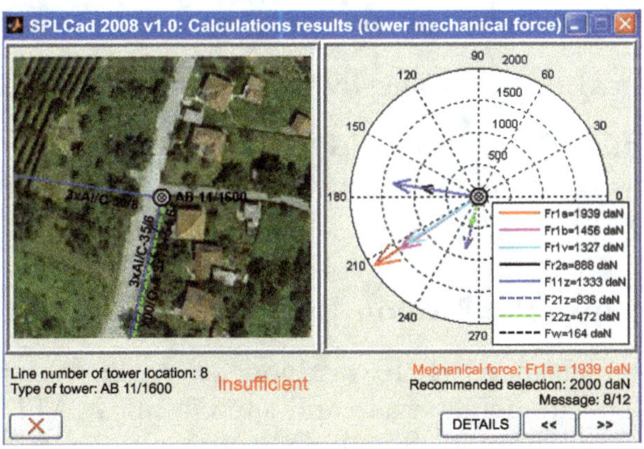


Fig. 2.38 – Dialog box with vector diagram of force of the strain of towers

and coordination. Automation of calculation and the forming of corresponding databases leads to the efficient and precise development of project documentation. SPLCAD can, with a little modification, also be used for designing cable lines.

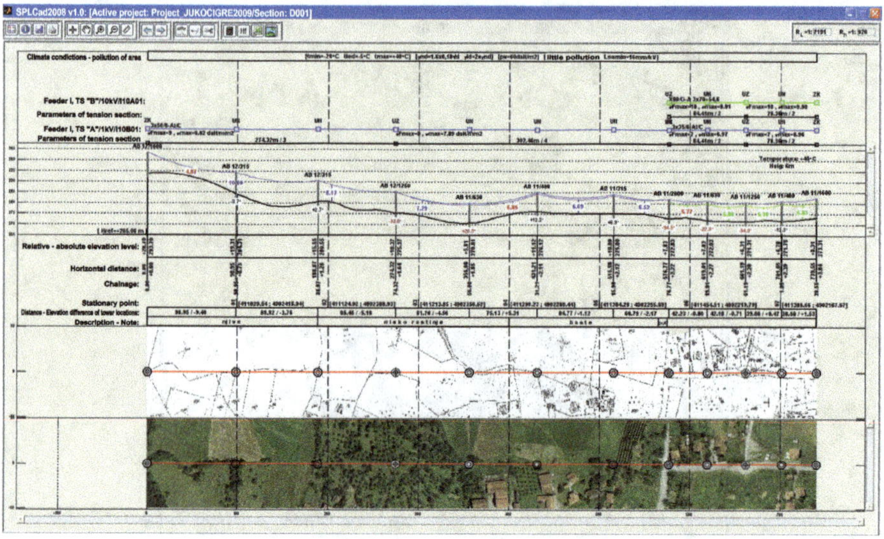

Fig. 2.39 – Longitudinal profile of section determined by calculation

## 2.1.6. Simulink® model of a turbine regulator in the "Kokin Brod" hydroelectric power plant

### 2.1.6.1 General data on the hydroelectric power plant

The "Kokin Brod" hydroelectric power plant (HPP) was constructed on the river Lim. The basic technical characteristics are provided in Table 2.6.

### 2.1.6.2 Modeling the "Kokin Brod" HPP

The aforementioned HPP was modeled within the framework of project [22]. The modeling was conducted using MATLAB® technical computing software and Simulink® software (Fig. 2.40). With the goal of verifying the results of the simulation, the corresponding measurements were taken. A comparative analysis of the numerically and experimentally obtained results for the turbine regulator, as well as for the hydraulic and mechanical section is displayed in the selected literature [23, 24].

The model consists of two closed loops. The first loop refers to the turbine regulator, hydraulic section (system of pipes and losses within the pipes), losses of power within the generator and turbines, as well as

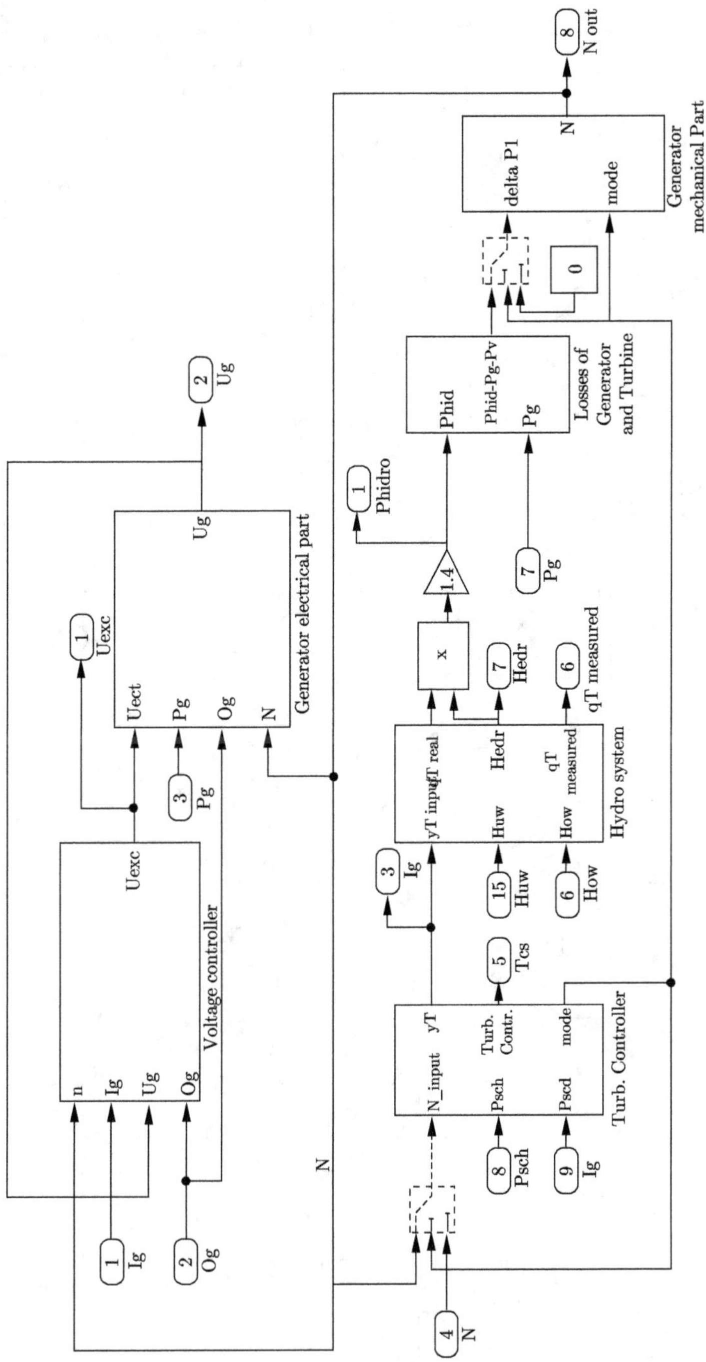


Fig. 2.40 – Complete model of the "Kokin Brod" HPP

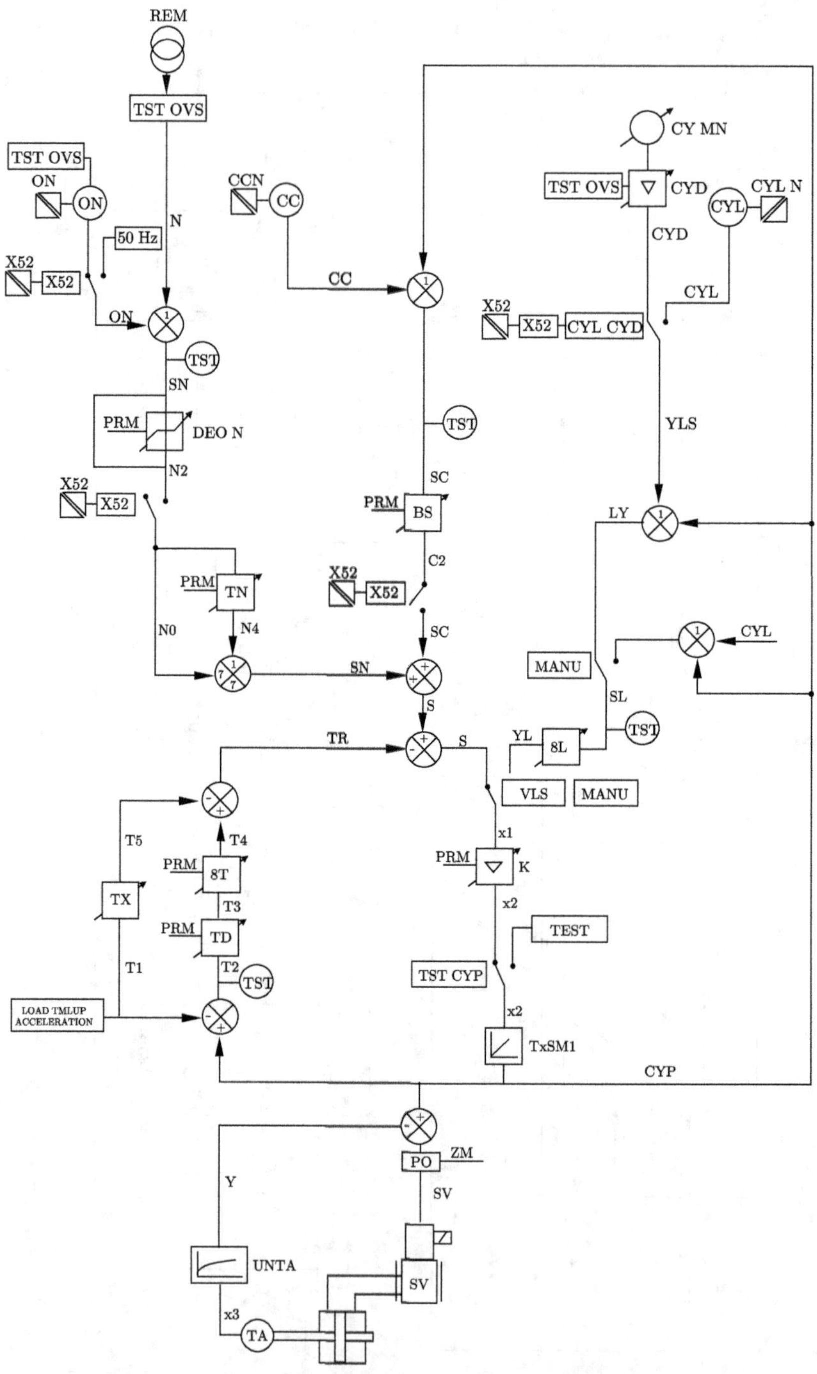


Fig. 2.41 – Electrical diagram of a turbine regulator

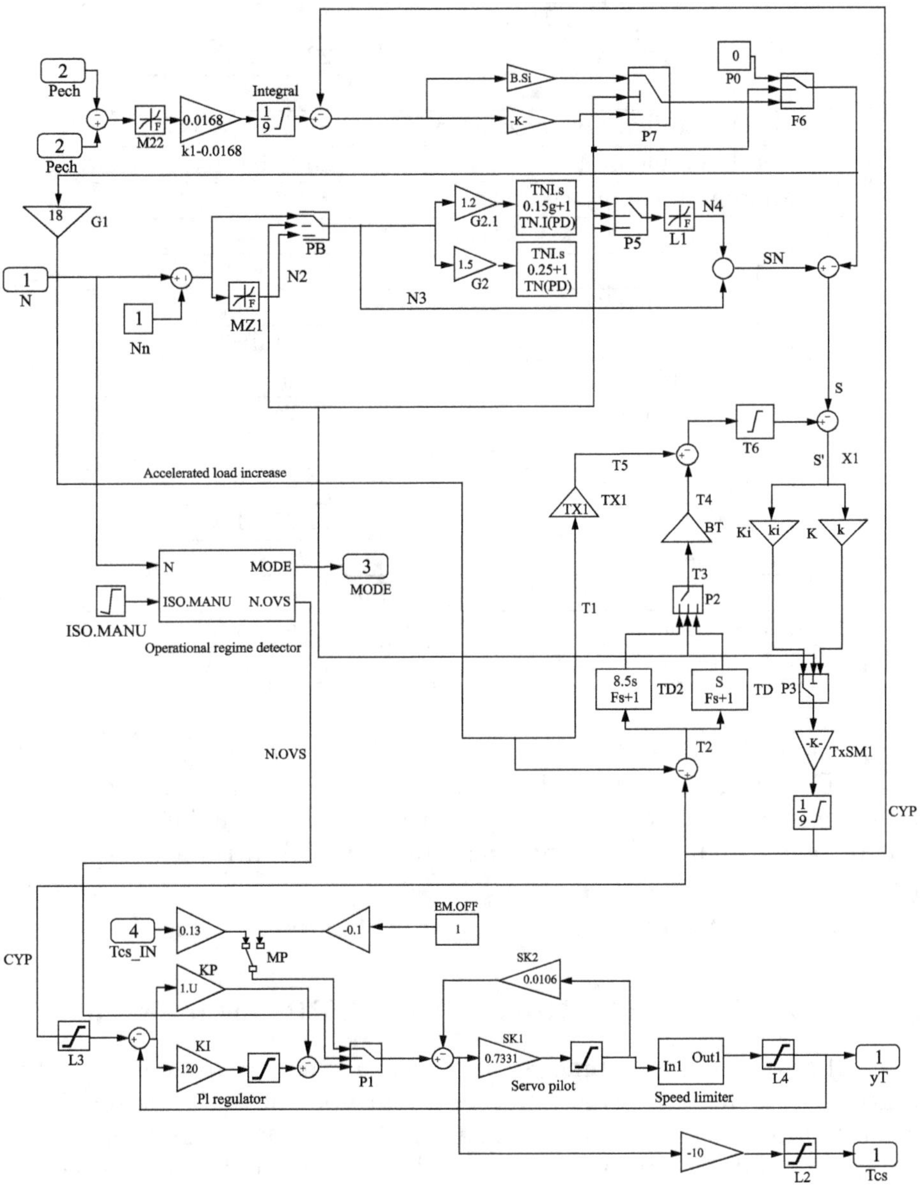


Fig. 2.42 – Detailed Simulink® model of turbine regulator

Table 2.6 – Technical characteristics of the "Kokin Brod" HPP

| *1. Data on the HPP* | |
|---|---|
| Installed capacity | 22.5 MW |
| Type | Dam |
| Total installed flow discharge | 37.4 $m^3/s$ |
| Total volume of accumulation | 250 106 $m^3$ |
| Maximum energy content | 202 GWh |
| *2. Turbine* | |
| Type | Francis 1.45/230 |
| Installed capacity | 11.25 MW |
| RPM | 375 $min^{-1}$ |
| Maximum drop | 72 m |
| Minimum drop | 36 m |
| *3. Generator* | |
| Apparent power | 12.5 MVA |
| Active power | 11.25 MW |
| Power factor | 0.9 |
| *4. Transformer Block* | |
| Rated capacity | 12.5 MVA |
| Transformation ratio | 121/6.3 kV |

the mechanical model of the generator. The return loop is according to speed. The second loop is made up of the voltage regulator and electrical model generator. The return loop is according to voltage. The principals of regulation of power systems are provided in the corresponding literature [25, 26].

A model of the turbine regulator is elaborated below.

#### 2.1.6.3 Display of the "Kokin Brod" HPP turbine regulator

The role of the turbine regulator is to keep the frequency of generator within the permitted limits or to turn off the entire system if the generator drops out of synchronization. The block which refers to the turbine regulator (Turb.Controller) is located in the lower right corner in Fig. 2.40. The turbine regulator inputs are:

- speed (N),
- signal for increasing active power ($P_{sch}$),
- signal for decreasing active power ($P_{SCD}$).

The turbine regulator outputs are:

- turning angle of blades on the turbine expressed in percents ($y_T$),
- MODE signal which provides information on the operational regime,
- turbine control signal ($T_{CS}$).

The turbine regulator has three operational regimes:

- normal regime,
- island regime,
- open circuit regime.

When the speed of the generator is within a range of +0.01 and −0.01 around the rated values the turbine is operating in normal mode. When the speed is outside of the range of ±0.01 for longer than 2 s, the operational regime detector activates the island mode. In the island regime the power is reduced down to the self-consumption of the power plant. Within the open circuit regime there is no load.

The Simulink® model of the turbine regulator is formed on the basis of the electrical diagram and description of individual blocks within the documentation. The electrical diagram of the turbine regulator is displayed in Fig. 2.41.

In Fig. 2.41 three branches can be seen:

- branch CN, which represents the input speed,
- branch CC, which represents the input order to increase or decrease the load (power),
- branch YL, which is used for starting up the power plant.

The meaning of the control points from Fig. 2.41 is provided in the corresponding documentation on the turbine regulator of the denoted HPP [27].

#### 2.1.6.4 Simulink® model of turbine regulator

A detailed Simulink® model of a turbine regulator is displayed in Fig. 2.42. Table 2.7 shows the basic Simulink® blocks which were used for modeling the turbine regulator. Details of all subsystems of the turbine regulator and the procedure for their modeling are provided in literature [22, 23].

Table 2.7 – Basic Simulink® blocks used for modeling a turbine regulator

| Block | Description |
|---|---|
| 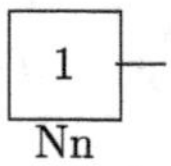 | Block which defines constant value. |
|  | Source of step function: start and end value of the function are defined, as is the time when the step will occur. |
| 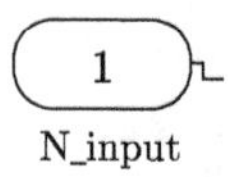 | Input variables from the block which is higher according to hierarchy or the input vector defined in MATLAB® technical computing software. |
| 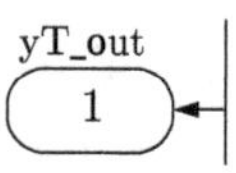 | Output simulated variables or signal which is sent to another subsystem or to MATLAB® technical computing software in the form of a vector - column. |
| 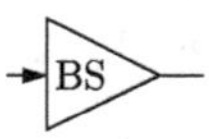 | Gain, the signal which enters into this block is multiplied by the value which is defined in the block. |
| 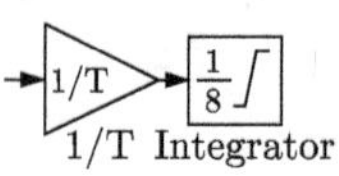 | Integrator which collects values of signals according to the value which is defined in the initial conditions. A limiter can be added to limit the output signal from the integrator. In combination with the gain an integrator with a defined time constant is achieved, where the value in the gain is equal to $1/T$, $T$ – time constant. Within the integrator block one can assign the initial condition of integration and the limits between which integration is conducted. |
| 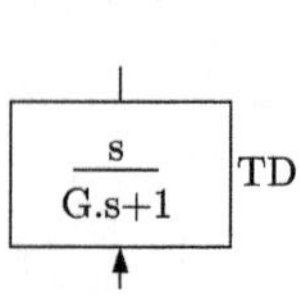 | ID regulator which is used in combination with a delay (part in the denominator) in order to decrease the input of instability created by the differentiator. This block has two types: the first type is without initial conditions and the second type is with initial conditions. The block without initial conditions is always used when the integrator is supposed to start from zero. The block with initial conditions is used in the case when the initial state of the integrator is supposed to have some starting value. In that case the initial input is set to the value from which integration begins. |

| Block | Description |
|---|---|
| 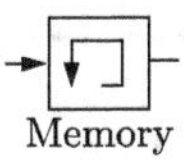 | The block which memorizes the value upon input. This block is used in breakers and operational regime detectors. |
| 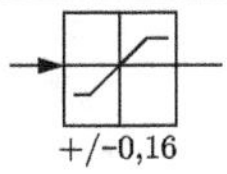 | Limiter which limits the input signal to a lower and upper boundary. |
| 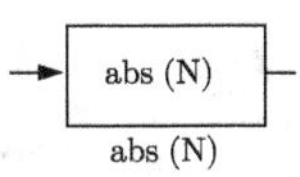 | Functional block which provides the absolute value from the input to the output, with this functional block other functions can be calculated ($x^2$, $\sqrt{x}$, polynomial function, etc.). |
| 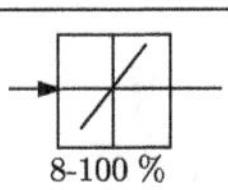 | Limiter of changes which limits the speed of increase and decrease of input signal. |
| 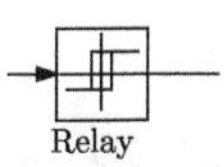 | Relay which is turned on/off depending on the assigned values. |
| 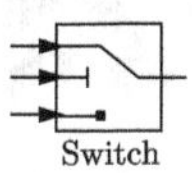 | Conditional switch which activates the upper input if the value of the middle input is greater than the assigned value and conversely activates the lower input. |
| 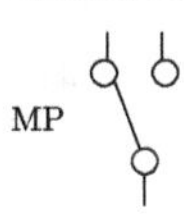 | Manual switch which is used at locations of manual transfer between two branches (without conditions). |

### 2.1.6.5 Comparative analysis of numerically and experimentally obtained results

Verification of the formed model is complete following the comparison of simulation results with the measurement results at the HPP. These results are displayed in detail within the provided literature [22, 23]. Figs. 2.43a and 2.43b show a portion of the results which relate to the dependencies $y_T$ and $T_{CS}$ in the function of time. The thick line represents the measured values and the thin line represents those obtained through simulation.

The illustrated modeling process shows only a portion of the possibilities which are offered by MATLAB® technical computing software and Simulink® software within computer-aided design in power engineering.

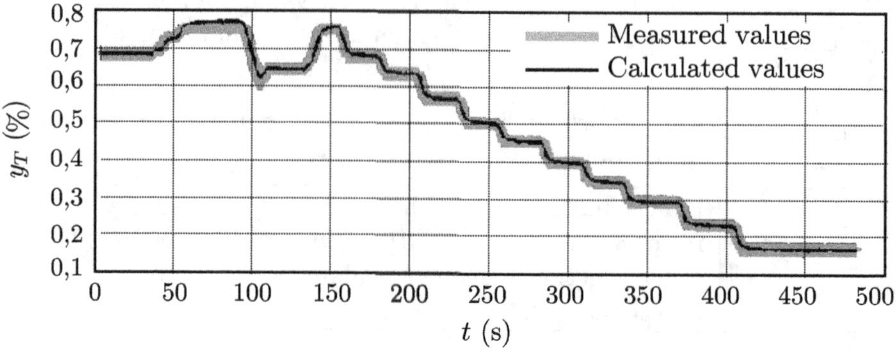


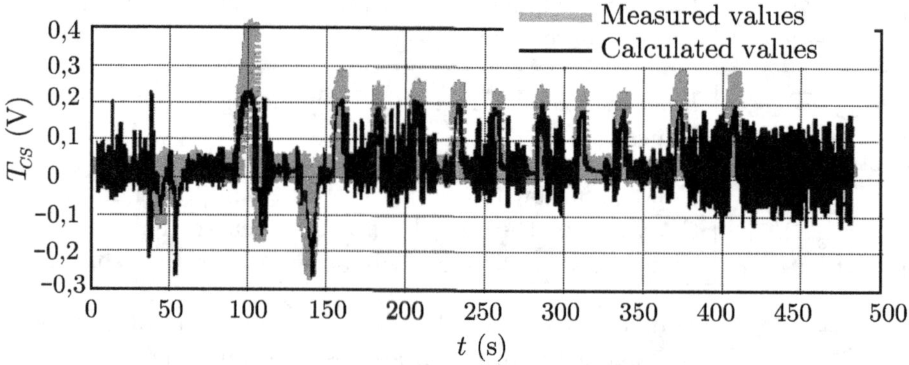


Fig. 2.43 - Numerically and experimentally determined dependencies $y_T$ (a) and $T_{CS}$ (b) in the function of time

## 2.2. EMTP / ATP

### 2.2.1. Introduction

The Alternative Transients Program (ATP) version of the Electromagnetic Transients Program (EMTP) is considered to be one of the most widely used universal program system for digital simulation of transient phenomena of electromagnetic as well as electromechanical nature in electric power systems [12]. With this digital program, complex networks and control systems of arbitrary structure can be simulated. ATP has extensive modeling capabilities and additional important features besides the computation of transients.

A partial overview of the problems which can be solved using the software tool EMTP/ATP are displayed in Section 1.6.5.3, Table 1.4.

The ATP program predicts variables of interest within electric power networks as functions of time, typically initiated by some disturbances. Basically, trapezoidal rule of integration is used to solve the differential equations of system components in the time domain. Non-zero initial conditions can be determined either automatically by a steady-state phasor solution or they can be entered by the user for simpler components.

ATP has many models including rotating machines, transformers, surge arresters, transmission lines and cables. Interfacing capability to the program modules TACS (Transient Analysis of Control Systems) and MODELS (a simulation language) enables modeling of control systems and components with nonlinear characteristics such as arcs and corona. Dynamic systems without any electrical network can also be simulated using TACS and MODELS control system modeling.

Symmetrical or unsymmetrical disturbances are allowed, such as faults, lightning surges and several kind of switching operations including commutation of valves. Frequency-domain harmonic analysis using harmonic current injection method (HARMONIC FREQUENCY SCAN) and calculation of the frequency response of phasor networks using FREQUENCY SCAN feature is also supported. The model-library of ATP at present consists of the following components:

- uncoupled and coupled linear, lumped R, L, C elements,
- transmission lines and cables with distributed and frequency-dependent parameters,
- nonlinear resistances and inductances, hysteretic inductor, time-varying resistance, TACS/MODELS controlled resistance,
- components with nonlinearities: transformers including saturation and hysteresis, surge arresters (gapless and with gap), arcs,
- ordinary switches, time-dependent and voltage-dependent switches, statistical switching (Monte-Carlo studies),
- valves (diodes, thyristors, triacs), TACS/MODELS controlled switches,
- analytical sources: step, ramp, sinusoidal, exponential surge functions, TACS/MODELS defined sources,
- rotating machines: 3-phase synchronous machine, universal machine model,
- user-defined electrical components that include MODELS interaction.

MODELS in ATP is a general-purpose description language supported by an extensive set of simulation tools for the representation and study of time-variant systems [28]:

- the description of each model is enabled using free-format, keyword-driven syntax of local context and that is largely self-documenting,
- MODELS in ATP allows the description of arbitrary user-defined control and circuit components, providing a simple interface for connecting other programs/models to ATP,
- as a general-purpose programmable tool, MODELS can be used for processing simulation results either in the frequency domain or in the time domain.

TACS is a simulation module for time-domain analysis of control systems. It was originally developed for the simulation of HVDC converter controls. For TACS, a block diagram representation of control systems is used. TACS can be used for the simulation of:

- HVDC converter controls,
- excitation systems of synchronous machines,
- power electronics and drives,
- electric arcs (circuit breaker and fault arcs).

Interface between electrical network and TACS is established by exchange of signals such as node voltage, switch current, switch status, time-varying resistance, voltage- and current sources.

Supporting routines are integrated utilities inside the program that support the users in conversion between manufacturers' data format and the one required by the program, or to calculate electrical parameters of lines and cables from geometrical and material data. Supporting modules in ATP are:

- calculation of electrical parameters of overhead lines and cables using program modules LINE CONSTANTS, CABLE CONSTANTS and CABLE PARAMETERS,
- generation of frequency-dependent line model input data (Semlyen, J.Marti, Noda line models),
- calculation of model data for transformers (XFORMER, BCTRAN),
- saturation and hysteresis curve conversion,
- Data Base Modularization (for $INCLUDE usage).

ATPDraw™ is a graphical, mouse-driven preprocessor to the ATP version of the Electromagnetic Transients Program (EMTP) on the MS-Windows platform [29]. The program is written in CodeGear Delphi 2007 and runs under Windows 9x/NT/2000/XP/Vista. In ATPDraw™ the user can construct an electrical circuit using the mouse and selecting components from menus, then ATPDraw™ generates the ATP input file in the appropriate format based on "what you see is what you get". The simulation program ATP and plotting programs can be integrated with ATPDraw™.

ATPDraw™ supports multiple circuit modeling that makes possible to work on more circuits simultaneously and copy information between the circuits. All kinds of standard circuit editing facilities (copy/paste, grouping, rotate, export/import, undo/redo) are available. In addition, ATPDraw™ supports the Windows clipboard and metafile export. The circuit is stored on disk in a single project file, which includes all the simulation objects and options needed to run the case. The project file is in zip-compressed format that makes the file sharing with others very simple.

Most of the standard components of ATP as well as TACS are supported, and in addition the user can create new objects based on MODELS or $Include (Data Base Module). Line/Cable modeling (KCLee, PI-equivalent, Semlyen, JMarti and Noda) is also included in ATPDraw™ where the user specifies the geometry and material data and has the option to view the cross section graphically and verify the model in the frequency domain. Special components support the user in machine and transformer modeling based on the powerful Universal Machine and BCTRAN components in ATP-EMTP [29]. In addition the advanced Hybrid Transformer model XFMR and Windsyn support is included.

ATPDraw™ supports hierarchical modeling by replacing selected group of objects with a single icon in an almost unlimited numbers of layers. Components have an individual icon in either bitmap or vector graphic style and an optional graphic background.

ATPDraw™ is most valuable to new users of ATP-EMTP and is an excellent tool for educational purposes. However, the possibility of multi-layer modeling makes ATPDraw™ a powerful front-end processor for professionals in analysis of electric power system transients, as well.

### 2.2.2. Examples of calculations of overvoltages caused by switching operations of the disconnector in metal-enclosed $SF_6$ gas insulated switchgears

The application of the software tool ATP is illustrated in the examples of calculating transient processes in enclosed, $SF_6$ gas insulated switchgears (GIS). The first example relates to the numerically and experimentally determined electromagnetic transient processes in the secondary circuits of the measurement transformers of the 123 kV three-phase enclosed, $SF_6$ GIS Karlsruhe-Oberwald. The second example illustrates the procedure for calculating the increase of potential of the metal enclosure of the 420 kV single-phase enclosed, $SF_6$ GIS within the "Višegrad" hydroelectric power plant (HPP).

#### 2.2.2.1 General considerations

Metal-enclosed switchgears with an operating voltage up to 800 kV are used for the transmission and distribution of electrical energy to cities, regions and industrial centers. The main advantages of these switchgears are:

- high level of reliability,
- low life-cycle expenses,
- small space requirement.

These advantages enable a location to be found close to the consumption centers. High-voltage air-insulated substations for a voltage of $U_n \geqslant 170$ kV require a great deal of space so they are built out in the open. However, in cities and industrial complexes space is limited and expensive. For this reason metal-enclosed $SF_6$ gas insulated switchgears (GIS) have emerged. These GIS consist of modular cells which contain all switching devices and measuring transformers. As $SF_6$ gas is a much better insulator than air, the space between modules is significantly less. This is why these GIS have small dimensions and can be place in buildings. GIS are constructed as single-phase and three-phase enclosed. With single-phase enclosed GIS each pole of the device is located in a separate metal enclosure, while with three-phase enclosed GIS all three poles are located in one enclosure. Three-phase enclosed GIS are used for a voltage of up to 170 kV.

The metal enclosure is most often made from aluminum and serves as the reservoir for the $SF_6$ gas. For the purpose of eliminating the occurrence of dangerous touch voltage, all metal parts of the GIS are grounded.

The dominant advantages of using metal-enclosed GIS are:

- small dimensions,
- hermetically sealed,
- lack of meteorological influences,
- lack of influences from elevation, because the GIS is kept at a constant pressure (2-4 times higher than atmospheric pressure),
- absence of corona interference,
- insulation is self-restoring,
- up to 20% of air in $SF_6$ gas does not significantly affect the insulation characteristics,
- vast deionization power,
- very small separation under the affects of arc,
- fast establishment of dielectric strength at the arc location,
- low arc voltage,
- pressure of $SF_6$ gas varies only slightly with changes in temperature.

Transient overvoltages caused by switching operations of the disconnector in metal-enclosed GIS are characterized by an exceptionally short rise time (from 5 ns to 20 ns). Due to that characteristic, these overvoltages are called Very Fast Transients (VFT) and can cause significant strain to the insulation elements of the GIS. These overvoltages also represent a potential source of interference which can affect the proper functioning of the control system and protection system [30-34]. One part of the generated overvoltage transfers through stray capacitances to the command-signal cables of the installed current and voltage transformers. These overvoltages travel to the secondary connections of the control system and protection system of the GIS, which requires them to be properly protected.

The disconnector, aside from its basic function of separating a portion of the circuit, also performs the role of switching on capacitive currents of unloaded busbars, conducting insulators, measuring devices and switching devices. The operational characteristics of the disconnector during the switching on of small capacitive currents within a metal-enclosed GIS in comparison to the classic open substations are [35, 36]:

- high gradient of electrical field between the arc and grounding enclosure during switching off operations,

- higher capacitive current due to greater specific capacitance towards the ground,
- lower wave impedance,
- greater gradient of transient voltage and current.

When the voltage between disconnector contacts crosses the dielectric strength of the arc gap, a re-ignition or pre-insertion ignition occurs which results in the flow of high-frequency current which causes the equalization of the voltage side and side under a load. Due to the slow contact speed (ca. cm/s) re-ignition occurs up to 100 times per operation. Re-ignition causes traveling waves the shapes of which are formed within the metal-enclosed GIS and with a frequency of up to several tens of MHz depending on the configuration of the substation. Depending on the voltage level, overvoltages can reach values which cause a flashover from the busbar to the grounded enclosure. Due to this it is important to discover the critical locations within the GIS where the highest overvoltages can occur in order to conduct the proper selection of protection.

The following cases are considered for the purpose of reviewing the strictest conditions which can occur during the switching on of small capacitive currents by the disconnector:

- switching on short sections in an open regime from the load side,
- switching on the long section of a busbar from the load side during which other components may be switched on, such as a circuit breaker or power transformer,
- switching on during phase opposition.

The increase of potential of the metal enclosure of the GIS is one of the phenomena found in practice during disconnector switching operations. This phenomenon, in literature known as Transient Ground Potential Rise (TGPR), manifests itself through the appearance of sparks between the enclosure and other grounded parts. The growth of potential of the metal enclosure reaches a value of ca. 100 kV and more and as such may damage individual components of the control system and protection system as well as endanger operating personnel of the substation.

Causes of growth of potential of the metal enclosure are:

- switching operations of the disconnector within GIS,
- faults within GIS such as short circuits,
- faults caused by GIS such as short circuits or lightning discharge.

The first two causes lead to a quick break of dielectric strength of the $SF_6$ gas. As a consequence of this break, a traveling wave is generated with an exceptionally short rise time $U_1$ (Fig. 2.44). For the case of switching operations by the disconnector, this traveling wave is divided into two components which travel from the disconnector, which is shown in Fig. 2.44.

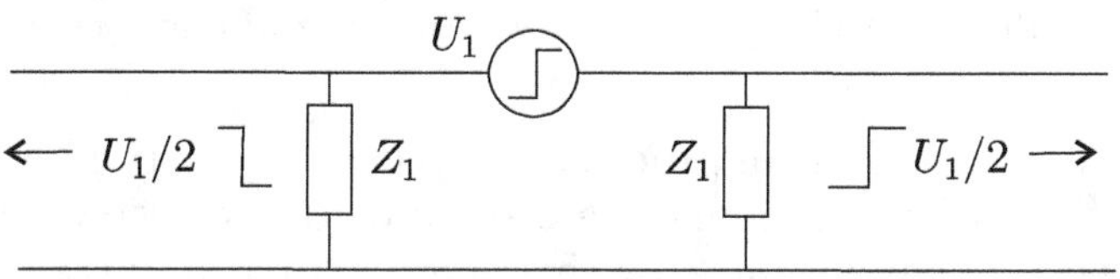


Fig. 2.44 - Equivalent circuit for illustrating the occurrence of VFT due to switching operations of a disconnector within GIS

The symbol $Z_1$ represents the wave impedance of the so-called internal system which is composed of the surface conductor and internal surface of the metal enclosure. The maximum value of overvoltage $U_1$ is equal to two times the amplitude of the phase voltage.

The grounding system and its couplings with equipment have been envisaged to the conduct current of short circuit industrial frequency without an increase of potential which could endanger human life. The impedance of the grounding system is mostly of an inductive character and with that increases along with the frequency. A frequency in the kHz domain will not condition a large increase of voltage of grounding components because the currents are negligible in comparison to a short circuit current. In the MHz domain, however, the impedance of the grounding system reaches high values so, despite of the low current values, the voltage difference between the grounding components reach a value of up to 100 kV [37-41].

The situations in which the disconnector within a metal-enclosed GIS can switch on small capacitive currents are:

- switching on a section of a metal-enclosed GIS (Fig. 2.45),
- switching on during phase opposition (Fig. 2.46).

The symbols in Fig. 2.45 have the following meanings: $U_1$ – voltage source, $Z_l$ – wave impedance of connected lines, $Z_m$ – wave impedance of metal-enclosed GIS, $D$ – disconnector.

The voltage situation shown in Fig. 2.45, which is most common in practice, refers to the use of a disconnector for switching on sections

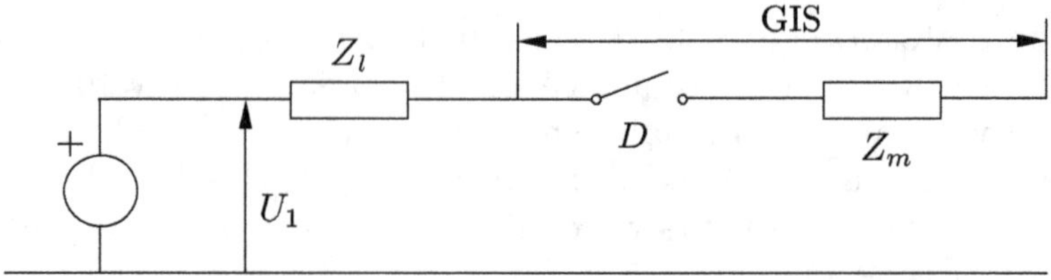


Fig. 2.45 – Switching on a section of a metal-enclosed GIS

of unloaded busbars. This in practice means that a portion of the metal-enclosed GIS is switched on using a disconnector from connecting lines, during which the capacitance of this section of the current circuit is dependent on the length of the busbars and the connecting equipment on it. The value of this capacitance can amount to several nF.

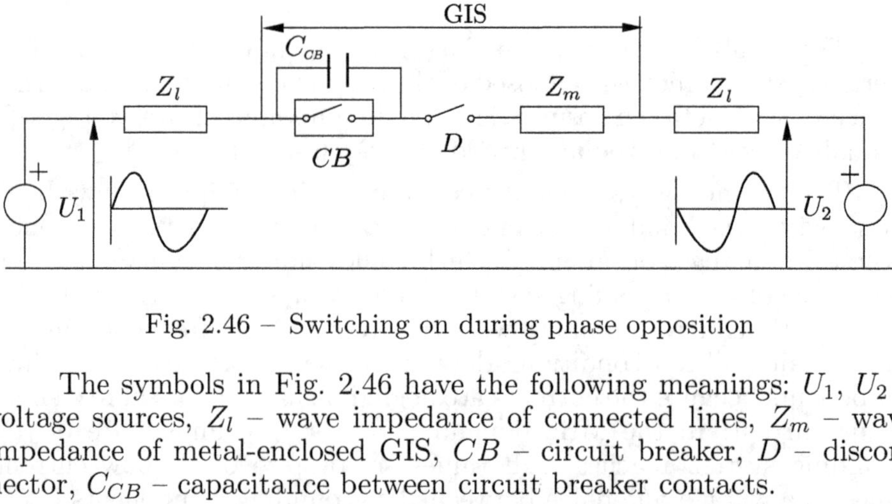


Fig. 2.46 – Switching on during phase opposition

The symbols in Fig. 2.46 have the following meanings: $U_1$, $U_2$ – voltage sources, $Z_l$ – wave impedance of connected lines, $Z_m$ – wave impedance of metal-enclosed GIS, $CB$ – circuit breaker, $D$ – disconnector, $C_{CB}$ – capacitance between circuit breaker contacts.

In the case which is illustrated in Fig. 2.46 two separate networks are switched on using a disconnector through parallel capacitances $C_{CB}$ between open breaker contacts. During the most unfavorable condition, the phases of these two networks can be moved 180° (phase opposition). Then between the disconnector contacts there is a voltage difference equal to two times the value of the amplitude of phase voltage.

Two examples are displayed below of the software tool EMTP/ATP in the calculations of transient overvoltages caused by switching operations of the disconnector in metal-enclosed GIS.

The first example shows numerically and experimentally determined electromagnetic transient processes in a 123 kV GIS caused by the switching on of the disconnector [42, 43]. The elements of a three-phase enclosed GIS have been modeled using the corresponding equivalent circuits with their parameters being determined from the project documentation and physical disposition of the switchgear. For the purpose of determining the transient processes in the secondary circuits of measuring transformers, special attention is given to the development of models for these elements. Equivalent circuits have been incorporated into the software tool EMTP/ATP. The effect of interphases-interferences is encompassed by the model which is of exceptional importance for the analysis of transient processes in a modern three-phase enclosed GIS. The results of the mathematical model were successfully confirmed through experiment in the example of the Karlsruhe-Oberwald 123 kV three-phase enclosed GIS.

The second example shows the numerical determination of increased potential of the metal enclosure of a 420 kV GIS caused by switching on the disconnector [44, 45]. Numerical simulations were executed using the software tool EMTP/ATP in the example of the 420 kV single-phase enclosed GIS of the HPP "Višegrad". Measures have been described for the effective lowering of the considered potential which are necessary to implement during the phase of design, installation and exploitation of these switchgears.

### 2.2.2.2 Electromagnetic transient processes in secondary circuits of measurement transformers in GIS

*a) Modeling of GIS*

For transient processes, the cylindrical character of GIS is represented by a high-frequency circuit with distributed parameters [46, 47]. Due to the skin effect, the current travels along the surface of the conductor and does not penetrate into its interior. For that reason it can be defined as an internal system, which consists of the surface conductor and internal surface of the metal enclosure, and an external system, which consists of the external surface of the metal enclosure and the surface of the soil. One part of the overvoltage which is generated within the system transfers through stray capacitances to the control and data lines of the installed current and voltage transformers.

These occurrences cause unwanted interference to the control and protection systems. The determination and lastly the improvement of the susceptibility of these systems in the presence of these transient overvoltages requires an exact knowledge of the transient behavior of the inner and outer GIS systems.

*a.1 The 123 kV $SF_6$ GIS Karlsruhe-Oberwald*

The investigated GIS Karlsruhe-Oberwald was installed in the city of Karlsruhe, Germany and represents a part of a 110 kV power system. The rated voltage of the equipment in this substation is 123 kV. Fig. 2.47 shows the location of the old outdoor substation which

Fig. 2.47 - Location of the new 123 kV three-phase enclosed GIS Karlsruhe-Oberwald (1) and the dismantled outdoor substation (2)

was dismantled due to the construction of the new three-phase enclosed $SF_6$ GIS [48]. Based on Fig. 2.47 one can clearly see the advantage of an GIS in the sense of required space. The displayed GIS takes up one fifth of the surface area of the dismantled outdoor substation while maintaining the same technical characteristics.

Fig. 2.48 shows the complete configuration of the GIS Karlsruhe-Oberwald in which the experiments were conducted. Within the displayed switchgear, a triple busbar system with a corresponding connection bay was installed. Each busbar system contains three phases enclosed in a shared metal enclosure. All disconnectors in the switchgear are equipped with a three-pole drive mechanism.

Fig. 2.49 shows the single-pole diagram of the investigated switchgear. Experimentally and numerically determined overvoltages caused

Fig. 2.48 - Complete configuration of the 123 kV three-phase enclosed GIS Karlsruhe-Oberwald

by switching operations of the disconnector were conducted on line bay $= E02$, the cross section and sketch of which are provided in Figs. 2.50a and 2.50b, respectively.

In order to predict the transient electromagnetic phenomena in the secondary circuits of voltage (T5) and current (T1) transformers, several network models of GIS-components and physical effects in the GIS were developed [42, 43]. Based on the developed models, simulations of transients in GIS due to disconnector operation were executed. The most significant models of the GIS elements are briefly described below.

*a.2 Arc model*

The operation of the disconnector was modeled using a modified Kopplin model which describes the arc resistance in a disconnector [42, 43]. This resistance represents a significant portion of the damping in the entire system of the GIS [49]. Normally the resistance is a frequency dependent parameter due to the skin effect. In this case of arc-discharge there exists a strong time-dependency according to temperature, diameter and losses of the discharge. Thus the time behavior of the spark's

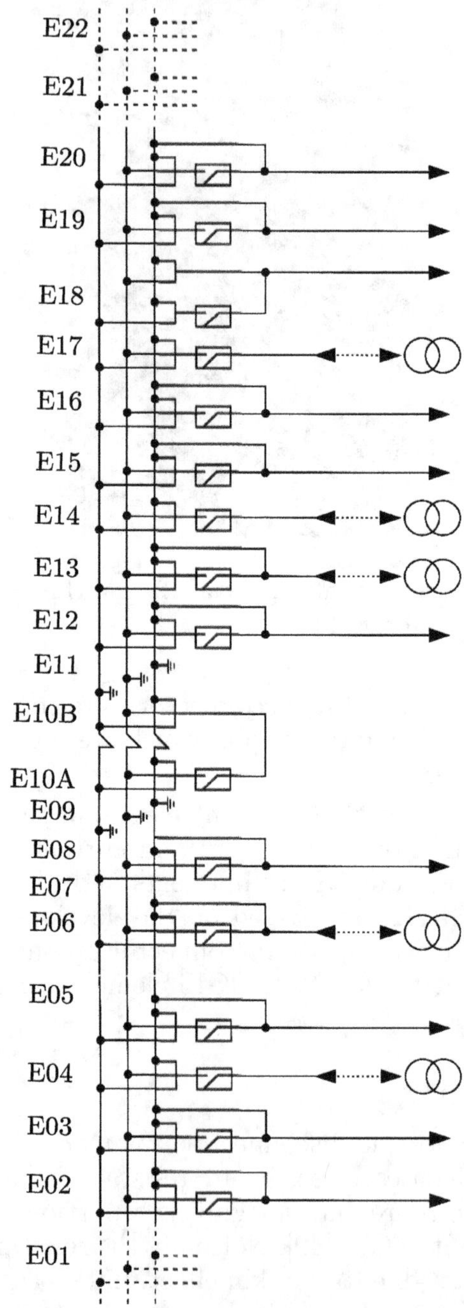


Fig. 2.49 - Single-pole diagram of the 123 kV three-phase enclosed GIS Karlsruhe-Oberwald

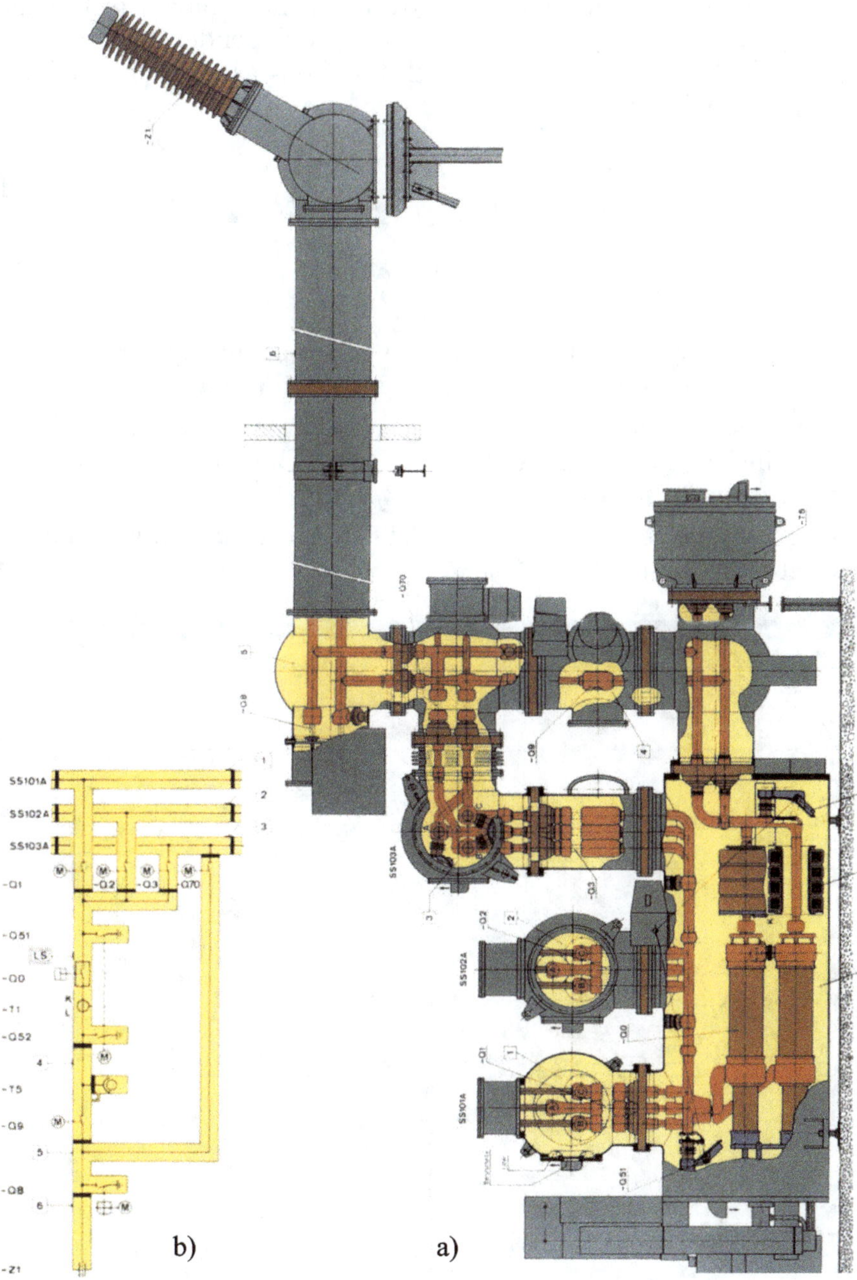


Fig. 2.50 – The analized line bay =E02; a) Cross section; b) Sketch

resistivity has to be evaluated correctly. The time behavior of the conductivity $g(t)$ is mainly influenced by the time dependent temperature function $\tau(t)$ of the arc-discharge. Both functions are shown by expressions (2.11) and (2.12):

$$\frac{1}{g}\frac{dg}{dt} = \frac{1}{\tau(g)}\left(\frac{ui}{P(g)} - 1\right) \tag{2.11}$$

$$\tau(g) = \tau_0\left(1 - e^{-\frac{g}{g_0}}\right) \tag{2.12}$$

where: $u$, $i$, $P$ – voltage, current and the power of the arc, respectively; $\tau_0$ – initial arc temperature, $g_0$ – initial arc conductivity.

This description of the physical arc-discharge process is valid from the beginning of the discharge up to its end. The model has been implemented into the software tool EMTP/ATP [12].

Fig. 2.51 shows a portion of an ATP file in which an arc model is incorporated [42, 43].

Section of ATP file - Notepad

File Edit Format View Help

```
C -------1---------2---------3---------4---------5---------6---------7---------8
BEGIN NEW DATA CASE
PRINTED NUMBER WIDTH, 13, 2,
  0.5E-9  3.0E-6        0         0  1.E-9   1.E-9
       1       1        1         1      1                       1
C -------1---------2---------3---------4---------5---------6---------7---------8
TACS HYBRID
91TR21A
90TR21
90TR28
C -------1---------2---------3---------4---------5---------6---------7---------8
99Q      58+TR21A                                                1.0PLUS1
99RT       = 0.3E-4/Q
C -------1---------2---------3---------4---------5---------6---------7---------8
99FUNK     = ABS(TR21A)
99VOR      = ABS(TR21-TR28)
99IQUAD    = FUNK*FUNK
99GD       = FUNK/VOR
99P        = 5.E+2+1.E+4*FUNK,
99GST      = IQUAD/P
99TAU      = 0.6E-6*(1.0-EXP(-5.*GD))
99CT2      = TAU .LE. 8.0E-9
99TAUX     = TAU*NOT(CT2)+8.0E-9*CT2
99T        = DELTAT/TAUX
99GDYN     = (GD+GST*T)/(1+T)
99CT1      = GDYN .LE. 1.E-6
99R        = INVRS(GDYN)*NOT(CT1)+1.E+6*CT1
C -------1---------2---------3---------4---------5---------6---------7---------8
99CTRL1    = TIMEX .LT. (1.E-9+2.4E-7)
98RM       = CTRL1*RT+NOT(CTRL1)*R
C -------1---------2---------3---------4---------5---------6---------7---------8
77TAUX          8.0E-9
77T                1.0
77RM            2.0E+5
77R             2.0E+5
77RT            2.0E+5
BLANK card ending all TACS cards
C -------1---------2---------3---------4---------5---------6---------7---------8
$VINTAGE, 1
-1C1    Q1                 3.E-6      215.6       21.33E-9   6.4         2
```

Fig. 2.51 – Section of ATP file with incorporated arc model

*a.3 Modeling of GIS components*

Due to the traveling nature of the transients the modeling of GIS makes use of electrical equivalent circuits composed by lumped elements and especially by distributed parameter lines, defines by surge impedances and traveling times. The quality of the simulation depends on the quality of the model of each individual GIS component. In order to achieve reasonable results even for very complex GIS structures highly accurate models for each internal equipment and also for components connected to the GIS are necessary. The equivalent circuits can be derived from the project documentation and from the internal physical arrangement. Table 2.8 shows the equivalent circuits which represent the main components of a typical GIS [42, 43, 46, 47].

The inner system, which consists of the high voltage bus duct and the inner surface of the encapsulation, has been represented thorougly by line sections modeled as transmission lines with distributed parameters. The phases and their interphase coupling have been investigated by applying the CABLE CONSTANTS SUBROUTINE and the method of modal components [12]. This method permits the calculation of each phase and its coupling to the other phases separately. Fig. 2.52 shows the calculation process using the method of modeled components.

Inhomogeneities in the inner systems are corners and Tee junctions. Table 2.8 shows sketches and models of these inhomogeneities. For the purpose of achieving greater precision of calculation in the field of high frequency, complex models of current and voltage transformers were developed (Table 2.8). Transients are transmitted to the secondary lines of the GIS by stray capacitances which result of the construction of the protection electrodes in the transformers.

The symbols provided in the equivalent circuits of the measurement transformers have the following meanings: $L_1$, $L_2$, $L_3$ – conductors of the internal system of the current transformer, $C_1$, $C_2$ – coupling capacitors, $L_{M1}$, $L_{M2}$ – conductors of the secondary circuits, $L_{E1}$ – shield electrode, $L_{E2}$ – ground conductor of the shield electrode, $R_B$, $L_B$, $C_B$ – resistance, inductance and capacitance of load, $DFK$ – pressure spring contact, $Lp$ – conductor of high-voltage electrode, $Cs$ – field-controlled capacitance, $C_k$ – coupling capacitor to the secondary winding of the voltage transformer.

*b) Experimental and calculation results*

Numerical simulations were conducted for the analyzed bay, a sketch of which is displayed in Fig. 2.50b. Busbar disconnectors Q1-Q3

Table 2.8 – Equivalent circuits of elements of a three-phase enclosed GIS

| Component | Equivalent circuit | Notes |
|---|---|---|
| Bus duct | 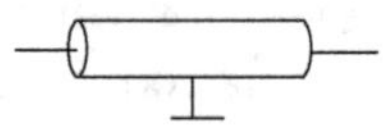 | Loss-free distributed parameter transmission line |
| Bushing (Capac. type) | 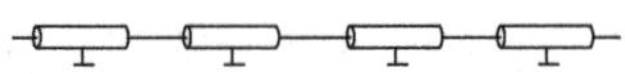 | $n$-number of equivalent shields (5-8) |
| Bushing (gas filled) | 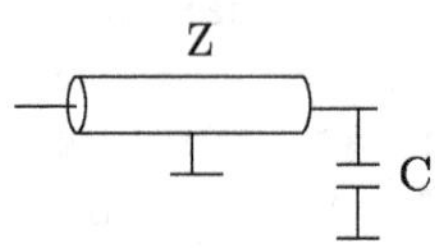  | $Z = 250\ \Omega$ <br> $C > 10$ pF |
| Spacer | 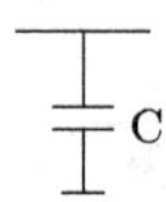  | $C > 4$ nF |
| Corner inhomo-geneity | 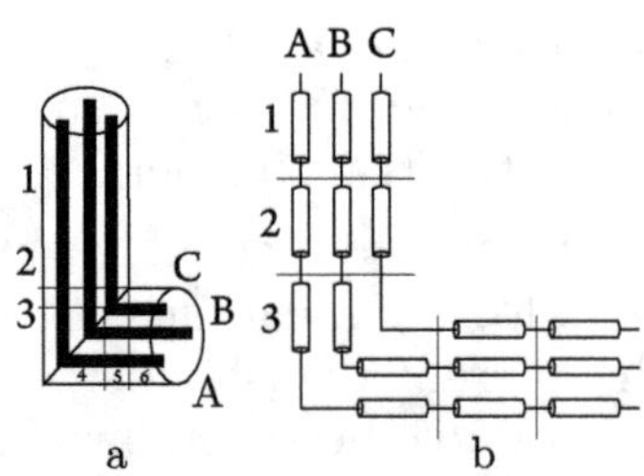  | |
| Tee-junction inhomo-geneity | 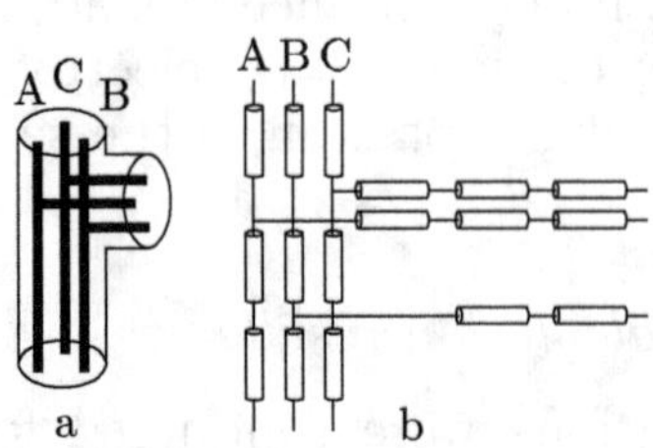  | |

| Component | Equivalent circuit | Notes |
|---|---|---|
| Closed disconnector | 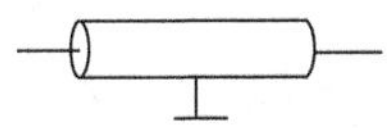 | |
| Open disconnector | 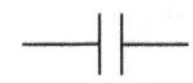 | |
| Disconnector during sparking | 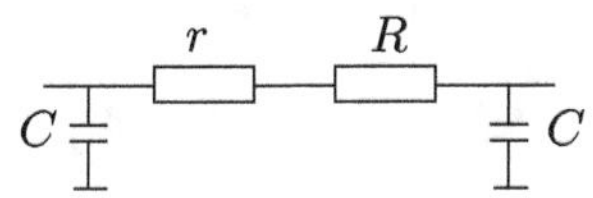  | $r = r(t)$;<br>$R < 10\ \Omega$<br>$C > 10$ pF |
| Open switch | 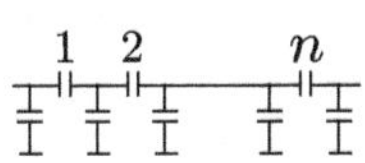  | $n$-number of breaking chambers |
| Current transformer | 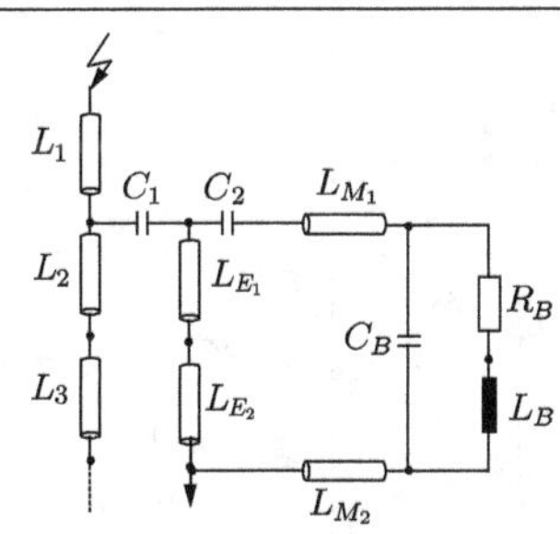  | $C_1 = 1$ pF<br>$C_2 = 300$ pF<br>$R_B = 20$ m$\Omega$<br>$L_B = 2,8\ \mu$H<br>$C_B = 20$ pF |
| Voltage transformer | 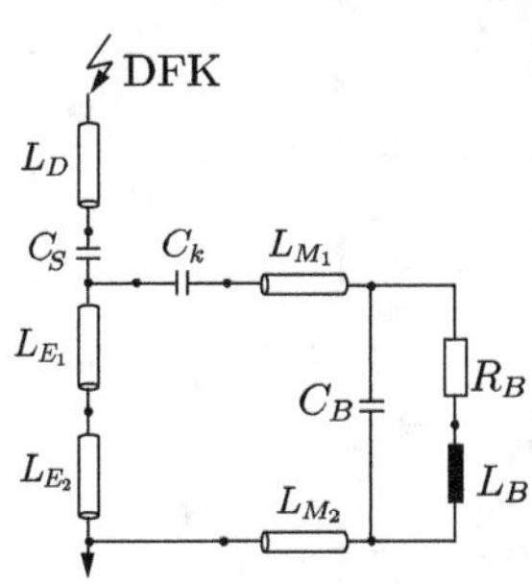  | $C_s = 9$ pF<br>$C_k = 62$ pF<br>$R_B = 180$ m$\Omega$<br>$L_B = 170$ mH<br>$C_B = 600$ pF |

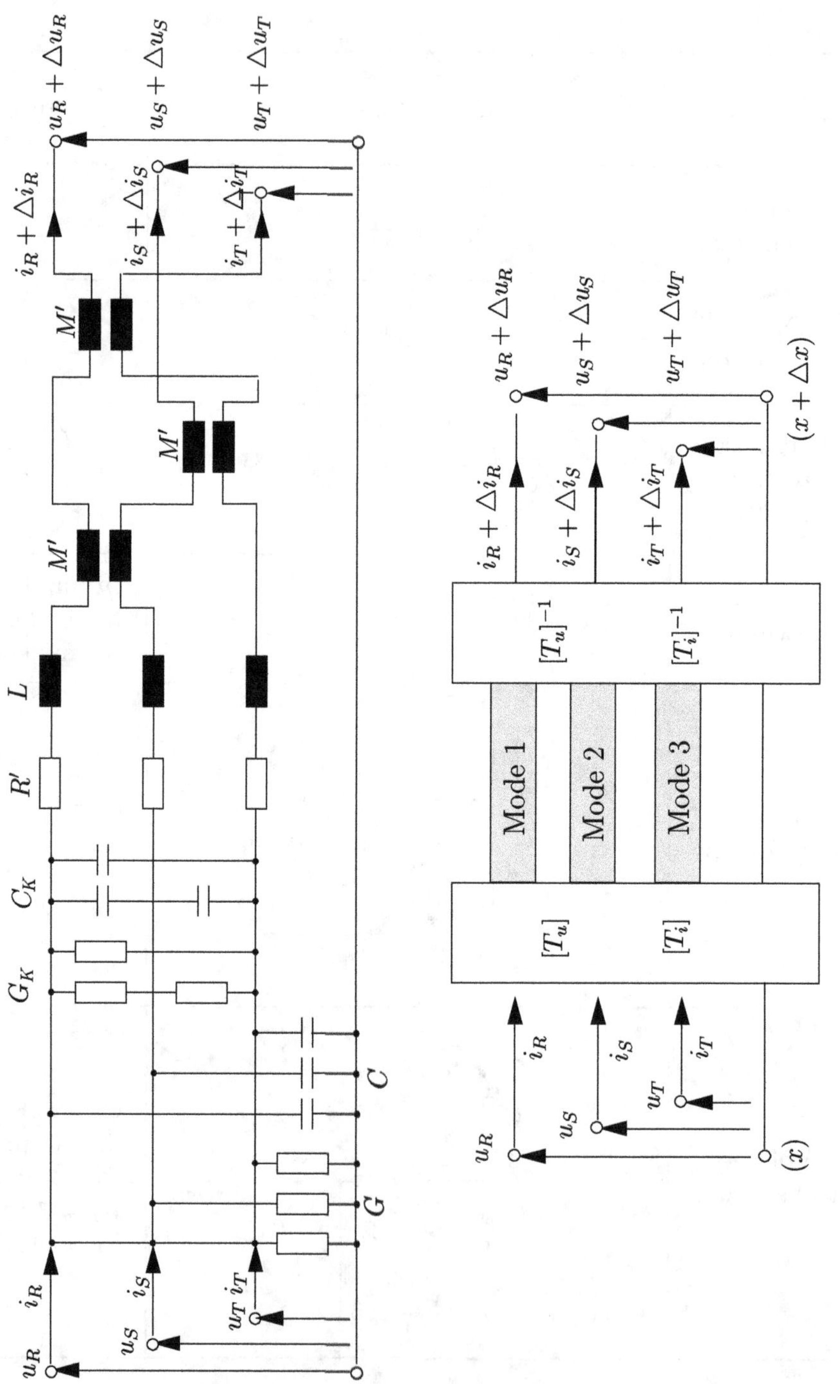


Fig. 2.52 – Calculation process using the method of modal components

and Q70, as well as circuit breaker Q0 were switched off. Transients caused by closing operation of the outgoing disconnector Q9 of the line bay were determined using the software tool EMTP/ATP and the corresponding models described in the previous section. A basic calculation time step of 0.20 ns was adopted which corresponds to the shortest element length which is 0.02 m. Only some of the several results of numerical simulations are displayed below.

Figs. 2.53a, 2.53b and 2.53c show the numerically determined transient overvoltage wave-shapes for three-phases at the terminal board of the secondary cable of the protection system. Fig. 2.54a shows the corresponding frequency spectrum of overvoltage provided in Fig. 2.53a.

The overvoltages resulting from the outgoing disconnector Q9 switch operation in the displayed line bay were determined experimentally. The transient overvoltages were measured at the secondary lines terminal board of the protection system using a Tektronix TD S744A digital oscilloscope with a 500 MHz sampling rate. Figs. 2.53d, 2.53e and 2.53f show the experimentally determined transient overvoltage wave-shapes determined at the secondary lines terminal board of the protection system. The corresponding frequency spectrum of overvoltage displayed in Fig. 2.53d is provided in Fig. 2.54b.

The displayed results of experimentally and numerically determined transient overvoltages within the internal system of GIS indicate a satisfactory level of accuracy for the applied model and simulation method. The quality of the model has also been confirmed by the agreement of the corresponding frequency spectrum provided in Figs. 2.54a and 2.54b. At a frequency of 20 MHz, the amplitudes of numerically and experimentally determined transient overvoltages amount to 3 V.

Based on the displayed results, it can be concluded that the main advantage of the proposed model is reflected in the ability to analyze electromagnetic transient processes in a modern three-phase enclosed GIS. Comparison of the experimentally and numerically obtained results indicates the applicability of the proposed model.

#### 2.2.2.3 Calculation of growth of potential of the metal enclosure of an $SF_6$ gas insulated switchgear caused by switching operations of the disconnector

*a) Modeling of GIS*

For transient processes, the cylindrical character of GIS is represented by a high-frequency circuit with distributed parameters [46, 47].

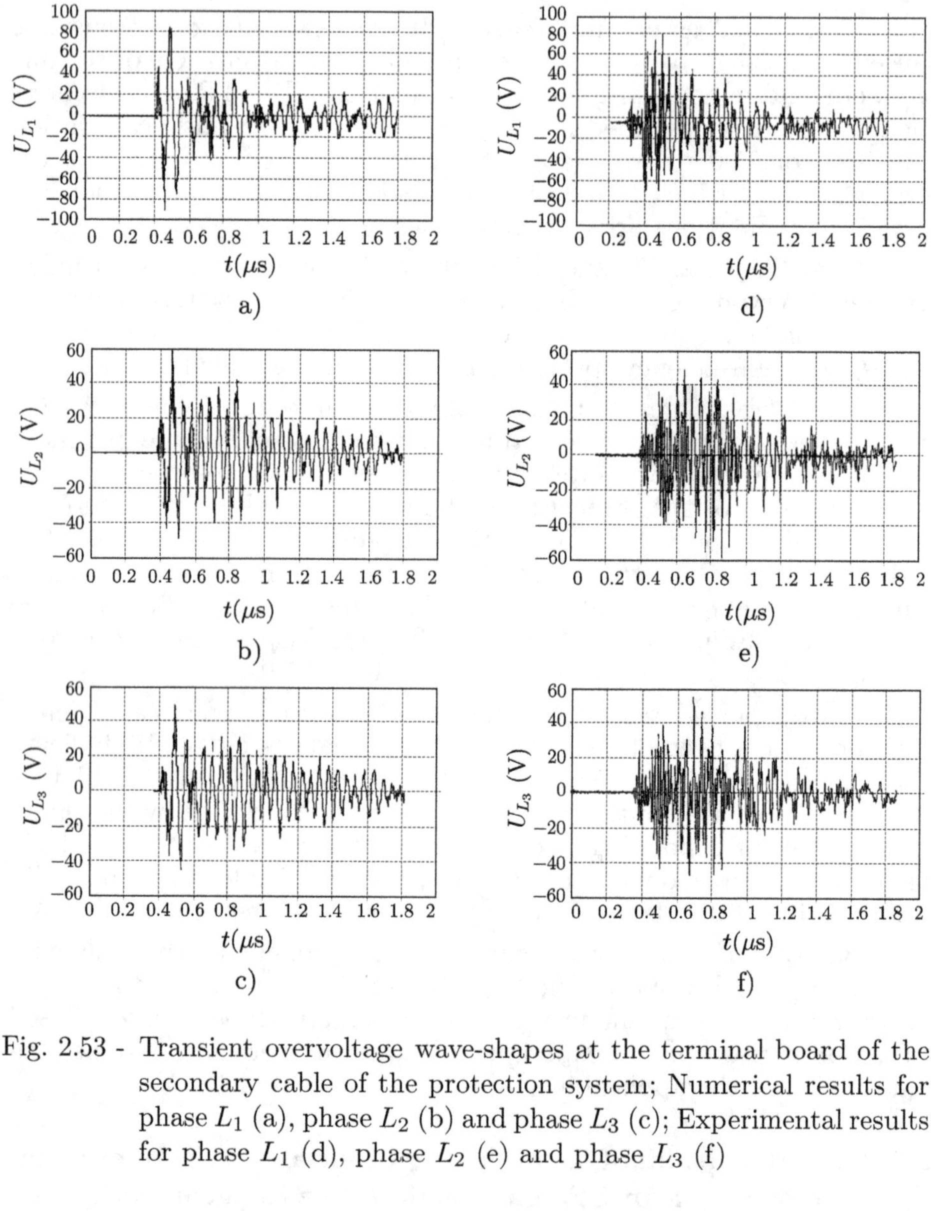


Fig. 2.53 - Transient overvoltage wave-shapes at the terminal board of the secondary cable of the protection system; Numerical results for phase $L_1$ (a), phase $L_2$ (b) and phase $L_3$ (c); Experimental results for phase $L_1$ (d), phase $L_2$ (e) and phase $L_3$ (f)

Due to the skin effect, the current travels along the surface of the conductor and does not penetrate into its interior. For that reason it can be defined as an internal system, which consists of the surface conductor and internal surface of the metal enclosure, and an external system, which consists of the external surface of the metal enclosure and the

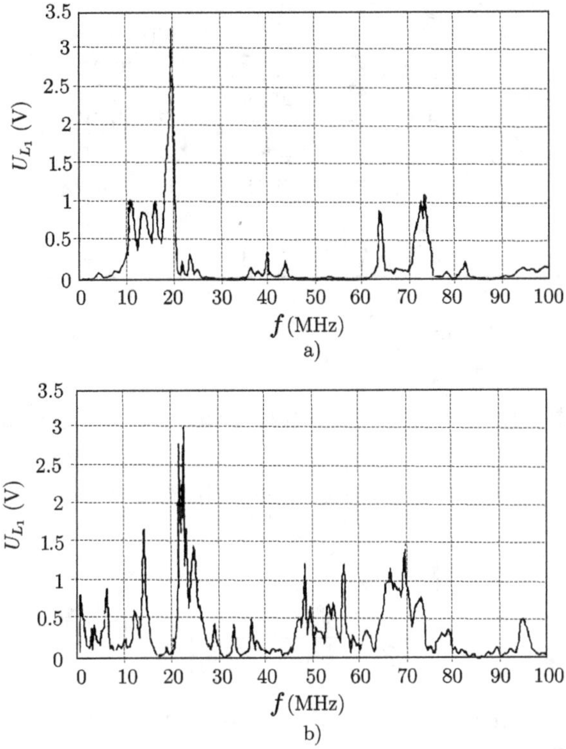


Fig. 2.54 - a) Frequency spectrum of numerically determined transient overvoltages provided in Fig. 2.53a; b) Frequency spectrum of experimentally determined transient overvoltages provided in Fig. 2.53d

surface of the soil (Fig. 2.55).

The symbols in Fig. 2.55 have the following meanings: $Z_1$ – wave impedance of internal system; $Z_2$ – wave impedance of external system; $Z_3$ – wave impedance of system of external conductor – soil.

For the model shown in Fig. 2.55, the following matrix of reflection and refraction can be defined:

$$S = \begin{bmatrix} s_{11} & s_{12} & s_{13} \\ s_{21} & s_{22} & s_{23} \\ s_{31} & s_{32} & s_{33} \end{bmatrix} \tag{2.13}$$

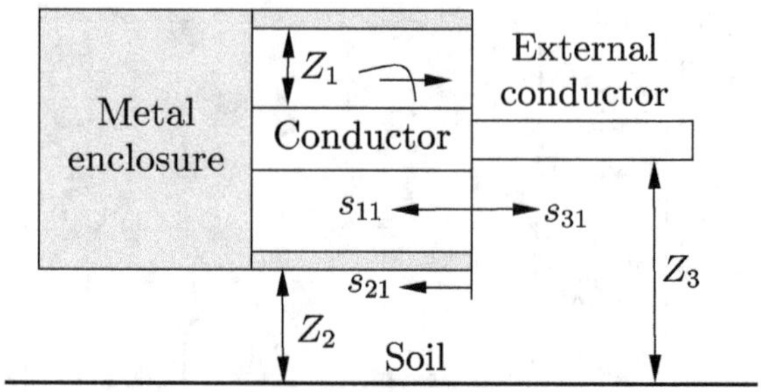


Fig. 2.55 – Display of internal and external systems of GIS

where the coefficient $s_{ij}$ is determined from the expression:

$$s_{ij} = \frac{2\,Z_i}{Z_1 + Z_2 + Z_3} \quad \text{for} \quad i \neq j \tag{2.14}$$

and:

$$s_{ij} = 1 - \frac{2Z_i}{Z_1 + Z_2 + Z_3}, \quad \text{for} \quad i = j \tag{2.15}$$

From the expressions (2.13)-(2.15) it can be concluded that the wave caused by switching operations of the disconnector in GIS is divided into three parts. The second part of the wave, shown through the coefficient $s_{21}$, is important from the aspect of calculation of growth of potential of the metal enclosure of GIS. The value of the coefficient $s_{21}$ is proportional to the amplitude of the potential of the metal enclosure.

The operation of the disconnector was modeled using a modified Kopplin model which describes the resistance of an arc in the disconnector [42, 43]. The $SF_6$ switchgear elements were modeled with equivalent circuits, defined by the wave impedance and travel time [42, 43, 46, 47]. The internal system of GIS was modeled using an equivalent circuit with distributed parameters. The module CABLE CONSTANTS is an integral component of the software tool EMTP/ATP and is used for modeling the internal system [12].

The SEMLYEN module was used for the calculation of conductor parameters. This module treats the conductor as an element with distributed parameters and enables the calculation of matrices of resistance and capacitance, as well as the wave impedance of the conductor at the given frequency. Fig. 2.56 shows a portion of an ATP file with a model of the conductor according to the SEMLYEN module.

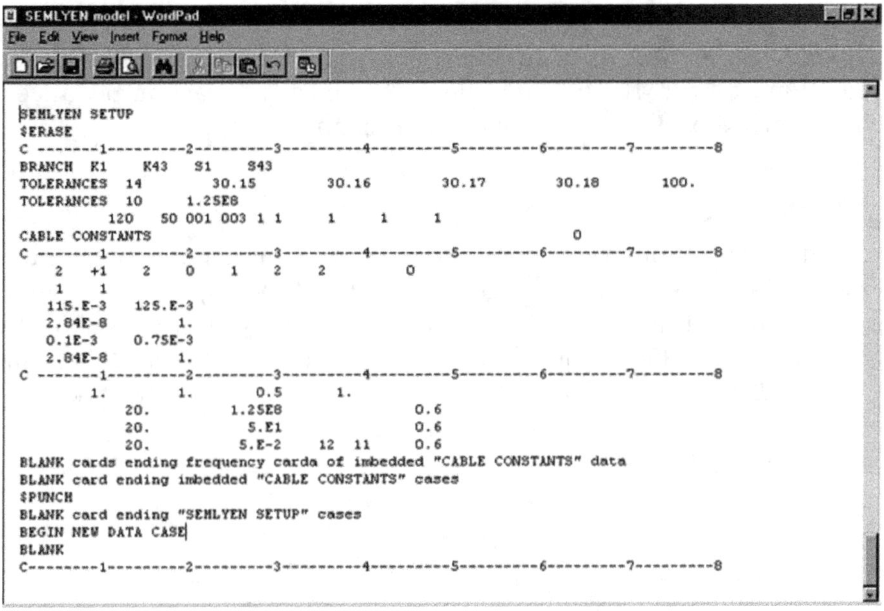


Fig. 2.56 - Portion of ATP file with model of conductor according to SEMLYEN module

Modeling of the external system consists of modeling the metal enclosure as a conductor with distributed parameters. For this reason the metal enclosure is divided into elements of determined lengths, while their wave impedance is determined from the expression:

$$Z = \sqrt{\frac{L'}{C'}} \tag{2.16}$$

where $L'$ and $C'$ represent the linear inductance and linear capacitance of the elements of the metal enclosure towards the soil. These parameters are defined with the expressions:

$$L' = \frac{\mu_0}{2 \cdot \pi} \cdot \ln \frac{2 \cdot h}{D} \tag{2.17}$$

$$C' = \frac{2 \cdot \pi \cdot \varepsilon_0}{\ln \dfrac{2 \cdot h}{D}} \tag{2.18}$$

where: $\mu_0$ – magnetic permeability of air, $\epsilon_0$ – dielectric constant of air, $h$ – height of element of metal enclosure in relation to soil, $D$ – median geometrical radius of element of metal enclosure.

The part which connects the metal enclosure with the grounding of GIS is also modeled as a conductor with distributed parameters. The application of a SEMLYEN module determines the wave impedance of the connecting part and the travel time of the wave.

*b) Calculation results*

Numerical simulations were executed on a model of a single-phase enclosed 420 kV line bay of the GIS of the HPP "Višegrad". The single-pole diagram and disposition of this bay are displayed in Figs. 2.57 and 2.58, respectively [50]. The dimensions of this bay in Fig. 2.58 equal $W \times H \times L = 4.5 \times 5.6 \times 12.6$ m, where $W$, $H$ and L represent the

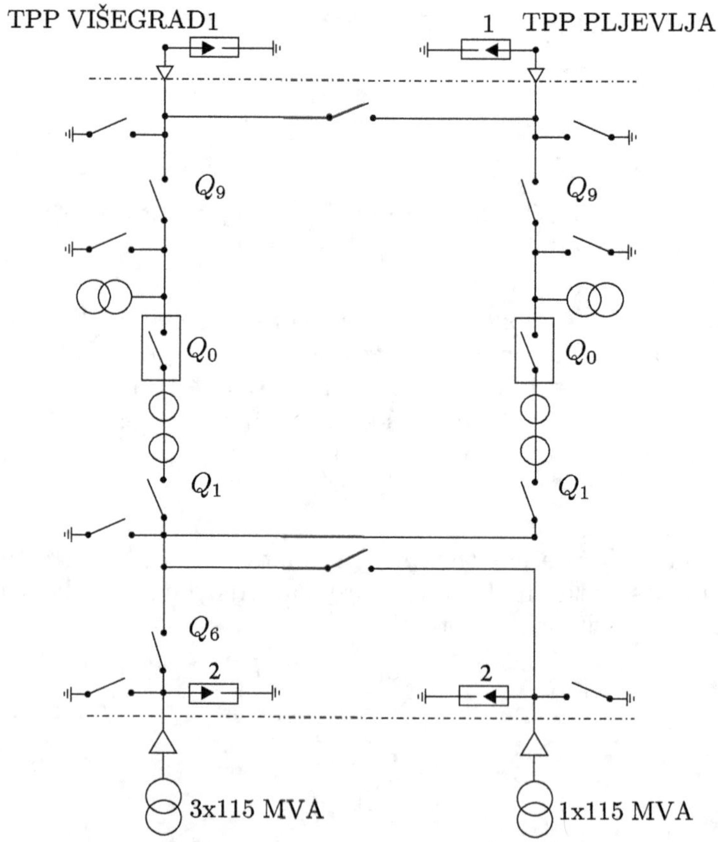


Fig. 2.57 - Single-pole diagram of the 420 kV single-phase enclosed GIS of the HPP "Višegrad"

width, height and length of the considered bay, respectively.

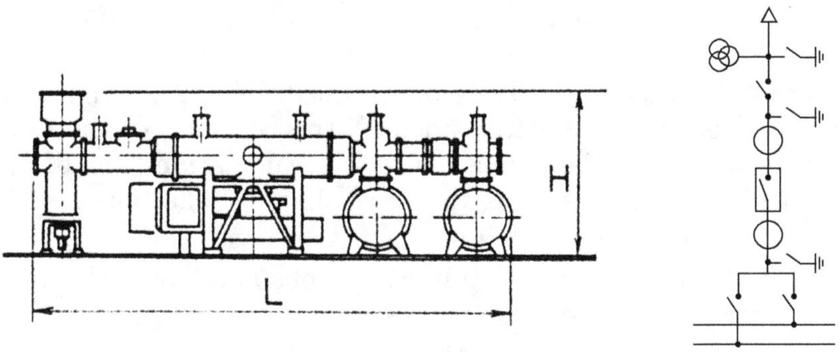


Fig. 2.58 - Disposition of 420 kV single-phase enclosed line bay of the GIS of the HPP "Višegrad" and corresponding single-pole diagram

Busbar disconnectors as well as the circuit breaker are switched off. The potential of the metal enclosure of the considered GIS, caused by the switching operation of the output disconnector of line bay Q9, is determined using the software tool EMTP/ATP and the corresponding model described in section 2.2.2.2 [12, 42-45, 51].

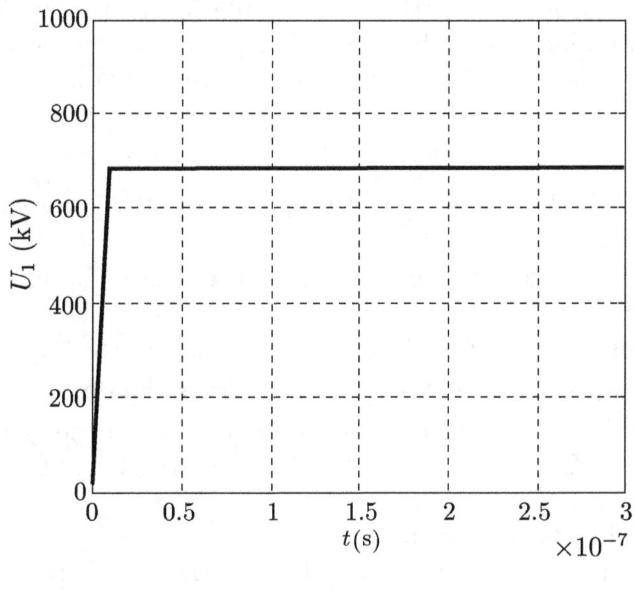


Fig. 2.59 - Voltage wave-shape between contacts of the output disconnector

Fig. 2.59 shows the voltage wave-shape between the contacts of the outgoing disconnector. The voltage amplitude has a maximum possible value of $U_1 = 2{\cdot}420{\cdot}\sqrt{2}/\sqrt{3} = 686$ kV, while the rise time amounts to 10 ns.

For the voltage wave-shape according to Fig. 2.59, the potential of the metal enclosure is determined at the coupling of the GIS and conducting insulator. Fig. 2.60a refers to a calculation time of up to 3 $\mu$s, while Fig. 2.60b illustrates the wave-shape of the potential in the time period from 240 ns to 360 ns.

Based on the graphs shown in Figs. 2.60a and 2.60b, the following conclusions can be made:

- the maximum value of the potential of the metal enclosure at the considered point amounts to 55 kV,
- the maximum value corresponds to the first maximum potential and is reached after 255 ns from the beginning of the transient process,
- the high frequency transient process on the metal enclosure of GIS lasts only a few $\mu$s.

The conclusions which were reached agree with the calculations presented in [41] according to which the potential of the metal enclosure of GIS has a relatively high amplitude but lasts only a few $\mu$s at frequencies around 30 MHz. The calculation results indicate somewhat lower amplitudes at the remaining points of the metal enclosure in relation to the amplitude of potential at the considered point.

*c) Measures for lowering potential of the metal enclosure of GIS*

Generally, there are two groups of measures which enable the lowering of the considered potential:

- measures which are taken in the phase of designing and installing GIS,
- measures which are taken in the phase of exploitation of GIS.

The first group of measures encompasses the optimization of:

- height of the metal enclosure in relation to the ground, which also determines the length of connecting parts to the grounding grid of GIS,
- number of connecting parts,
- characteristics of the ground below the metal enclosure of GIS.

The second group of measures includes:

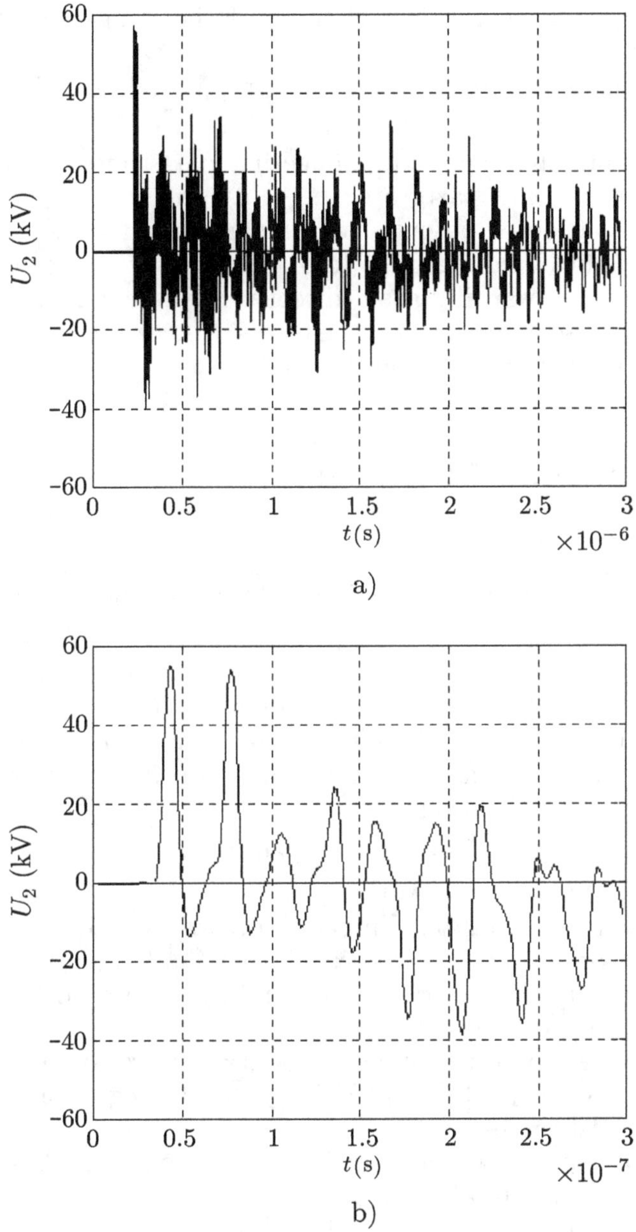


Fig. 2.60 - Wave-shape of potential of metal enclosure at the coupling of the GIS and conducting insulator; a) Time period from 0 to 3 $\mu$s; b) Time period from 240 ns to 360 ns

- informing operating personnel of the dangers of this phenomenon,
- warning operating personnel directly before switching operations of disconnectors.

Decreasing the length of connecting parts and the distance between two connecting parts are the most effective measures for decreasing the considered potential which can be undertaken in the phase of designing and installing GIS. Informing operating personnel of the dangers as well as warning them directly before conducting switching operations of disconnectors are measures which are necessary during exploitation of these substations.

## 2.3. MS Excel / MS Access

### 2.3.1. Introduction

The integrated software tool MS Excel combines the following [52, 53]:

- creating working tables,
- working with databases in tabular form,
- creating diagrams based on data from working tables,
- solving problems from the areas of business, science and engineering applications.

MS Excel features the following advantages:

- during creation of working tables:
    - formatting text in cells in a way which is supported by other applications within the Windows operating system,
    - capability of displaying and printing a portion of a table up to a level of detail,
    - capabilities of importing data from various databases (MS Access, dBase and MS SQL) and integration with MS Excel tables,
    - capabilities of exporting data from MS Excel into relational databases,
    - working with created tables,
- during creation of diagrams:
    - MS Excel supports at least 128 graphic formats, of which 24 are three-dimensional,
    - capability of dynamic linking of data,

- when solving problems from the areas of business, science and engineering applications:
  - existence of efficient add-on programs for analysis, such as:
    - * Goal Seek – program for back calculation,
    - * Solver – program for "what if" analysis which enables working with several variables and finding optimal solutions,
    - * Analysis ToolPak – program for conducting complex and sophisticated statistical analysis.
  - forming Gantt charts in the process of project management,
  - simple integration of MS Excel with other applications (MS Word, MS PowerPoint).

MS Access is a powerful database management system which is supported by modern technologies on the software market [54, 55]. It is an objectively oriented database management system (DBMS), which means that the entire databases is composed of objects which have defined characteristics, or attributes, as well as assigned properties which determine their structure, appearance and behavior. MS Access contains two cores: the first core is known as the Jet Database Engine (JDE) and the second core is the Microsoft Data Engine (MSDE).

MSDE is a key element of open DBMS. MS Access can be considered the front-end layer in the exploitation of large database systems such as SQL (Structured Query Language) Server, Oracle, SyBase and DB2. In such environments, the structures of data (tables, views, rules of accuracy, standard data procedures) are located on a network server, while the forms, reports and other elements of the application are located on the computers of clients. Interaction is conducted through a standard Open Systems Interconnections (OSI) layer of communication. MSDE is the bearer of this work, providing complete compatibility with SQL Server.

MSDE contains various advanced technologies necessary for the exploitation of databases, such as:

- dynamic inclusion of data,
- dynamic maintenance of databases.

This model provides the user with all the advantages of SQL Server, such as:

- automatic transformation of data during import and export,

- exploitation of Online Analytical Processing (OLAP) structures intended for specific data analyses,
- queries in natural language instead of complex SQL commands,
- parallel execution of queries.

MS Access enables the creation of MS Access Project model applications which become the front-end for the future client-server system. Access encompasses visual tools for creation of databases based on DaVinci technology which enables access to structures in SQL Server. If there is a "local" application which is based on a Jet database and which needs to be transferred into a client-server ambient, MS Access Upsizing Wizard is used to convert the application into the equivalent on SQL Server.

The primary programming language of MS Access is Visual Basic for Applications (VBA). MS Access integrates a unique environment for programming (Visual Basic Environment – VBE) which also functions in the other package programs.

In the first part of this section some of the possibilities for using databases in designing high-voltage substations were presented. The characteristic values of fault current, as an integral part of the criteria for selection of high-voltage equipment, are calculated using the programming language VBA and Visual Basic (VB). Various types of databases with elements of high-voltage equipment have been developed using MS Excel and MS Access. The basic principles of database architecture and the formation of filtering criteria which are based on the application of MS Query and VBA are discussed. The main elements for the proper creation of relational databases in MS Access are illustrated. The application of databases is presented within the examples of selection of high-voltage circuit breakers and disconnectors.

Modern design in power engineering entails the automation of calculations using macros. In the second part of the section the process for forming a macros as a VBA procedure is presented. The application of macros is illustrated in the examples of calculation of the total electricity and thermal impulse of injected current in the tested structure and automation of work with databases.

In the third part of the section the use of applications from the Microsoft Office program package is demonstrated in designing supply of telecommunications equipment for a control-commutation center. The capabilities of MS Excel when working with and processing data, as well as the dynamic linking of data between MS Excel and MS Word, have been presented in the example of calculation of a portion of elements of a system for supply of telecommunications equipment.

## 2.3.2. Application of databases in designing high-voltage substations

### 2.3.2.1 Criteria for selection of high-voltage equipment

The selection of high-voltage equipment is an integral component of designing power system substations. This selection is conducted on the basis of criteria which include indicative values of the network at the location of installation of the equipment and the rated, i.e. allowed values of the equipment [10].

The characteristic values of fault current represent a component of the criteria for selection of high-voltage equipment. As stated in section 2.1.4.1, these values can be calculated using various methodological procedures [10, 11] and software tools [5, 12-14]. The criteria for selection of high-voltage circuit breakers and disconnectors are displayed below. A review of the criteria for selection of the entire high-voltage equipment is provided in literature [10].

The process for selection of high-voltage equipment can be completely automated by forming corresponding databases. Data, instead of in the form of printed catalogs, is kept in logically organized files. A properly structured database enables the data which fulfills the corresponding filtering criteria to be efficiently located and separated.

Circuit breakers are selected on the basis of critical conditions in the network at the installation location. These conditions relate to the type of fault and configuration diagram of the network. After confirming the mentioned critical conditions for operation of the circuit breaker, it is necessary to calculate the characteristic values of fault current and other indicative values of the network and compare them with the rated, i.e. permitted values for the corresponding circuit breaker. The circuit breaker may be installed in the substation only if all criteria are sufficient for the selection of characteristic values.

Table 2.9 provides an overview of the characteristic values of the network and circuit breaker, as well as the necessary relation of these values for selection of the circuit breaker.

The symbols in Table 2.9 have the following meanings: $U_{ms}$ – maximum value of phase to phase voltage of the network, $U_n$ – rated voltage of circuit breaker, $I_{r_{\max}}$ – maximum operational current through the network branch in which the circuit breaker will be placed, $I_n$ – rated current of circuit breaker, $I_i$ – breaking current at circuit breaker installation location, $I_{in}$ – rated breaking current of circuit breaker,

Table 2.9 - Overview of characteristic values of the network and circuit breaker, as well as necessary relation of these values for selection of the circuit breaker

| Network value | Circuit breaker value | Selection criterion |
|---|---|---|
| $U_{ms}$ | $U_n$ | $U_{ms} \leqslant U_n$ |
| $I_{r\max}$ | $I_n$ | $I_{r\max} \leqslant I_n$ |
| $I_i$ | $I_{in}$ | $I_i \leqslant I_{in}$ |
| $I_{um}$ | $I_{un}$ | $I_{um} \leqslant I_{un}$ |
| $A$ | $A_d = I_t^2 \cdot t$ | $A \leqslant A_d$ |

$I_{um}$ – impulse current at location of circuit breaker installation, $I_{un}$ – rated switching on current of circuit breaker, $A$ – thermal impulse for short circuit at the circuit breaker installation location, $A_d$ – permitted thermal impulse for circuit breaker, $I_t$ – permitted thermal current during time period $t$.

The selection of a disconnector is conducted on the basis of the criteria defined in Table 2.10. The selection of a disconnector is considered proper if all of the conditions from the aforementioned table are fulfilled.

Table 2.10 - Overview of characteristic values of the network and disconnector, as well as necessary relation of these values for selection of the disconnector

| Network value | Disconnector value | Selection criterion |
|---|---|---|
| $U_{ns}$ | $U_n$ | $U_{ns} \leqslant U_n$ |
| $I_{r\max}$ | $I_n$ | $I_{r\max} \leqslant I_n$ |
| $I_{um}$ | $I_{un}$ | $I_{um} \leqslant I_{un}$ |
| $A$ | $A_d = I_t^2 \cdot t$ | $A \leqslant A_d$ |

The symbols in Table 2.10 have the following meanings: $U_{ns}$ – rated voltage of substation, $U_n$ – rated voltage of disconnector, $I_{r_{\max}}$ – maximum operational current through the branch of the network in which disconnector will be placed, $I_n$ – rated current of disconnector, $I_{um}$ – impulse current at installation location of disconnector, $I_{un}$ – rated impulse current of disconnector, $A$ – thermal impulse for short circuit at installation location of disconnector, $A_d$ – permitted thermal impulse for disconnector, $I_t$ – permitted thermal current during time period $t$.

### 2.3.2.2 Calculation of characteristic values of fault current

For selection of a circuit breaker and disconnector, according to the criteria in Tables 2.9 and 2.10, it is necessary to calculate the characteristic values of fault current. Within the example of a single-pole diagram of a portion of the power system displayed in Fig. 2.26, section 2.1.4.2, this calculation is illustrated for the case of a three-phase short circuit on busbars C, switches 1, 3', 3", 4', 4", 5 and 6 are switched on. The faults on the 220 kV side of the network are switched off in 0.2 s. The considered substation is located within a network with an efficient grounding neutral point.

Details of the classic process are displayed in sections 2.1.4.2 and 2.1.4.3. This encompasses the forming of a corresponding equivalent circuit and determining the characteristic values of fault current without explicit application of any software tools including the application of the software tool MATLAB® /Simulink® in the automation of the calculation of characteristic values of fault current. The aforementioned processes relate to the calculation of characteristic values of fault current outside of a database program.

Two processes for calculation of characteristic values which are executed within a database program are illustrated below. According to the first process, the corresponding calculation is executed using the programming language VBA. This language is an integral part of the software tool MS Excel, the capabilities of which for working with databases are illustrated in section 2.3.4.2.

The application of MS Access in automatically defining the parameters of an equivalent circuit are illustrated in another process. Based on the parameters entered in this way, using the programming language VB, the calculation of characteristic values of three-phase short circuit current is conducted.

*a) Calculation process using MS Excel and VBA*

The programming language VBA is an integral part of MS Excel. Through VBA a program is formed which, for the given parameters and elements of the system, enables the calculation of three-phase short circuit current.

Figs. 2.61a-2.61c show a document developed in MS Excel which encompasses a single-pole diagram (Fig. 2.61a), system parameters (Fig. 2.61b) and input data on the system elements (Fig. 2.61c). During the analysis of calculation results it is necessary to note that, only

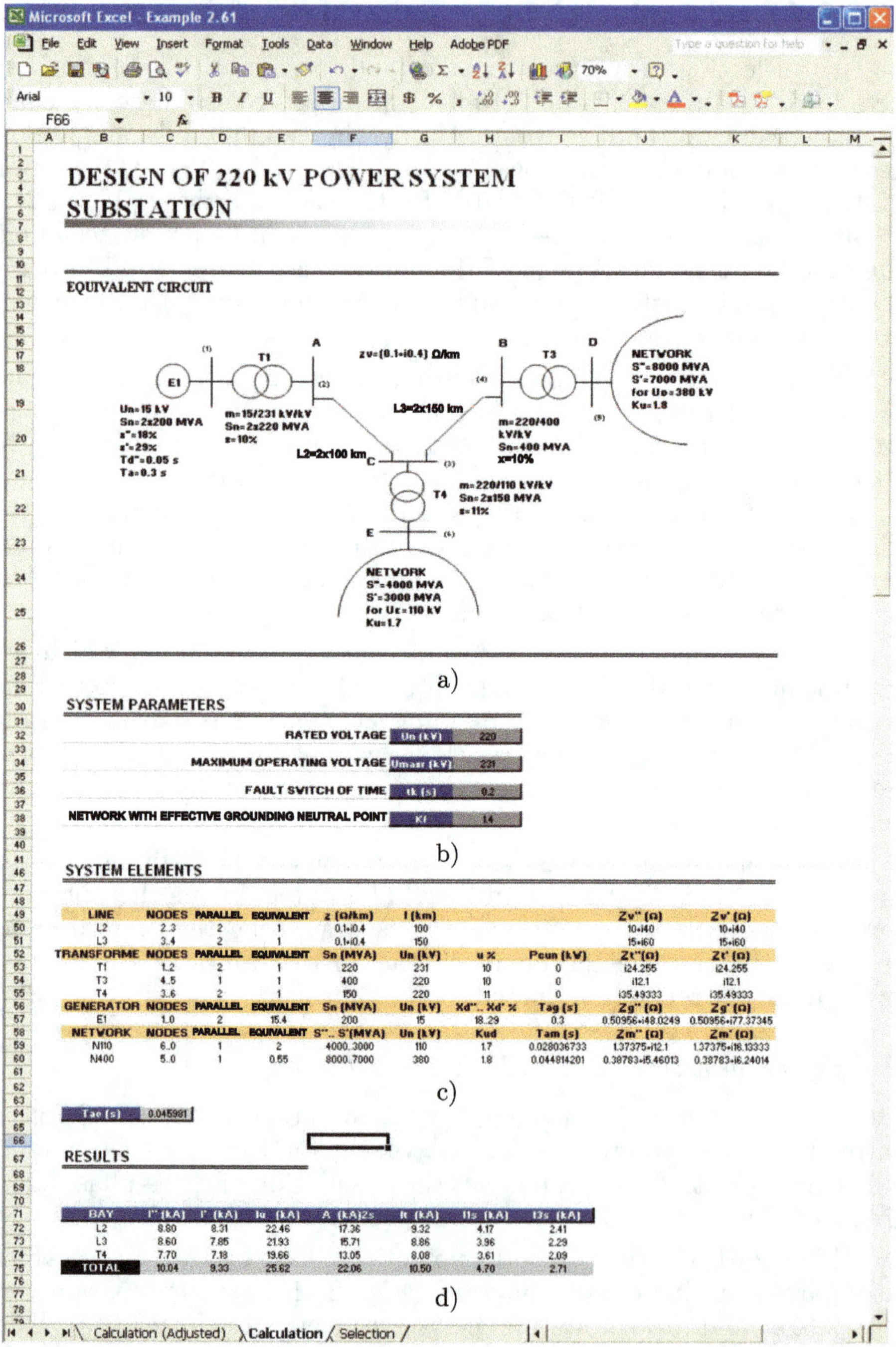
Microsoft Excel - Example 2.61
DESIGN OF 220 kV POWER SYSTEM SUBSTATION
EQUIVALENT CIRCUIT
E1
Un=15 kV
Sn=2x200 MVA
x"=18%
x'=29%
Td"=0.05 s
Ta=0.3 s
T1
m=15/231 kV/kV
Sn=2x220 MVA
x=10%
A
B
C
D
E
zv=(0.1+i0.4) Ω/km
L3=2x150 km
L2=2x100 km
T3
m=220/400 kV/kV
Sn=400 MVA
x=10%
T4
m=220/110 kV/kV
Sn=2x150 MVA
x=11%
NETWORK
S"=8000 MVA
S'=7000 MVA
for Un=380 kV
Ku=1.8
NETWORK
S"=4000 MVA
S'=3000 MVA
for Un=110 kV
Ku=1.7
a)
SYSTEM PARAMETERS
RATED VOLTAGE Un (kV) 220
MAXIMUM OPERATING VOLTAGE Umax (kV) 231
FAULT SWITCH OF TIME tk (s) 0.2
NETWORK WITH EFFECTIVE GROUNDING NEUTRAL POINT Kf 1.4
b)
SYSTEM ELEMENTS
LINE | NODES | PARALLEL | EQUIVALENT | z (Ω/km) | l (km) | | | Zv" (Ω) | Zv' (Ω)
L2 | 2.3 | 2 | 1 | 0.1+i0.4 | 100 | | | 10+i40 | 10+i40
L3 | 3.4 | 2 | 1 | 0.1+i0.4 | 150 | | | 15+i60 | 15+i60
TRANSFORME | NODES | PARALLEL | EQUIVALENT | Sn (MVA) | Un (kV) | u % | Pcun (kW) | Zt"(Ω) | Zt' (Ω)
T1 | 1.2 | 2 | 1 | 220 | 231 | 10 | 0 | i24.255 | i24.255
T3 | 4.5 | 1 | 1 | 400 | 220 | 10 | 0 | i12.1 | i12.1
T4 | 3.6 | 2 | 1 | 150 | 220 | 11 | 0 | i35.49333 | i35.49333
GENERATOR | NODES | PARALLEL | EQUIVALENT | Sn (MVA) | Un (kV) | Xd".. Xd' % | Tag (s) | Zg" (Ω) | Zg' (Ω)
E1 | 1.0 | 2 | 15.4 | 200 | 15 | 18..29 | 0.3 | 0.50956+i48.0249 | 0.50956+i77.37345
NETWORK | NODES | PARALLEL | EQUIVALENT | S".. S'(MVA) | Un (kV) | Kud | Tam (s) | Zm" (Ω) | Zm' (Ω)
N110 | 6.0 | 1 | 2 | 4000..3000 | 110 | 1.7 | 0.028036733 | 1.37375+i12.1 | 1.37375+i18.13333
N400 | 5.0 | 1 | 0.55 | 8000..7000 | 380 | 1.8 | 0.044814201 | 0.38783+i5.46013 | 0.38783+i6.24014
c)
Tae (s) 0.045981
RESULTS
BAY | I" (kA) | I' (kA) | Iu (kA) | A (kA)2s | It (kA) | I1s (kA) | I3s (kA)
L2 | 8.80 | 8.31 | 22.46 | 17.36 | 9.32 | 4.17 | 2.41
L3 | 8.60 | 7.85 | 21.93 | 15.71 | 8.86 | 3.96 | 2.29
T4 | 7.70 | 7.18 | 19.66 | 13.05 | 8.08 | 3.61 | 2.09
TOTAL | 10.04 | 9.33 | 25.62 | 22.06 | 10.50 | 4.70 | 2.71
d)
Calculation (Adjusted)
Calculation
Selection

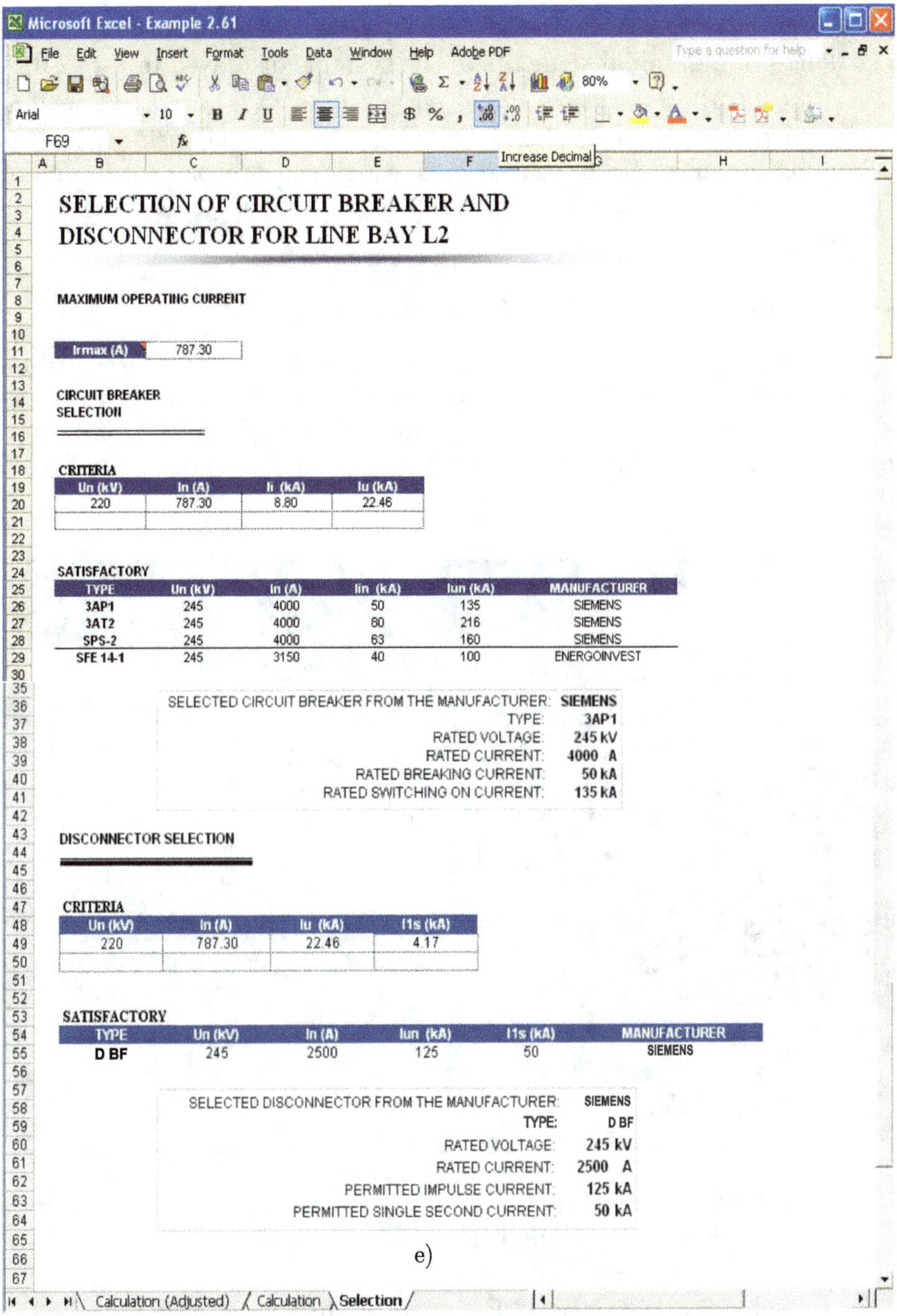


Fig. 2.61 - Document created in MS Excel using a program written in VBA; a) Single-pole diagram; b) System parameters; c) Data on system elements; d) Calculated characteristic values; e) Selection of circuit breaker and disconnector in line bay L2

in this case, the calculation was done while also recognizing the active resistances of individual elements of the systems displayed in Fig. 2.61c.

Fig. 2.61d shows the calculation results of characteristic values of the current of a three-phase short circuit obtained using the programming language VBA. The results which refer to bay L2 represent the characteristic values of current of a three-phase short circuit on busbars C.

*b) Calculation process using MS Access and VB*

Automation of the process for calculating the characteristic values of current of a three-phase short circuit and selection of high-voltage equipment was done using the program MS Access and the programming language VB. The dialog box of the formed program is provided in Fig. 2.62.

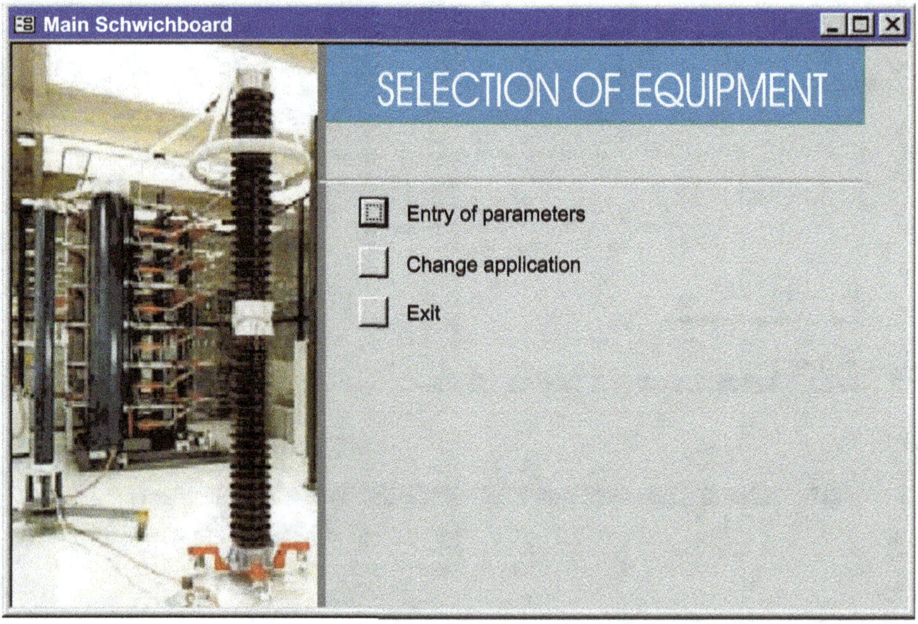


Fig. 2.62 - Dialog box for calculation of short circuit current and selection of high-voltage equipment

By selecting the option "Entry of parameters" from Fig. 2.62, a dialog box appears for entering in data. The dialog box contains fields for entering in the parameters and elements of the considered high-voltage substation. This dialog box, following the entry of input data for the considered portion of the power system, is displayed in Fig. 2.63.

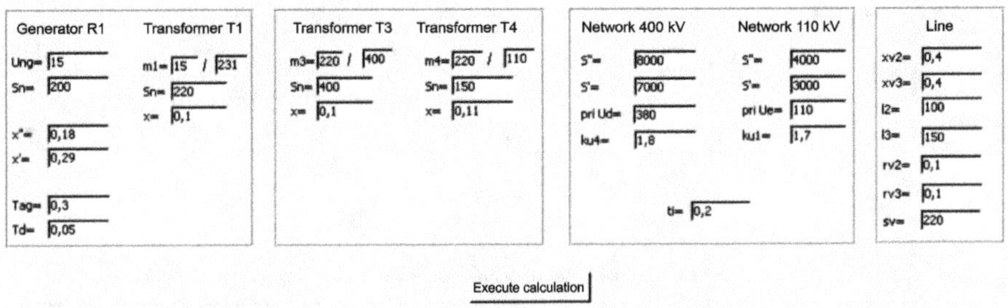


Fig. 2.63 - Dialog box with entered data for the portion of the power system from Fig. 2.61a

By selecting the icon "Execute calculation" from Fig. 2.63, the calculation is executed for characteristic values of the current of a three-phase short circuit on busbars C using the program written in Visual Basic. The calculation results are shown in Fig. 2.64 and partially in Table 2.11. Based on the determined calculation results and the criteria for selection of high-voltage equipment provided in Tables 2.9 and 2.10, filtration of the database is conducted which will be further explained in section 2.3.3.2.

By selecting the corresponding icon (Circuit breaker, Disconnector, Current tr.) in Fig. 2.64, a database search is executed.

*c) Collective overview of calculation results for a fault current determined using various calculation techniques*

Table 2.11 shows the characteristic values of subtransient current of a three-phase short circuit on busbars C, as well as the corresponding impulse current, determined using various calculation techniques.

The symbol $I''_{KC}$ in Table 2.11 refers to the effective value of subtransient components of a three-phase short circuit current on busbars C. Considering that the calculation of maximum current value $I''_{KC}$ (14.295 kA in Fig. 2.64) is performed by MS Access, for the purpose of comparing the results, Table 2.11 shows its effective value ($14.295/\sqrt{2} = 10.1$ kA).

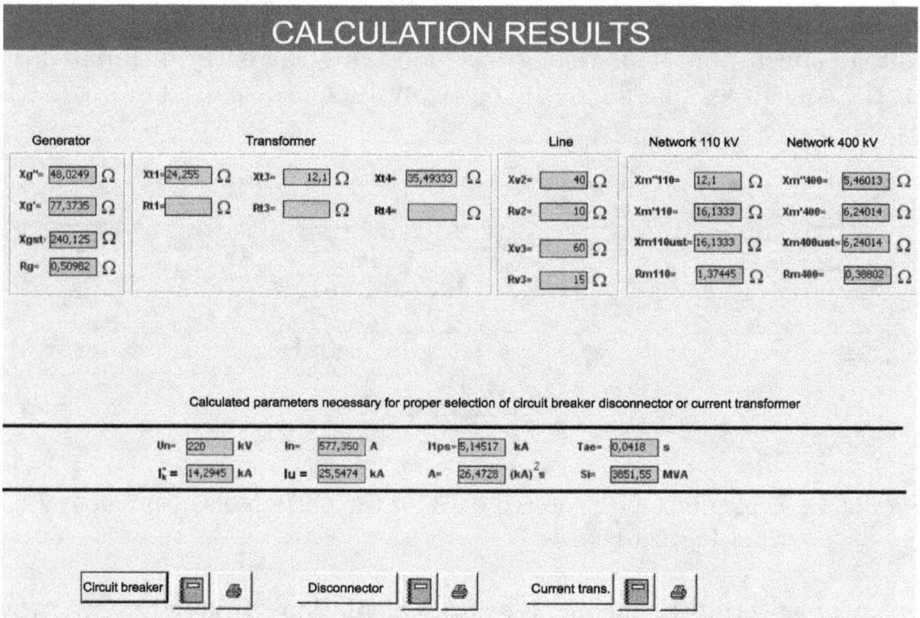

Fig. 2.64 - Results of calculation for a three-phase short circuit current on busbars C

Based on the calculation results shown in Table 2.11, it can be concluded that practically the same calculation results were obtained in the first, second and fourth examples. The lower values obtained in the third example are in this case the consequence of recognized active resistances of the substation elements.

### 2.3.3. Application of databases in selection of high-voltage equipment

#### 2.3.3.1 Definition of database types

A database is an organized collection of related information for a specific purpose. For example, a catalog of circuit breakers from a certain manufacturer represents a database which holds information on various types of circuit breakers which this manufacturer produces. The information refers to the characteristics of the circuit breakers and is necessary for proper selection during the design of high-voltage substations. The common denominator of this information is that it all refers

Table 2.11 - Collective overview of calculation results of three-phase short circuit current and the corresponding impulse current on busbars C on the single-pole diagram from Fig. 2.61a

| Example of application | $I''_{KC}$ (kA) | $i_{uC}$ (kA) |
|---|---|---|
| 1.Classic process (section 2.1.4.3) | 10,0 | 25,7 |
| 2. MATLAB® /Simulink® (section 2.1.4.3) | 10.2 | 26.0 |
| 3. MS Excel – VBA | 8.8 | 22.5 |
| 4. MS Access – VB | 10.1 | 25.5 |

to circuit breakers.

An electronic database is a location in a computer's memory where data is stored. The data, instead of in the form of a printed catalog, is kept in files. Logical units in which data is divided into in an electronic database are called tables. One table should contain just one type of data in order to later avoid the problems related to and increasing list. It is important that the data in the database be organized in such a way that it can be easily located and taken from the database for further processing, and it should also be easy to add new data or edit existing data.

A relational database is based on the principle of the division of information into collections of logically connected data which are kept in separate tables within the framework of the file. Tables are the basic object in relational databases and represent the active base of the system for storing and obtaining information.

Distribution of data between tables (relational databases) in comparison to the storage of all data in one table, enables improvements in working with databases such as:

- increased flexibility,
- simplicity in work,
- effortless control.

The basic elements of every relational database are the tables in which occurrences are defined as records with an identifier (key). The complex relationship between tables is determined through queries tables which can be used as query criteria and to view complex entities. The user interface is provided through several types of forms and reports. Finally, the database processes and automation are defined by macros and modules through which the database is promoted in the user application.

### 2.3.3.2 Example of the application of MS Excel

*a) Principles of a properly structured sheet in MS Excel*

Despite the basic function of MS Excel consisting of work with tables, the structure which is made up of lines and columns can also be used to form databases. Within MS Excel databases are called sheets which are a range of cells with two or more lines and at least one column. Each line represents a record and each column is called a field with each record containing several fields.

In order to form a database it is necessary to create its structure, which entails the defining of its content. The type of field which is entered – data, corresponds to the types of data which are entered into the working table.

To create an effective database there are certain rules which must be followed [5]:

- the sheet should not be divided by inserting empty columns or lines,
- each level of information should contain a separate column,
- empty space between lines is not allowed.

Fig. 2.65 shows a properly structured sheet with data on high-voltage circuit breakers.

The sheet shown in Fig. 2.65 is clear and well organized which makes it easier to enter data and use that data at a later time. The layout of fields and the format are consistently applied and each record has been completed with the circuit breaker type, rated voltage Un, rated current In, rated switching off current Iin, rated switching on current Iun and name of the manufacturer.

*b) Filtering of the sheet*

Working with databases includes the following processes:

- defining the database,
- editing records and fields,
- moving records,
- adding records and fields,
- deleting records and fields,
- sorting, i.e. arranging records in the database on the basis of one, two or several fields,
- filtering the database,

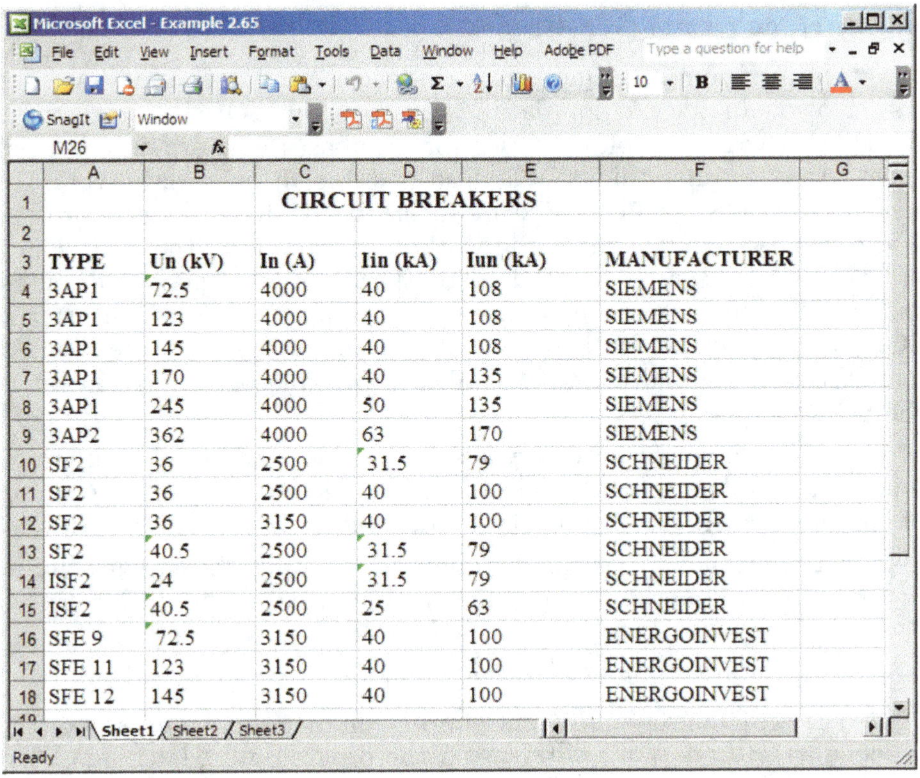

| | A | B | C | D | E | F |
|---|---|---|---|---|---|---|
| 1 | | | CIRCUIT BREAKERS | | | |
| 2 | | | | | | |
| 3 | TYPE | Un (kV) | In (A) | Iin (kA) | Iun (kA) | MANUFACTURER |
| 4 | 3AP1 | 72.5 | 4000 | 40 | 108 | SIEMENS |
| 5 | 3AP1 | 123 | 4000 | 40 | 108 | SIEMENS |
| 6 | 3AP1 | 145 | 4000 | 40 | 108 | SIEMENS |
| 7 | 3AP1 | 170 | 4000 | 40 | 135 | SIEMENS |
| 8 | 3AP1 | 245 | 4000 | 50 | 135 | SIEMENS |
| 9 | 3AP2 | 362 | 4000 | 63 | 170 | SIEMENS |
| 10 | SF2 | 36 | 2500 | 31.5 | 79 | SCHNEIDER |
| 11 | SF2 | 36 | 2500 | 40 | 100 | SCHNEIDER |
| 12 | SF2 | 36 | 3150 | 40 | 100 | SCHNEIDER |
| 13 | SF2 | 40.5 | 2500 | 31.5 | 79 | SCHNEIDER |
| 14 | ISF2 | 24 | 2500 | 31.5 | 79 | SCHNEIDER |
| 15 | ISF2 | 40.5 | 2500 | 25 | 63 | SCHNEIDER |
| 16 | SFE 9 | 72.5 | 3150 | 40 | 100 | ENERGOINVEST |
| 17 | SFE 11 | 123 | 3150 | 40 | 100 | ENERGOINVEST |
| 18 | SFE 12 | 145 | 3150 | 40 | 100 | ENERGOINVEST |

Fig. 2.65 - Example of a properly structure sheet with data on high-voltage circuit breakers

– using the built-in function of the database.

The process for filtering a sheet is described below. This is the process for taking data from a sheet which fulfills certain assigned criteria. MS Excel contains several tools which enable sheet filtering:

– conditional formatting,
– automatic filter,
– advanced filter,
– use of the program MS Query,
– use of the programming language VBA.

The first three filtering methods are described in detail within the provided literature [5]. The use of the program MS Query and the programming language VBA are displayed below.

*b.1 Use of the program MS Query*

The program MS Query is included in MS Excel and serves for the creation of queries for databases using the SQL language. It is possible to write complex queries for databases and draw data from them. Among other things, it is also possible to draw data from MS Excel workbooks in which there are sheets with data. The entire process is executed without opening the workbook containing the data which practically serves as a protection for the database against unwanted changes. The program is equipped with graphical tools which use the SQL language in the background so it is not necessary to be familiar with the language, although it is always possible to view and change the SQL commands.

Further on it is necessary to assign the criteria defining which data will be taken from the selected table. In the case of the selection of circuit breakers, those criteria are defined in Table 2.9.

Fig. 2.66 shows a completed query for filtering a sheet with circuit breakers, as well as the results which can be used further in MS Excel.

*b.2 Use of the programming language VBA*

The selection of equipment can be conducted on the basis of the independent forming of a filter using the programming language VBA which is packaged with MS Excel. This programming language enables the coding of special purpose functions which can later be used in MS Excel in the same way as built-in functions [56, 57].

By using VBA the following was performed:

- complete calculation of three-phase short circuit current in MS Excel,
- selection of a workbook,
- assignment of filtering criteria,
- selection of elements which fulfill the assigned conditions.

The selected elements may then be further used in MS Excel for forming the corresponding equipment specifications. The selection of elements is illustrated below.

*c) Example of application*

Through the application of the programming language VBA, a program was created which, for the assigned system parameters, calculates the current of a three-phase short circuit and, on the basis of the

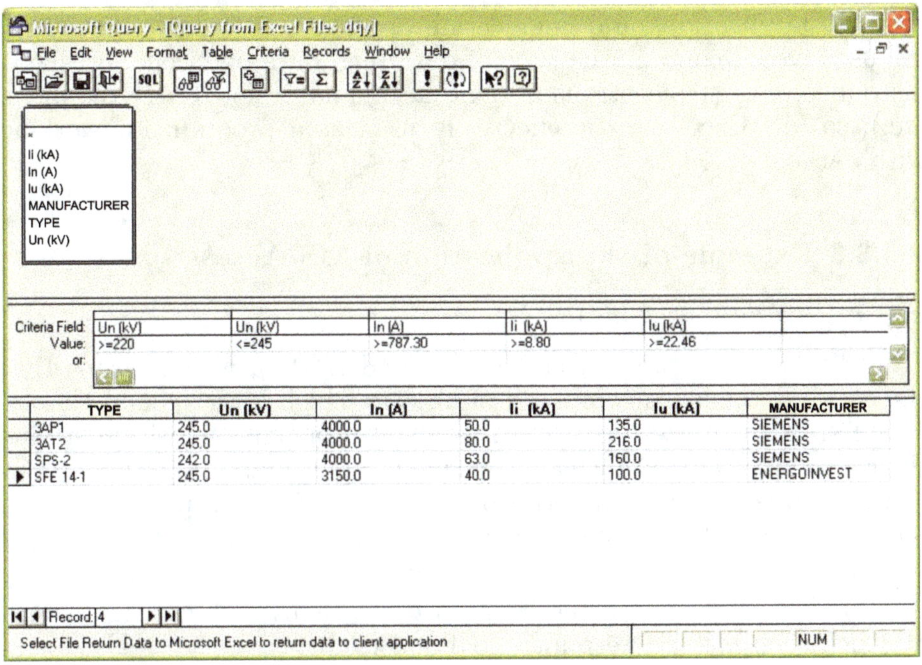

Fig. 2.66 - Completed query for selection of circuit breakers created using the program MS Query

assigned criteria, enables the selection of high-voltage equipment. Figs. 2.61a-2.61d show the appearance of this document, while the result of the proposed circuit breakers and disconnectors in line bay L2 are as illustrated in Fig. 2.61e [16].

The selection of the remaining elements of the substation (current and voltage transformers, surge arresters, busbars, etc.) is conducted in the same manner, by carrying out the displayed procedures.

Finally, it is important to note that the sheets formed in MS Excel do not represent a relational database. An MS Excel worksheet has only two dimensions, which practically means that one sheet can manage only one type of data. It is not possible to establish a link between two sheets so when updating data in one table, data in another table is automatically updated on the basis of the defined relations, which is the case with relational databases. In this case it would be necessary to manually update data in all tables, which is practically impossible if the number of tables is large, during which there is also the risk of errors. However, when designing high-voltage substations, business applications and engineering applications, most often sheets are used which contain one type of data (e.g. data on circuit breakers, disconnectors, current transformers, accumulator batteries, rectifiers, etc.), so MS Excel can be effectively used as a program for working with databases [58].

#### 2.3.3.3 Example of the application of MS Access

*a) Creating databases in MS Access*

MS Access is an objectively oriented DBMS, which means that the entire database is composed of objects which have defined characteristics, or attributes, as well as assigned properties which determine their structure, appearance and behavior. For example, the table properties include the table description, table subject and arrangement of records in the table, alphabetically or chronologically, based on one or several fields.

The process of creating a database can be divided into eight steps, of which each has a determined goal and result:

- determining what users expect from the database and what data is necessary to provide the corresponding result,
- planning distribution of data,
- analysis of fields of each table,
- determining a unique field which will eliminate the existence of two identical records within the table,
- determining how the tables are interconnected,
- checking the design and going through the procedure with users,
- creating the table and entering in data,
- analyzing and optimizing the database.

The aforementioned steps in creating a database in MS Access are provided in the specialized literature [54, 55]. Below a short review is provided for the most important elements for the proper creation of relational databases on the basis of which the selection of high-voltage equipment is conducted.

*a.1 Normalization*

Normalization is the process of restructuring files with data. The end goal of normalization is to decrease data in the database to the simplest structure and minimization of redundancy of data, i.e. organization of data fields for the purpose of achieving an efficient and flexible method of data storage.

Normalization has a complex mathematical background which contains specific conditions called normal forms. Each normal form represents an obstacle which the database must cross before going to the next step. The higher the level of the normal form, the more restrictive the test.

Reaching the third normal form is considered sufficient for the database. The first normal form eliminates duplicate data. The second normal form requires that all data from the table relate directly to the subject of the table, which is normally indicated by the primary key. The third normal form eliminates fields which can be obtained from other fields.

*a.2 Determining the primary key*

In a relational database system it is very important to group and find connected data which is located in the tables. In order to achieve that, each record in the table must be unique in some way. A field, or fields which contain a unique value represent a primary key. MS Access does not allow the existence of multiple identical values of a primary key, nor does it allow null values of a primary key. The field or fields of a primary key must contain a valid unique value.

*a.3 Types of dependencies*

Tables can be connected in three ways: one-to-several, one-to-one, several-to-several. The way which is used depends on the number of records of each table which will have the same values. In order to connect tables, each must have a primary key, a field which contains a unique value in each record. The primary key may contain two or more fields, which when combined provide a unique value for each record.

The most commonly used dependency is one-to-several, when one record of a table can have several corresponding records in another table. The first table is often called the "parent" and the second is called the "child". A one-to-one dependency is a type of matrix in which each record of one table corresponds to one record of another table. Neither table is a "parent". The key fields in both tables are primary keys. One of the uses of this kind of dependency is storing additional information which is rarely accessed in the first table.

A several-to-several dependency is not allowed in a relational database. Several records of one table have the same values of key fields as several records of another table. In order for this kind of dependency to be applied in MS Access it is necessary to create a third table, which is called a junction table, between these two tables. In this way a several-to-several dependency can be broken down into two one-to-one dependencies.

All that is necessary for defining the dependency between two tables is to indicate which fields of the tables are shared. In a one-to-one dependency, the field of the “parent” table is called the primary key and must be the primary key of that table or a unique index. The field in the second table is called the non-primary key and is not required to contain a unique value. Locating data is, however, quicker when there is an index for the non-primary key.

With a one-to-one dependency both fields are primary keys or unique indexes. A several-to-several dependency is actually a one-to-several dependency in which a third table is created so that its primary key is a combination of the primary keys of the shared fields of both tables.The table of connections becomes page "one" of the dependencies for both tables.

*a.4 Referential integrity*

Referential integrity is an optional system of rules which provide proper dependencies and unaltered state of the database during entry, updating and removal of data. The basic rule of referential integrity is that for each record of another table (page "more") there must be just one corresponding record in the "parent" table (page "one"). The rules of referential integrity which can be applied in MS Access are:

- a record cannot be entered into another table if the record does not exist in the "parent" table,
- a record in the "parent" table cannot be removed if there is a record which it is connected to in another table,
- a record in another table cannot be altered in such a way that the non-primary key does not have a corresponding value in the "parent" table,
- the value of the primary key in the "parent" table cannot be changed until there is a connected record in another table.

The Relationships window in MS Access contains all of the tools necessary for forming dependencies and connecting tables, as well as for setting the rules of referential integrity.

*b) Example of application*

*b.1 Determining the purpose of the database*

Determining the purpose of the database is the first step in creating a new database. The main purpose of the database called Substations relates to the calculation of short circuit currents and selection of high-voltage equipment [16]. In order to achieve this, the database must contains forms for data entry on the basis of which the calculation of short circuit currents are calculated, as well as forms for displaying high-voltage equipment which fulfills the selection criteria. Additionally, this can also include sub-forms for displaying additional information on high-voltage equipment, as well as reports for the purpose of printing the obtained calculations and equipment specifications.

*b.2 Distribution of data*

The second step represents determining the division of data between the tables. This is one of the more important tasks in creating a database. In the considered case, information on manufacturers and types of equipment is repeated. For the purpose of decreasing redundancy, this information is stored in separate tables.

The data which is used as selection criteria is located in an individual table. Other, conditionally less important pieces of data, such as the dimensions of the circuit breaker or disconnector, are also stored in separate tables.

The types of dependencies between the tables Selection of circuit breaker and Circuit breaker characteristics as well as the table Disconnectors and Disconnectors characteristics are one-to-one, with application of the rules of referential integrity. Data which is not necessary during selection of equipment is stored in separate tables, separated from the main portion of information. This enables data to be searched through faster. All other types of dependencies are one-to-one and also use the rules of referential integrity. This completes the process of creating tables and defining the dependencies between the tables.

Fig. 2.67 shows the Relationships window after establishing dependency lines for circuit breakers, disconnectors and current transformers.

By selecting the "Circuit breaker" icon in Fig. 2.64, a database search is initiated. On the basis of the assigned selection criteria defined in Table 2.9, the user, as a result of the search, is provided with the

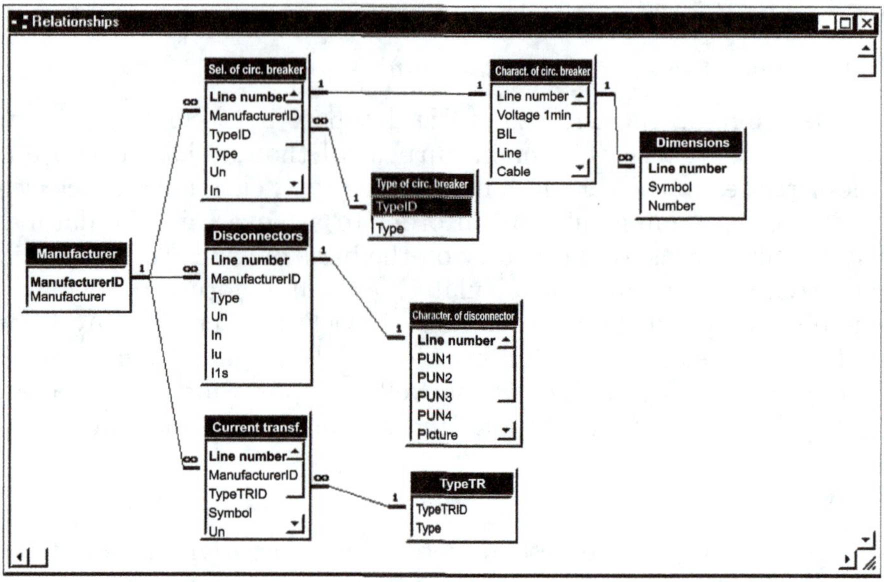


Fig. 2.67 - Relationships window after establishing dependency lines for circuit breakers, disconnectors and current transformers

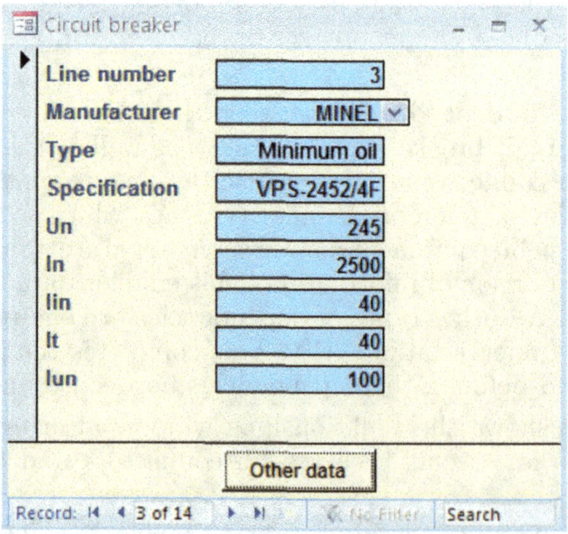


Fig. 2.68 – Recommended circuit breaker using MS Access

recommended circuit breaker (Fig. 2.68) with corresponding characteristics (Fig. 2.69).

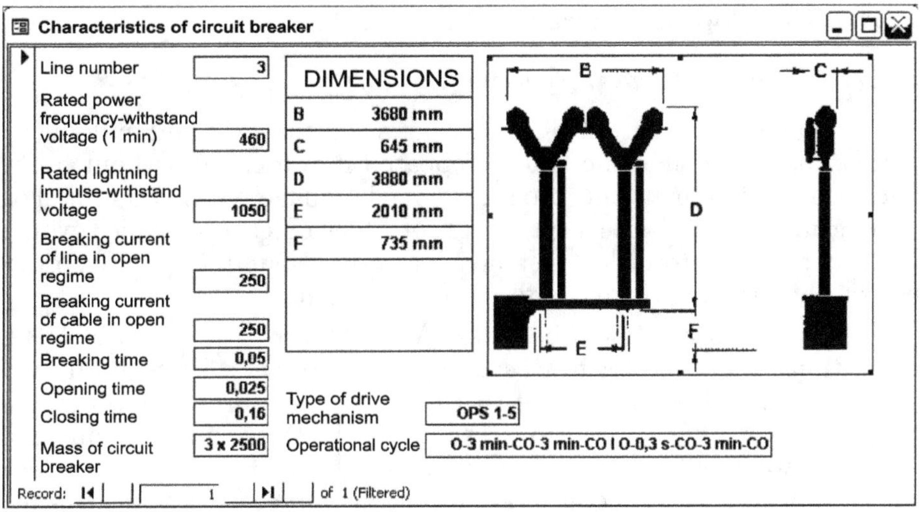


Fig. 2.69 – Characteristics of the recommended circuit breaker

The same procedure may also be carried out for the selection of other elements of high-voltage equipment.

Based on the displayed results the following can be concluded.

- characteristic values of fault current, necessary for proper selection of high-voltage equipment, can be calculated using existing tools or a user-developed program,
- implementation of properly structured and appropriately filtered databases enables the efficient selection of high-voltage equipment and development of corresponding specifications,
- the illustrated examples indicate that, for the purposes of designing high-voltage substations, it is possible to use databases developed using various programs. The efficiency of MS Excel is sufficient when working with smaller databases, as well as with databases taken from other programs. MS Access, as an objectively oriented relational DBMS, has the advantage when working with more complex applications,
- the application of database programs and software tools for drafting technical documentation enables the automation of computer design of high-voltage substations, whereby increased productivity is achieved [59].

## 2.3.4. Application of MS Excel macros for design in power engineering

### 2.3.4.1 Developing macros in MS Excel

*a) Defining macros and their basic function*

A macro is a program which executes certain tasks and in doing so ensures that each time the task is repeated it is done so without fault. Macros are written in the programming language VBA [56, 60]. For forming macros, MS Excel has the tool Macro Recorder which works similarly to a tape recorder. Instead of recording sound, Macro Recorder records the pressing of keys and mouse functions.

A macro is a VBA procedure. A procedure is a group of VBA instructions which perform a certain task or provide a result. In MS Excel it is possible to create two types of procedures. Subprograms are procedures which perform a certain task. With subprograms the beginning word is "Sub" and the final words are "End sub". A macro is a type of routine procedure. Functions are procedures which return a result. VBA enables the forming of special purpose functions which can be used as built-in MS Excel functions. Functions begin with the word "Function" and end with the words "End function".

A good characteristic of macro recording in MS Excel relates to the automatic assignment of comments from the program, i.e. information about the recording process: name and date of macro, who recorded the macro and keyboard shortcuts, if any, which are assigned to the macro. When recording a macro, Macro Recorder automatically inserts procedure lines so it can be easily read. Typical activities which are preferred to be automated using a macro are:

- formatting and printing reports,
- assistance in filling out Excel forms,
- merging data from several workbooks into a main workbook,
- displaying data in a diagram,
- assigning keyboard shortcuts for commonly used commands,
- application of a certain AutoFormat to a collection of cells,
- application of the AutoFit command,
- forming a special purpose application for a table.

*b) Recording macros using Macro Recorder*

Before beginning the recording of a macro it is necessary to decide where the macro will be placed and how it will be recorded. MS Excel offers three options for placing the macro:

– active workbook,
– personal workbook,
– new workbook,

If a macro is placed in an active workbook it can only be executed when that workbook is open. If a macro is placed in a personal workbook it will be available in all workbooks, regardless of which workbook is open. If a macro is placed in another workbook, it is necessary to open that workbook whenever it is to be used.

A macro can be recorded absolutely or relatively. If a macro is recorded relatively, the macro will always be executed from the current position of the cell indicator. If a macro is recorded absolutely, it will always be executed in the same collection of cells as it was when being recorded. If the purpose of the macro does not require the modification or selection of a collection of cells, then it is not necessary to be concerned with whether the macro is recorded relatively or absolutely.

In order for the macro to be recorded it is necessary to perform the following:

– select Tools > Macro > Record New Macro in order to open the Record Macro dialog box,
– enter in the name of the macro into the Macro Name text box. The name can consist of letters, numbers and underscores, but may not contain spaces. The name must begin with a letter and may not be longer than 64 characters,
– when necessary, in the Store Macro In field, either select or enter in the location where the macro should be placed. For a new macro the active workbook option is selected as the default. When recording several macros in a row, one may notice that the Store Macro In field displays the same selection as the previous recording. When MS Excel is reopened the default This Workbook option is restored,
– if the assignment of a keyboard shortcut is desired, it is necessary to enter in the letter which will be a part of the shortcut into the Shortcut Key text box. Keyboard shortcuts for macros have an advantage over MS Excel's built-in shortcuts. Because MS Excel already uses a majority of the keyboard combinations Ctrl + <letter> for keyboard shortcuts, it is better to use a combination of Ctrl + Shift + <letter>,
– in the Description text box a short description of the function of the macro can be entered. This description will be displayed within the dialog box during the execution of the macro. The

description helps when recording a large number of macros for the purpose of reminding one of which tasks individual macros perform,
- press the OK button to close the Record Macro dialog box. The word Recording will be displayed on the status line along with the Stop Recording tool palette which is necessary in order to stop the recording of the macro,
- for a macro to be recorded relatively it is necessary to click on the Relative Reference button on the Stop Recording tool palette,
- perform all the actions which are part of the task which is being recorded,
- click on the Stop Recording button on the Stop Recording tool palette.

It can very easily occur that the Macro Recorder remains turned on. It will continue to record until it is given the command to stop working. For this reason caution is needed. If the user forgets to turn off the Macro Recorder on time, it is not necessary to record the entire macro over. One can simply remove the surplus recorded actions by editing the macro.

The workbook consists of objects such as worksheets, sheets with diagrams and sheets with modules. A module is an object which holds VBA code inside of it. All objects of the workbook together represent the project. By recording a workbook, one is essentially recording a project which contains all objects of the workbook. If a macro is placed in the active workbook, then it is necessary to record that workbook after recording the macro. If the macro is placed in the personal workbook, then it is necessary to record the same before closing the program by selecting the option Yes which is offered by MS Excel. Failing to do so will result in the loss of all macros which were located in that workbook.

*c) Executing a macro*

In MS Excel there are several ways to execute a macro:
- by selecting the macro from the list in the Macro dialog box,
- by activating the keyboard shortcut,
- by clicking on the button on the tool palette,
- by placing it as a menu item,
- by linking it to a graphical object.

The method which is selected depends on the type of problem which the macro will solve, as well as the competency and experience of the user. The steps for executing and editing a macro are described in the selected literature [56, 60].

### 2.3.4.2 Examples of the application of macros

*a) Application of a macro in determining the total electricity and thermal impulse of injected current in the tested structure*

An analysis of the behavior of a roof covering under the effects of lightning discharge current was conducted on the basis of experiments in laboratory conditions. The injection of impulse current into the tested structure in laboratory conditions provided the oscillatory dampened form which is given in the following analytical expression [5]:

$$i(t) = 3225 \cdot \sin(\omega \cdot t) \cdot e^{\delta \cdot t} \tag{2.19}$$

where: $i$ – current (A), $t$ – time ($\mu$s), $\omega = 2\pi/T$ – frequency, $T$ – period ($\mu$s).

For a period of 50 $\mu$s and a damping factor of $\delta = -0.0054735$ it is necessary to calculate the total quantity of lightning discharge electricity according to the expression:

$$q = \int_0^{T_t} \text{abs}(i) \cdot dt \tag{2.20}$$

where $T_t = 350$ $\mu$s.

The defined integral can be determined using the extended trapezoidal rule of integration:

$$\int_{x_1}^{x_N} f(x) \cdot dx = h \cdot (0.5 * f_1 + f_2 + f_3 + \cdots + f_{N-1} + 0.5 * f_N) + \text{ error} \tag{2.21}$$

where: $h$ – calculation step, $f_i$ ($i = 1 - N$) – values of function for equidistant values $x_1 - x_N$.

A calculation step value of $h = 1$ $\mu$s is adopted for the calculation and the quantity of electricity is stated in As.

The basic purpose of a macro which is reflected in the automation of calculation has also found its application in this example [61]. By using a macro, a problem can be solved in a quicker and more elegant way, which also includes the possibility of automatically changing the value of amplitude, period and damping factor of injected current. A portion of the procedure which enables the drawing of diagrams of time dependent injected current has been added to the program.

Creating a macro begins with the selection of the option Tools > Macro > Record New Macro which opens the Record Macro dialog box. In the field intended for the name of the macro the name "Thermalimpmacro" is entered. Considering that this macro does not belong the group of general purpose macros, but rather its use is related to a specific example from a unique workbook, it would be most appropriate to place the macro in the active workbook.

In the description field the function of the macro is briefly described. Due to the nature of the problem which is being solved with this macro and the precisely defined steps, it is best to record this macro absolutely. Upon closing the Record Macro dialog box the Stop Recording tool palette appears whereby the recording of the macro is activated. The process begins by entering in the formula for current form into cell B2 (Fig. 2.70):

$$\mathrm{B2} = \$\mathrm{E}\$5 * \mathrm{SIN}(\mathrm{A2} * 2 * \mathrm{PI}()/\$\mathrm{E}\$8) * \mathrm{EXP}(-\$\mathrm{E}\$11 * \mathrm{A2})$$

In cells E5, E8 and E11, data is inserted on the amplitude, period and damping factor of the injected current (Fig. 2.70). In all formulas, current is expressed in amperes and time in microseconds. The symbols of the cells in the formula are addressed relatively which enables the copying of the formula through the entire column containing current samples. The remaining current samples are obtained using the AutoFill option, which means that cell B352, for the last current sample, will contain the formula:

$$\mathrm{B352} = \$\mathrm{E}\$5 * \mathrm{SIN}(\mathrm{A352} * 2 * \mathrm{PI}()/\$\mathrm{E}\$8) * \mathrm{EXP}(-\$\mathrm{E}\$11 * \mathrm{A352})$$

In all formulas it is possible to use the preset MS Excel functions such as the sine function, exponential function and function for obtaining Pi. By using the ABS function the absolute values of the current samples are obtained, which are located in the cells of the third column (column C). Column D contains the square values of the current samples which are used for calculating thermal impulse.

Using the trapezoidal rule of integration, on the basis of the assigned data, the total amount of electricity and thermal impulse are

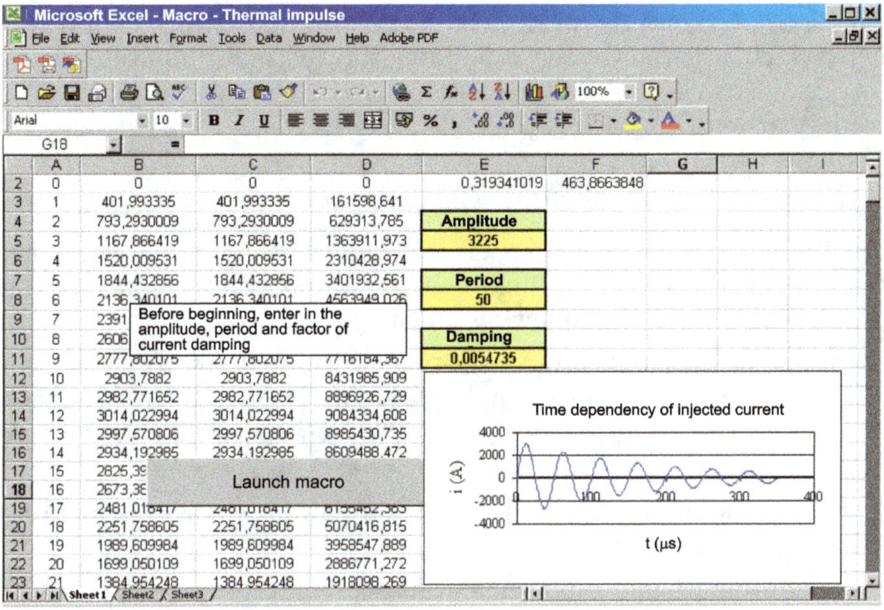


Fig. 2.70 - Worksheet of the workbook with the macro "Thermalimpmacro"

calculated. The formula for the total electricity is placed in cell E2 and has the form:

$$\mathrm{E2} = (0.5 * (\mathrm{C2} + \mathrm{C352}) + \mathrm{SUM}(\mathrm{C3} : \mathrm{C351})) * 0.000001$$

Thermal impulse is calculated on the basis of the formula in cell F2:

$$\mathrm{F2} = (0.5 * (\mathrm{D2} + \mathrm{D352}) + \mathrm{SUM}(\mathrm{D3} : \mathrm{D351})) * 0.000001$$

The formulas from cells E2 and F2 contain the preset MS Excel function SUM which adds together the values from the selected group of cells.

Once the diagram has been created it is necessary to discontinue recording of the macro by activating the Stop Recording button. For the execution of this macro it is best to select the method of assigning the macro to a graphical object due to the clear visibility of the object and capability of entering in a description of the function or method for using the macro. The option Forms and the Button tool can assign an arbitrarily sized button to the macro which will initiate the reproduction of the macro.

Fig. 2.70 shows the worksheet of the workbook which contains the macro which when activated calculates the thermal impulse and total electricity, as well as draws a diagram of the time dependency of the injected current.

*b) Application of macros in working with databases*

Table 2.12 shows the capacitance values of individual equipment items from various types of high voltage switchgears. The symbols for the switchgear have the following meanings:

- A – metal-enclosed, single-phase insulated 400 kV switchgear,
- B – metal-enclosed, single-phase insulated 110 kV switchgear,
- C – typical 121 kV disposition of switchgear 400 kV / 121 kV for outdoor installation.

Using the option for working with a database it is possible to:

- sort the database,
- automatically filter the database,
- set the criteria which enables the capacitances of individual elements to be found,
- add data into an existing database,
- determine the sum of capacitance according to type of switchgears, as well as total capacitance of all elements of all switchgears,
- remove entered subtotals, i.e. generate the previous database.

The application of a macro is shown below in the process of calculating sums of capacitance according to the type of switchgear, as well as the total capacitance of all elements from the existing sheet [61]. In the example two macros are used. The macro named "Subtotmacro" serves to calculate the sums of capacitance, while the macro named "Generdatabase" enables the defining of a database on the basis of which the adding together of capacitances is executed.

Within the macro "Subtotmacro" it is necessary to record the sorting of the sheet according to type of switchgear and in the Data-Subtotal option select the field Capacitance in the column on the basis of which data is added together. The macro also encompasses the formatting of individual cells through which subtotals and the sum of all capacitances within the sheet are extracted.

In the macro "Generdatabase" the selection of the option Remove All from the Data-Subtotal dialog box is recorded whereby the initial

Table 2.12 - Values of capacitance of individual equipment items for various types of high voltage switchgears

| Type of switchgear | Element | Voltage (kV) | Capacitance (pF) |
|---|---|---|---|
| A | Power transformer | 400 | 3230 |
| A | Capacitive voltage transformer | 400 | 4000 |
| A | Inductive voltage transformer | 400 | |
| A | Current transformer | 400 | 50 |
| A | Circuit breaker | 400 | 300 |
| A | Disconnector closed | 400 | 100 |
| A | Disconnector open | 400 | |
| A | Cable terminal | 400 | 400 |
| B | Power transformer | 110 | 3000 |
| B | Capacitive voltage transformer | 110 | 16000 |
| B | Inductive voltage transformer | 110 | 100 |
| B | Current transformer | 110 | 23 |
| B | Circuit breaker | 110 | 200 |
| B | Disconnector closed | 110 | 100 |
| B | Disconnector open | 110 | |
| B | Cable terminal | 110 | 200 |
| C | Power transformer | 121 | 3230 |
| C | Capacitive voltage transformer | 121 | 4400 |
| C | Inductive voltage transformer | 121 | 300 |
| C | Current transformer | 121 | 700 |
| C | Circuit breaker | 121 | 500 |
| C | Disconnector closed | 121 | 60 |
| C | Disconnector open | 121 | 40 |
| C | Cable terminal | 121 | |

database is regenerated. Both macros are executed by clicking on the assigned button or inserted graphical object.

Fig. 2.71 shows the worksheet after the execution of the macro "Subtotmacro".

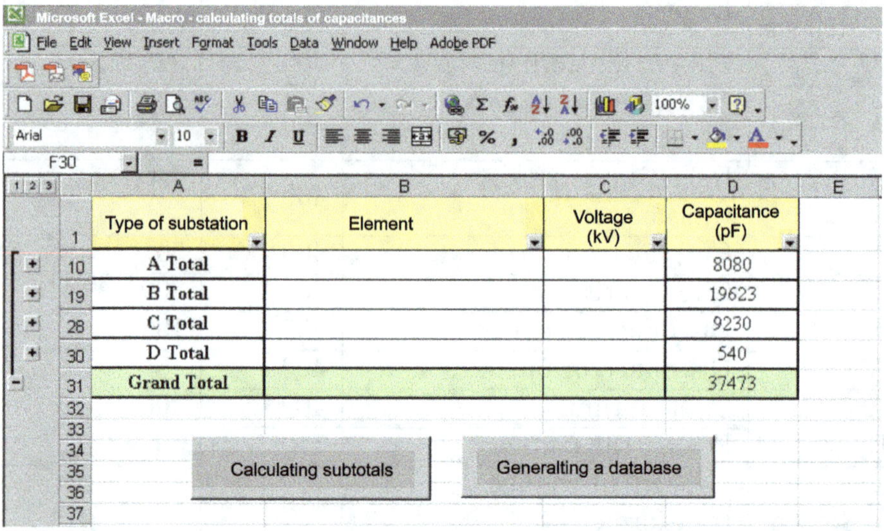


Fig. 2.71 – Worksheet following the execution of the macro "Subtotmacro"

### 2.3.5. Application of MS Excel in designing the power supply of telecommunications equipment

This section presents the technical description of power supply devices of a control-commutation center with a direct voltage of −48 V. The selection of devices for power supply and back-up power supply was conducted according to the technical instructions [62]. The technical calculations and corresponding selection of the aforementioned devices were completely automated using the programs MS Excel and Visual Basic (VB), which is illustrated in the examples of the selection of accumulator batteries and rectifiers [63-67].

#### 2.3.5.1 Technical description of power supply with direct voltage

A brief overview is provided below of the technical description for power supply of devices of a control-commutation center with a direct voltage of −48 V. The telecommunications (TC) equipment which is supplied with direct voltage is located in special purpose cabinets, envisaged for internal installation. The cabinets are located within the existing structure. To obtain the necessary direct voltage, two cabinets have been envisaged which are outfitted with rectifiers, distribution panels, battery terminals and control panels.

Supplying the devices with direct voltage is achieved using eight parallel-connected rectifiers with four in each cabinet. Each rectifier is a D48 / 120 WBRU6 – FGE18 GR60. The maximum current of one rectifier with a direct voltage of −48 V amounts to 120 A. For the needs of possible future expansion of capacity, in each cabinet a space for installation of one more rectifier with the same characteristics has been envisaged.

Each rectifier unit is outfitted with a microprocessor controller which monitors the operation of the rectifier and communicates with the central control unit which is located on the door of the rectifier cabinet. The output characteristic of the rectifier is a V-I type with current limited to 120 A. All parameter settings of the rectifier and rectifier installation have been digitalized and are set using the system menu.

The central unit is in constant communication with the rectifier units and monitors the condition of all circuit breakers/fuses, and provides the necessary alarms and notifications on the LCD display placed on the door of the rectifier cabinet. Additionally, in the case of a drop of network voltage, the central unit provides a warning regarding the situation which has occurred and constantly monitors the voltage of the accumulator batteries. If the voltage in the accumulator batteries falls below the previously programmed values, the battery circuit breakers are automatically switched off for the purpose of protecting the accumulator batteries from deep discharge.

The power supply system is intended to operate parallel to the accumulator batteries, rectifiers and devices, which means that the rectifiers are connected parallel and maintain the constant voltage of the batteries and telecommunications installation. In the case of a drop from the network, the installation is supplied with power from the accumulator batteries.

The necessary back-up power supply is secured from two accumulator batteries manufactured by FIAMM, type SMG1000. Each battery consists of 24 cells with a voltage of 2 V per cell and a nominal capacity of 1000 Ah for 10 hours of current discharge. These batteries have been envisaged for vertical installation in two levels. The batteries are connected to the rectifier cabinet through the battery circuit breakers/fuses produced by Siemens, MW4 with a rated current of 630 A, which are installed at the foot of the rectifier cabinet. The accumulator batteries are connected to the outputs from the rectifier cables 3×P/F 150 mm$^2$ per pole.

The accumulator batteries must fulfill the special requirement of being hermetically sealed and being equipped with a regulated valve. The aforementioned batteries are VR type (VR – valve regulated), with gelified electrolyte, so they fulfill the required conditions.

In order for them to be properly maintained, an ambient temperature of 20° C is necessary. The necessary conditions are provided by air conditioners.

For the purpose of creating the highest quality back-up system, batteries were ordered with an entirely sufficient capacity, even for the option of additional cabinets with equipment and connection to a larger number of users in a mobile telephony system.

The structure in which the telecommunications equipment is located is equipped with a diesel – generator set (genset).

### 2.3.5.2 Technical calculations

Automation of technical calculations was achieved entirely using MS Excel [63]. This process was implemented during the drafting of project documentation for supplying power to the control-commutation centers of mobile telecommunications [64-66]. Fig. 2.72 shows the worksheet in MS Excel. The processes for selection of accumulator batteries and rectifiers are illustrated below.

*a) Calculation and selection of stationary accumulator batteries*

The basic parameters used for determining the necessary capacity of the accumulator batteries, as the back-up source of power supply, are:

- $I_{max}$ – maximum current load of TC equipment and invertor at a direct voltage of −48 V, in accordance with item 4.1. of the technical instructions [62] (A),
- $T$ – time of required autonomy in supplying power to TC equipment, in accordance with item 4.2. of the technical instructions [62] (h),

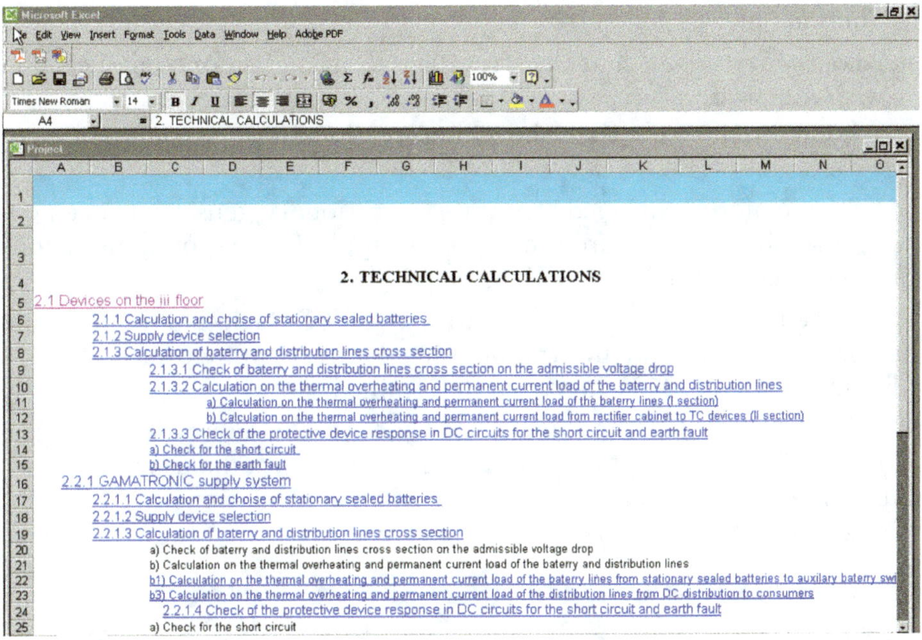


Fig. 2.72 – Worksheet in MS Excel

- $K_1$ – 1.15 – factor of increasing capacity due to sulphatization of pannels and possibly lower ambient temperature with classic batteries. In batteries with pressure which is regulated by a valve, so-called hermetically sealed batteries, this factor is not taken into consideration, i.e. it equals 1,
- $K_2$ – factor of increased capacity of accumulator batteries which are dimensioned for autonomy of less than 10 hours. This factor is defined by the supplier of accumulator batteries depending on the envisaged autonomy.

The capacity of accumulator batteries Q (Ah) is determined based on the expression:

$$Q = \frac{I_{\max} \cdot T \cdot K_1}{K_2} \tag{2.22}$$

The required autonomy is provided by at least two accumulator batteries with a total capacity which must be greater than Q.

The required time autonomy for supplying power to TC equipment during a full load, in accordance with item 4.2 of the technical instructions [62], amounts to:

- 2 hours, in the case that the structure (location) is not equipped with a genset,
- 1 hour, in the case that the structure (location) is equipped with a genset.

In the aforementioned structure, a stationary genset has been envisaged so the time required for reserve supply for telecommunications equipment is adopted as $T = 1$ h.

Based on the data on consumptions of TC devices obtained from the manufacturer, and for the envisaged direct voltage of −48 V, a value for Imax of 725 A has been determined.

By substituting the numerical values in the expression (2.22), a capacity of $Q = 954$ Ah is obtained for the accumulator batteries. This value is determined on the basis of the calculations in MS Excel (Fig. 2.73).

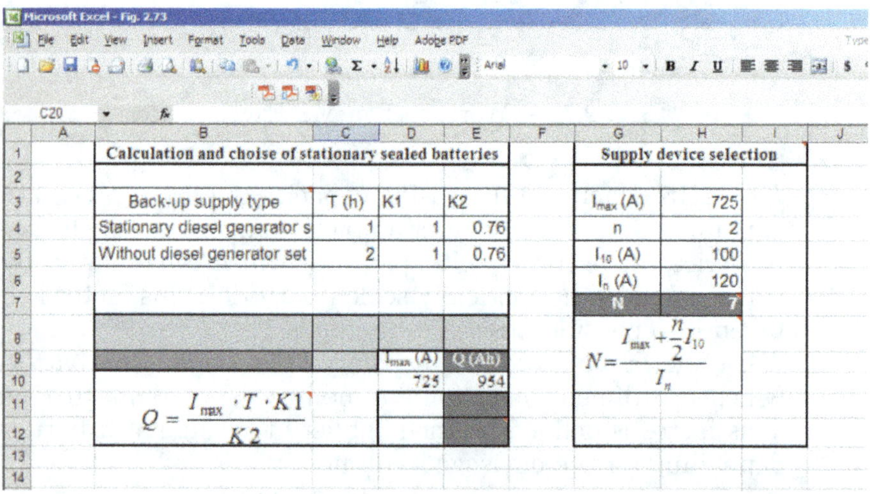


Fig. 2.73 – Calculation of accumulator battery capacity

On the basis of the calculated capacity, the following selection was made:

- two accumulator batteries manufactured by FIAMM, type SMG1000 were adopted and feature the following characteristics:
- each battery has 24 cells, with a voltage per cell of 2 V, i.e. a total of 48 V,

- the capacity per cell amounts to 1000 Ah for ten hours of current discharge,
- the batteries are valve-regulated lead acid (VRLA) with gelified electrolyte,
- the batteries are installed vertically in two levels on metal holders resistant to damage in the event of an earthquake.

Based on the calculated and adopted values of the capacity of the accumulator batteries, it can be concluded that the total capacity (capacity of both batteries) is significantly greater than the necessary capacity, which also enables the use of the selected batteries in the case of possible future expansion of capacity of the telecommunications devices.

In the meantime, the process of automation of the selection of system power supply elements has been updated with the development of a corresponding software tool based on the application of the program VB [67]. The application of this software in the example of selection of the considered accumulator batteries is illustrated below.

Fig. 2.74 shows the dialog box of the aforementioned software tool.

By activating the Stationary Batteries Catalog option and then the Calculate button, the dialog box displayed in Fig. 2.75 appears. After entering in the corresponding input data, the calculation is executed for the necessary capacity of the accumulator batteries.

Based on the calculation results, a FIAMM accumulator battery was selected, type SMG1000, the basic characteristics of which are displayed in Fig. 2.76.

*b) Selection of power supply device*

The selection of rectifier units is conducted in such a way to provide the greatest possible security and flexibility in their use during all exploitational states. The selection of rectifier units in relation to their rated current load is determined depending on:

- maximum current load under a direct voltage of −48 V, in accordance with item 4.1 of the technical instructions [62] (A),
- capacity and number of used accumulator batteries,
- whether or not the facility is equipped with a genset.

In the case that the facility (location) does have a stationary genset, the total number of rectifier units is determined on the basis of

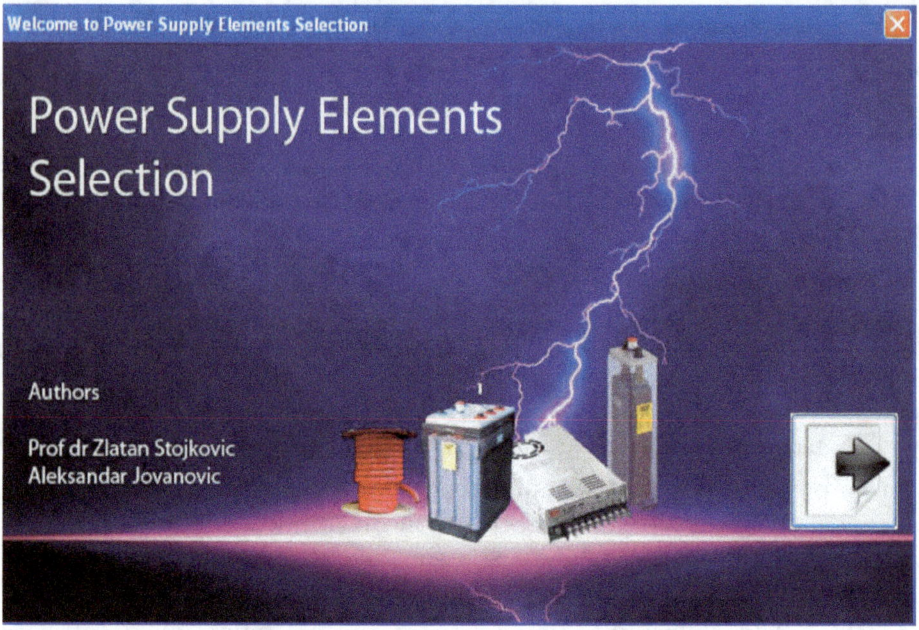


Fig. 2.74 - Dialog box of the software tool for automation of the selection of system power supply elements using the program VB

the expression:

$$N = \frac{I_{\max} + \frac{n}{2} \cdot I_{10}}{I_n} \tag{2.23}$$

In the case that the facility (location) does not have a stationary genset, the total number of rectifier units is determined on the basis of the expression:

$$N = \frac{I_{\max} + \frac{n}{2} \cdot I_{10}}{I_n} + 1 \tag{2.24}$$

The symbols in expressions (2.23) and (2.24) have the following meanings:

- $N$ – total number of rectifier units,
- $I_{\max}$ – maximum load under direct voltage of −48 V, in accordance with item 4.1 of the technical instructions [62] (A),

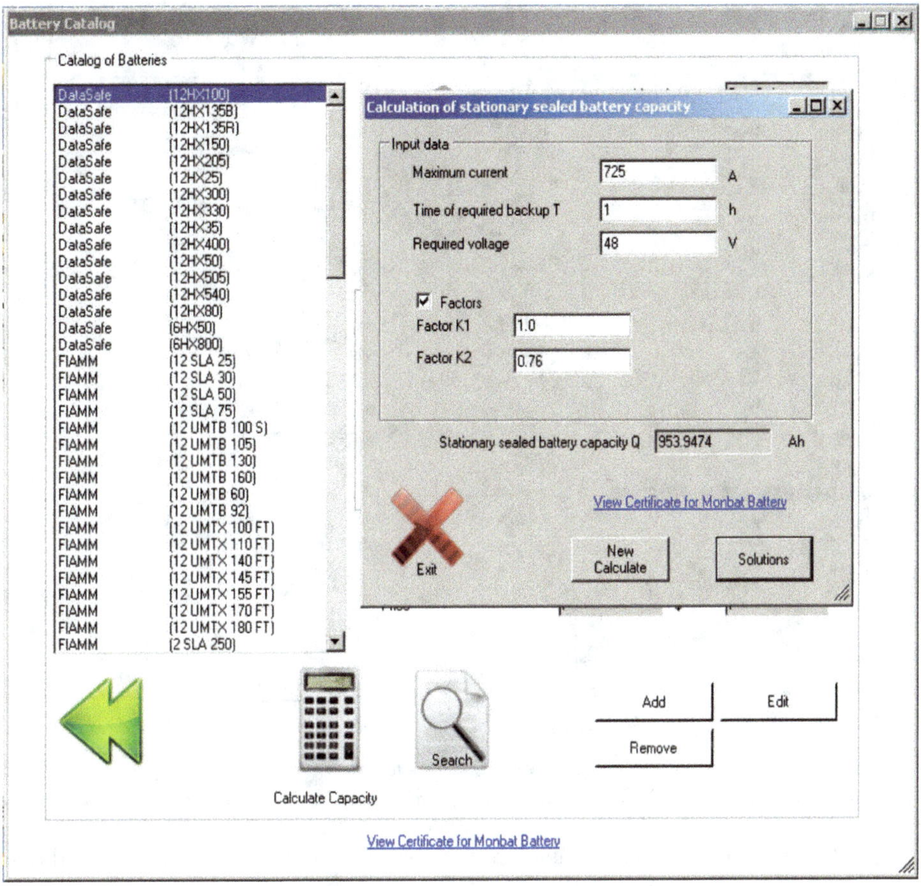


Fig. 2.75 - Dialog box of the software tool for calculating the capacity of accumulator batteries

- $I_n$ – rated output current of the selected rectifier unit in (A),
- $n$ – total number of envisaged accumulator batteries for a rated voltage of $-48$ V,
- $I_{10}$ – ten-hour current charging of one of the envisaged accumulator batteries (A).

If the obtained result for the number of rectifiers is not a complete number it is rounded up to the next highest complete number. During this selection it is assumed that the envisaged accumulator batteries are of the same capacity.

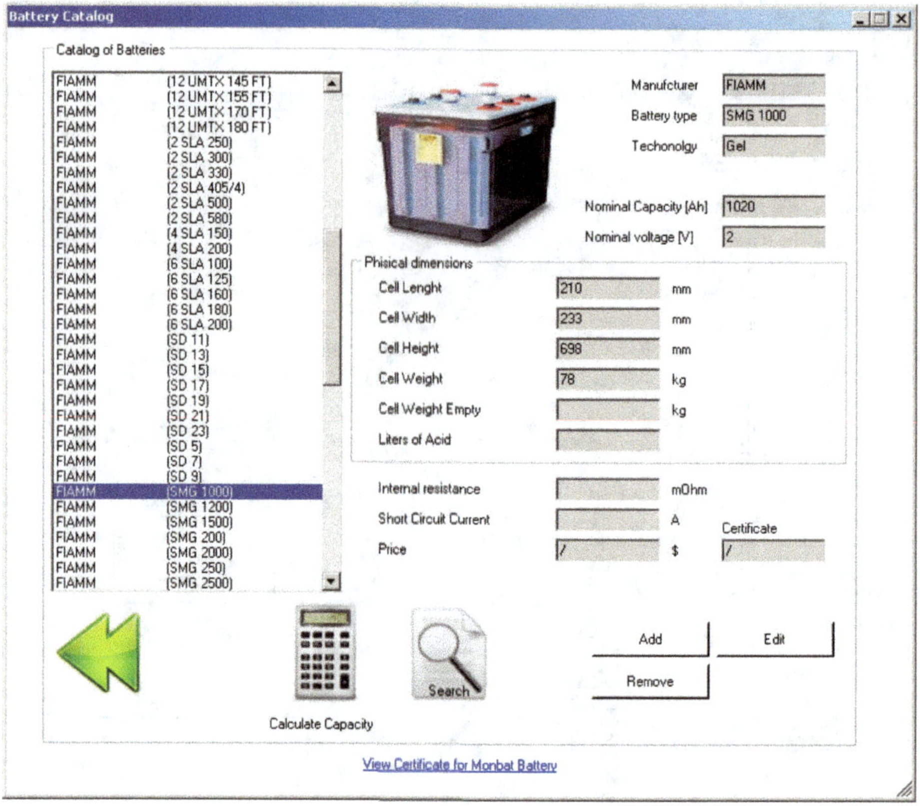


Fig. 2.76 - Dialog box with basic characteristics of the selected accumulator battery FIAMM SMG1000

The calculation and selection of a rectifier can be conducted in a similar manner both through the balance power and output power of the rectifier unit.

In the considered case the facility (location) does have a stationary genset, so the total number of rectifier units was determined by applying expression (2.23). For supplying power to equipment with direct voltage, a D48 / 120 WBRU6 – FGE18 GR60 type rectifier was selected. The maximum current of one rectifier under a direct current of −48 V amounts to 120 A.

The numerical values of the input data in expression (2.23) are: $I_{\max} = 725$ A , $n = 2$, $I_{10} = Q/t = 1000/10 = 100$ A, $I_n = 120$ A.

The total number of rectifier units envisaged for the capacity of the TC devices is obtained on the basis of this input data and expression (2.23). For supply power to equipment with direct voltage, a total of $N = 7$ rectifiers were envisaged (Fig. 2.73). For the requirement of possible expansion of capacity, a place for the installation of additional rectifiers of the same characteristics has been envisaged.

## 2.4. AutoCAD

### 2.4.1. Introduction

The program AutoCAD is a powerful software tool for computer-aided design. Some of the advantages of this program are [68-70]:

- the possibility of drafting technical documentation in various fields,
- spatial (3D) modeling which includes realistic displays of surfaces and bodies, as well as the capability of calculating physical characteristics of models,
- programming using the programs AutoLISP, Visual LISP and VBA, through which the automation of the process of drafting technical documentation is enabled,
- dynamic linking of data, which enables two-way communication between AutoCAD and other software tools,
- accessing external databases.

This section presents some of the aspects of advanced use of AutoCAD in computer-aided design of power system substations and structures. The first example illustrates an effective technique of 3D modeling of some of the most complex structures, such as a turbogenerator. The calculation of the physical characteristics of a 3D model is executed in the example of the inertia moment of a U-profile busbar. The developed software tool for designing lightning protection for general and special purpose structures was realized using the programs AutoLISP, Visual LISP and VBA. By using this software tool, calculations were made for zone of protection from lightning discharge for a residential structure, special purpose structure, substation and overhead line.

### 2.4.2. Spatial (3D) model of a turbogenerator

AutoCAD can generate three types of 3D objects: wire-frame models, surface and full body. Wire-frame models look like models made from wire. These models have no actual surface nor do they have the traits of solid bodies. However, they are very useful for forming the shape which can then be transformed into surfaces or full bodies. Surfaces, as opposed to wire-frame models, can cover up objects in the background. They are especially useful for forming irregularly shaped objects. Full bodies are shapes which with their volume consume space. Full bodies can be combined and in this way the most complex objects are obtained along with information about their physical traits.

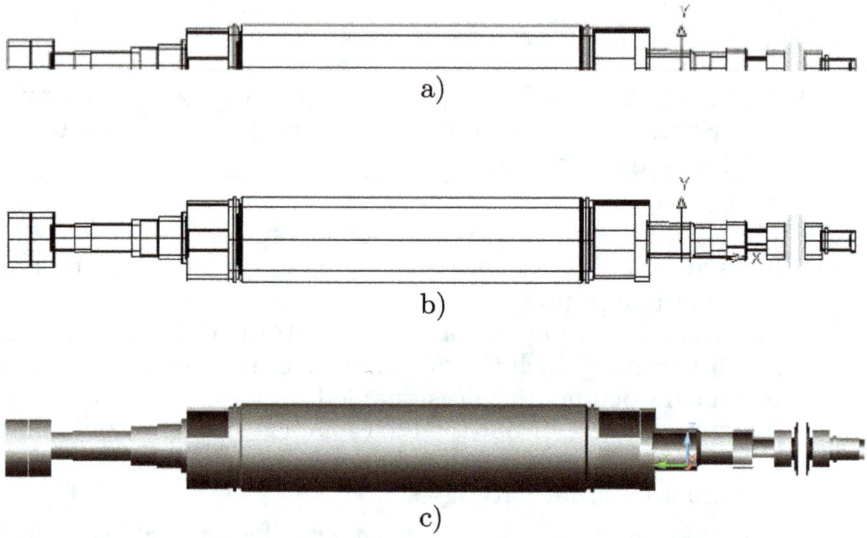


Fig. 2.77 - a) Contour of a portion of a turbogenerator shaft with rotor in the XY plane; b) Contour of turbogenerator shaft with rotor formed using the Revolve command; c) Model of a turbogenerator shaft with rotor after shading

Three-dimensional modeling using AutoCAD enables the efficient drafting of technical documentation for the most complex structures, as well as the importing of such models into some of the specialized programs for calculating in a 3D coordinate system [71]. The following example displays the details of forming some elements of a turbogenerator, as well as a complete model of a turbogenerator observed from various points [59, 72]. A detailed description of the commands for developing a 3D model in AutoCAD can be found in the reference guides [68-70].

The first step in forming a 3D model of a shaft with a rotor consists of drawing the contours in the $XY$ plane. Fig. 2.77a shows the contour of a section of the turbogenerator shaft with rotor in the $XY$ plane. The Revolve command enables the forming of an axis of a symmetrical object in relation to the assigned axis and defined angle. Using this command on the contour from Fig. 2.77a in relation to the central axis and an angle of 360° forms the contour of a turbogenerator shaft with rotor (Fig. 2.77b).

AutoCAD enables the drawing of only planar 2D sketches in the $XY$ plane. For drawing full bodies the Solids tool palette is used along with rotation of the coordinate system with the goal of creating a sketch with various sides. This manipulation with the coordinate system is performed with the UCS (User Coordinate System) command, whereby the user coordinate system is defined.

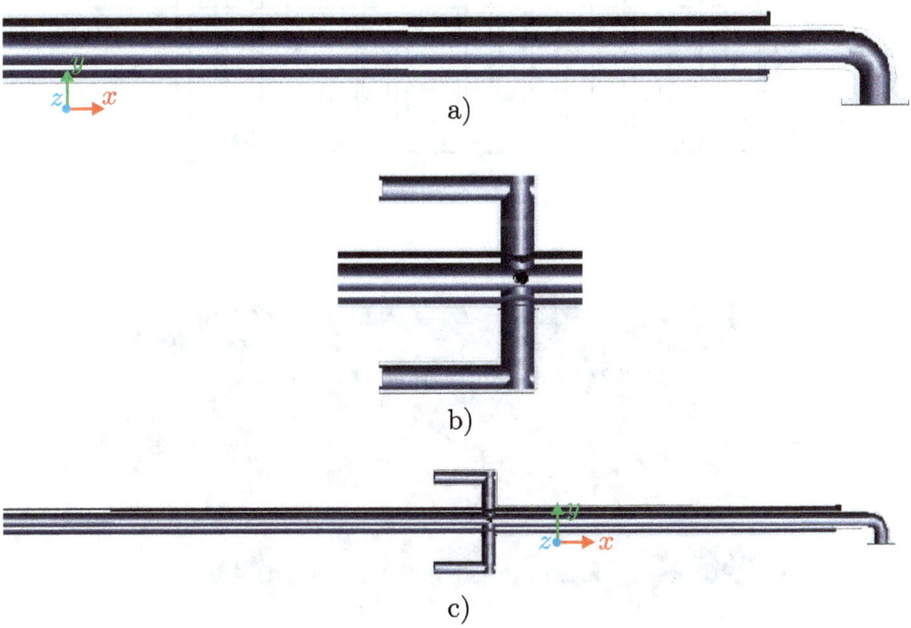


Fig. 2.78 - a) Contour of portion of turbogenerator shaft; b) Contour with connecting elements; c) Complete element

For the purpose of visualization of the model, it can be shaded or rasterized. Shading is based on one implied source of light which illuminates the model. AutoCAD automatically places this source of

a)

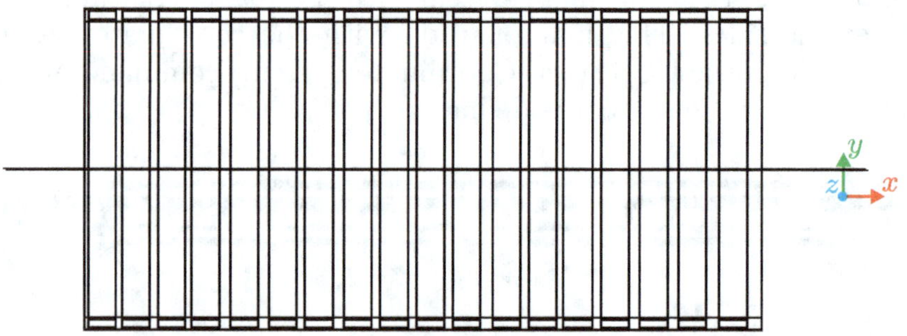


b)

c)

Fig. 2.79 - a) Initial element; b) Contour formed using Revolve command; c) External part of turbogenerator after shading

light behind the observer. The command for shading is Shade and includes various options. Fig. 2.77c shows the model of the turbogenerator shaft with rotor which is shaded using the command Shade → Gouraud Shaded. This option forms a smooth shading between individual surfaces, which provides a realistic picture especially for oval surfaces. The second option which is used in visualization of 3D models is rasterization (Rendering). This option enables the realistic display of 3D drawings by forming shadows, transparent objects, as well as adding a background and mapping a 2D picture on the surface of a 3D model.

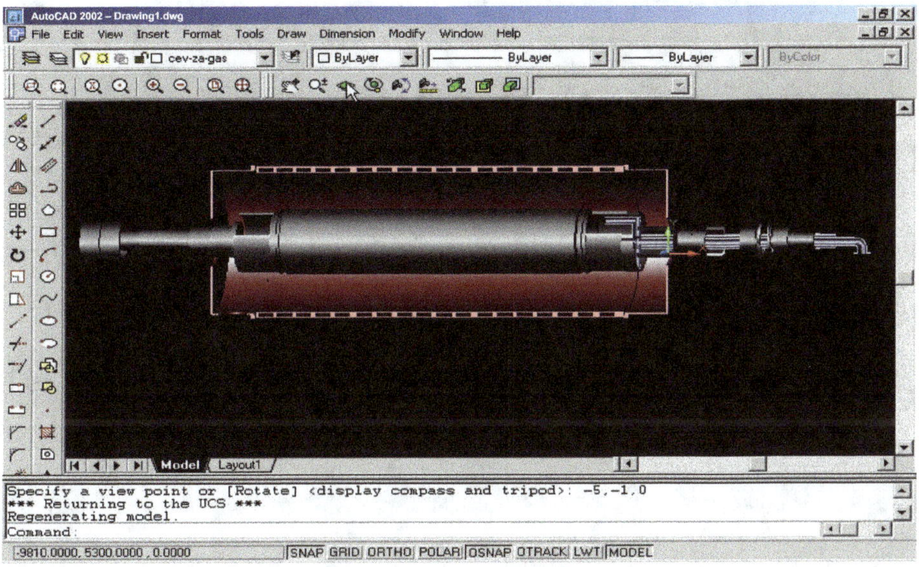


Fig. 2.80 - Cross section of turbogenerator with included layers shown in Figs. 2.77c, 2.78c and 2.79c

The steps for forming elements of the turbogenerator shaft are illustrated below. By using the Layer command, a layer is formed for a portion of the turbogenerator shaft (Fig. 2.78a). To form the contour displayed in Fig. 2.78b, the commands Explode and Extrude were utilized. The command Explode enables the separation of an axial symmetric object on the plane with which it borders. By using the Extrude command it is possible to expand the profile along the ortogonal axis or along some assigned path. Fig. 2.78c shows the complete detail which results from joining the contours given in Figs. 2.78a and 2.78b.

Figs. 2.79a, 2.79b and 2.79c show the process of forming the external portion of the turbogenerator. The initial element is provided in

Fig. 2.79a. By using the Revolve command on the initial part, a contour is formed as in Fig. 2.79b. By selecting the command View → Shade → Gouraud Shaded, shading is implemented whereby the external element of the turbogenerator obtains the shape as in Fig. 2.79c.

Fig. 2.80 provides the cross section of a turbogenerator which resulted from the inclusion of the layers shown in Figs. 2.77c, 2.78c and 2.79c. Fig. 2.81 illustrates the layer which contains the stator winding of the turbogenerator.

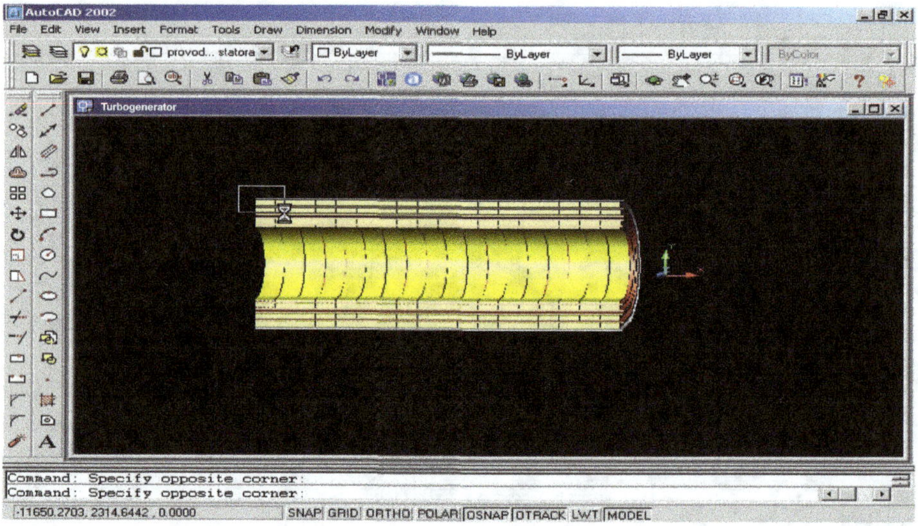


Fig. 2.81 – Display of layer with stator winding of the turbogenerator

The result of the complete process of 3D modeling of the turbogenerator is displayed in Fig. 2.82. The forming of this model utilized all the advantages of AutoCAD in organizing drawings using various layers, colors and lines [6]. The displayed model can be observed from various points, which is realized using the command Vpoint. The result of the application of this command is provided in Fig. 2.83.

### 2.4.3. Calculation of the moment of inertia of busbars

For the purpose of mechanical calculation of busbars, it is necessary to determine the moment of inertia for the given configuration and profile of the conductors. The mechanical characteristics of the corresponding profiles of the busbars are determined according to the

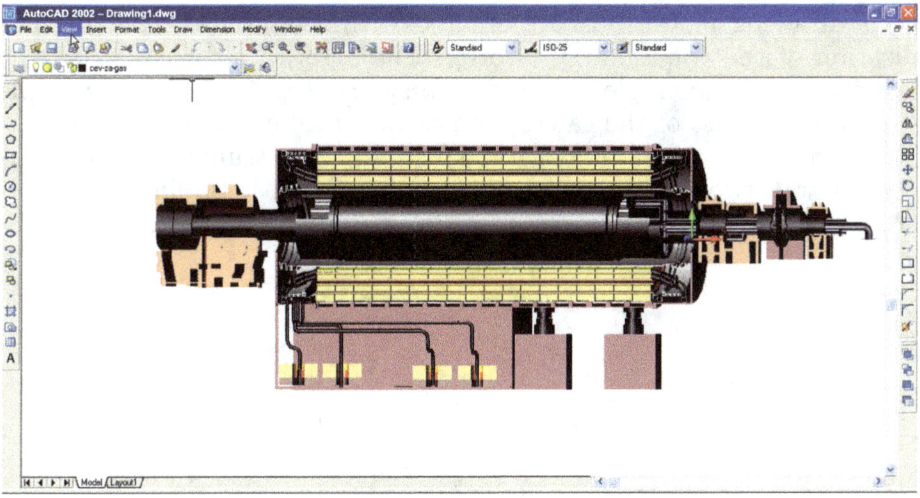


Fig. 2.82 – 3D model of a turbogenerator

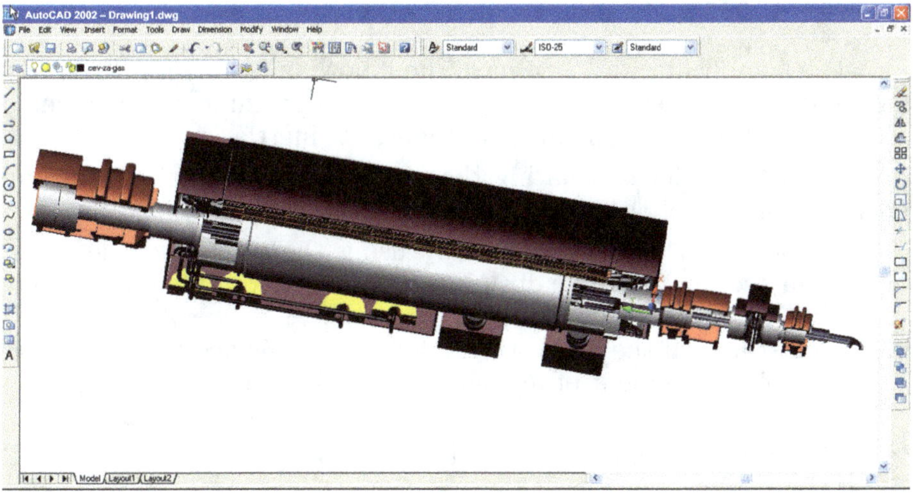


Fig. 2.83 – 3D model of a turbogenerator observed from a different point

expression provided in [10]. However, instead of performing such a calculation, the moment of inertia can be directly determined on the basis of the drawing of the busbars created within AutoCAD and the application of the command Massprop. The algorithm for calculation is described in the example of U-profile busbars (Fig. 2.84). This example is an integral part of the calculation of the trunking system generator-transformer in the "Bajina Bašta" HPP [73]. The trunking system was created with two aluminum conductors per phase. U-profiles were used with the dimensions 120 × 45 × 10 mm (profile U12).

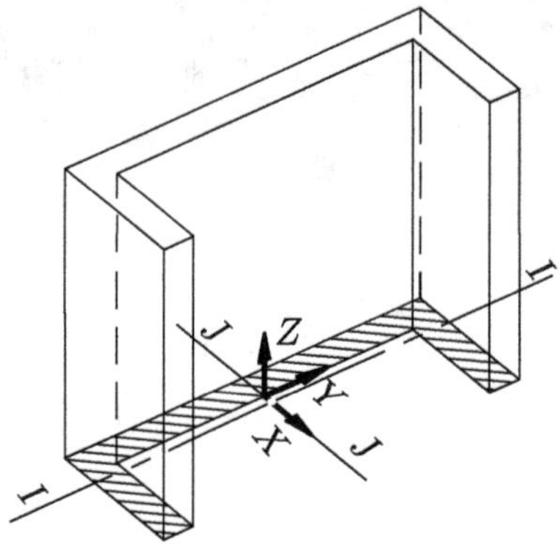


Fig. 2.84 – 3D model of a U-profile busbars

The symbols in Fig. 2.84 have the following meanings: $X, Y, Z$ – world coordinate system (WCS); $I - I$ – axis of minimum moment of inertia; $J - J$ – axis of maximum moment of inertia.

The process for calculating the moment of inertia of the busbars generally consists of the following steps:

- form a 3D model of busbars using the commands described in section 2.4.2; for forming a cross section of the 3D model (Section Plane) select the XY plane,
- distinguish the cross section for which it is necessary to calculate the moment of inertia (shaded cross section in Fig. 2.84),
- use the Massprop command on the defined cross section.

The results of applying the aforementioned algorithm for a U12 profile are displayed in Fig. 2.85.

When drawing it is necessary to use the proper units. In this case the units are provided in centimeters. For this reason the moment of inertia taken from Fig. 2.85 (principal moment) amounts to $I = 30.527\ \text{cm}^4$.

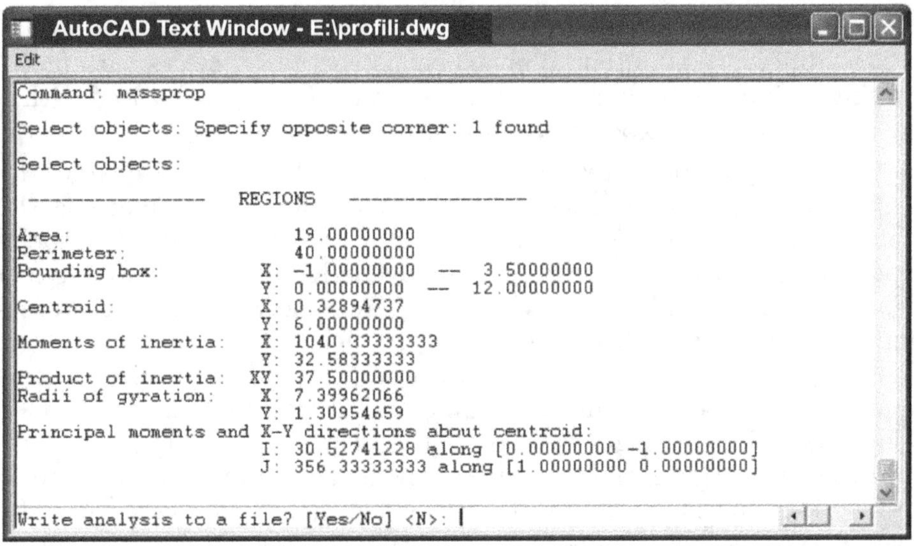


Fig. 2.85 - Calculation results of the considered U12 profile using the Massprop command

The displayed process enables the calculation of the moment of inertia of complex profiles without the direct application of the Stein theorem by the user. It is necessary to transfer the user coordinate system (UCS) to the determined emphasis and apply the Massprop command.

## 2.4.4. Designing lightning protection for general and special purpose structures

### 2.4.4.1 General considerations

The term lightning protection encompasses complex protective measures against a direct lightning strike and its secondary effects, which provide security to people and animals, equipment and materials from explosion, fire and destruction. The result of a direct lightning strike is the destruction of the structure and facilities, ignition of flammable and explosive materials and injury to people and animals. The effects of a direct lightning discharge are suffered most by solitary and tall buildings and structures. During a direct lightning discharge, secondary effects of the discharge may also occur. These effects are the result of electromagnetic induction and the manifestation of an increased difference of potential in steel constructions, devices with metal housings, pipelines, cranes and other equipment which is located within the area and is not exposed to a direct discharge.

For purpose of protecting structures from lightning discharge, the proper installations are designed and implemented. Implementing the aforementioned protection has a social and economic aspect. The social aspect relates to the protection of humans and animals, and is based on legal regulations and norms. The economic aspect entails the protection of material goods from the damaging effects of lightning discharge. Lightning discharges on structures without the proper lightning protection are the cause of significant damages and losses in commerce. It is also known that damages caused by the disruption of the production process can be significantly greater than the actual physical damage or destruction caused to the production structures themselves [74]. The goal of constructing the prescribed lightning protection installations is to reduce such damages down to the smallest possible measure.

A lightning protection installation for protection of structures from lightning discharge consists of an external and internal lightning protection installation. The regulations which relate to the process of designing, implementing, maintaining, inspecting and verifying lightning protection installations for general purpose structures are provided in the corresponding standards [75-77]. Analysis of the grounding system and the calculation of impulse characteristics of a grounding grid are an integral part of designing lightning protection installations [5].

#### 2.4.4.2 Theoretical assumptions for calculation of the protection level and protected zone

A structure is considered to be protected from direct lightning discharge if the probability of discharge within the vicinity of the lightning protection installation is less than the technically acceptable values. There is no absolutely certain protection from a direct lightning discharge which would be economically acceptable. Due to this there are protected zones which are defined in which, with a high probability, structures can be considered protected from a direct lightning discharge. Along with protection for a direct lightning discharge, structures must also be equipped with protection from the inductive effects resulting from lightning discharge into the lightning protection installation of the protected structure or in the vicinity of the protected structure.

Fig. 2.86 shows the algorithm for determining the level of protection of a lightning protection installation. The adopted frequency for lightning strikes is determined from the expression:

$$N_c = 3 \cdot 10^{-3}/(C_1 \cdot C_2 \cdot C_3 \cdot C_4) \tag{2.25}$$

where the coefficients $C_1$, $C_2$, $C_3$ and $C_4$ are determined according to Tables 2.13, 2.14, 2.15 and 2.16.

Table 2.13 – Values of coefficient $C_1$

| $C_1$ – type of structure construction | | | |
|---|---|---|---|
| Structure construction | Metal roof | Combined roof | Flammable roof |
| Metal | 0.5 | 1 | 2 |
| Combined | 1 | 1 | 2.5 |
| Flammable | 2 | 1.5 | 3 |

Table 2.14 – Values of coefficient $C_2$

| Structure content | $C_2$ |
|---|---|
| Without value or unknown | 0.5 |
| Little value or mostly flammable | 1 |
| Greater value or especially flammable | 2 |
| Extremely valuable, irreparable if damaged, very flammable or explosive | 3 |

Table 2.15 – Values of coefficient $C_3$

| Purpose of structure | $C_3$ |
|---|---|
| Unoccupied | 0.5 |
| Mainly unoccupied | 1 |
| Difficult evacuation or danger from panic | 3 |

According to vulnerability of the structure to lightning discharge, all structures are divided into five protection levels. Table 2.17 provides the levels of protection, calculated efficiencies of lightning protection, as well as the amplitude of discharge current, according to which the striking distance is determined which corresponds to the observed level

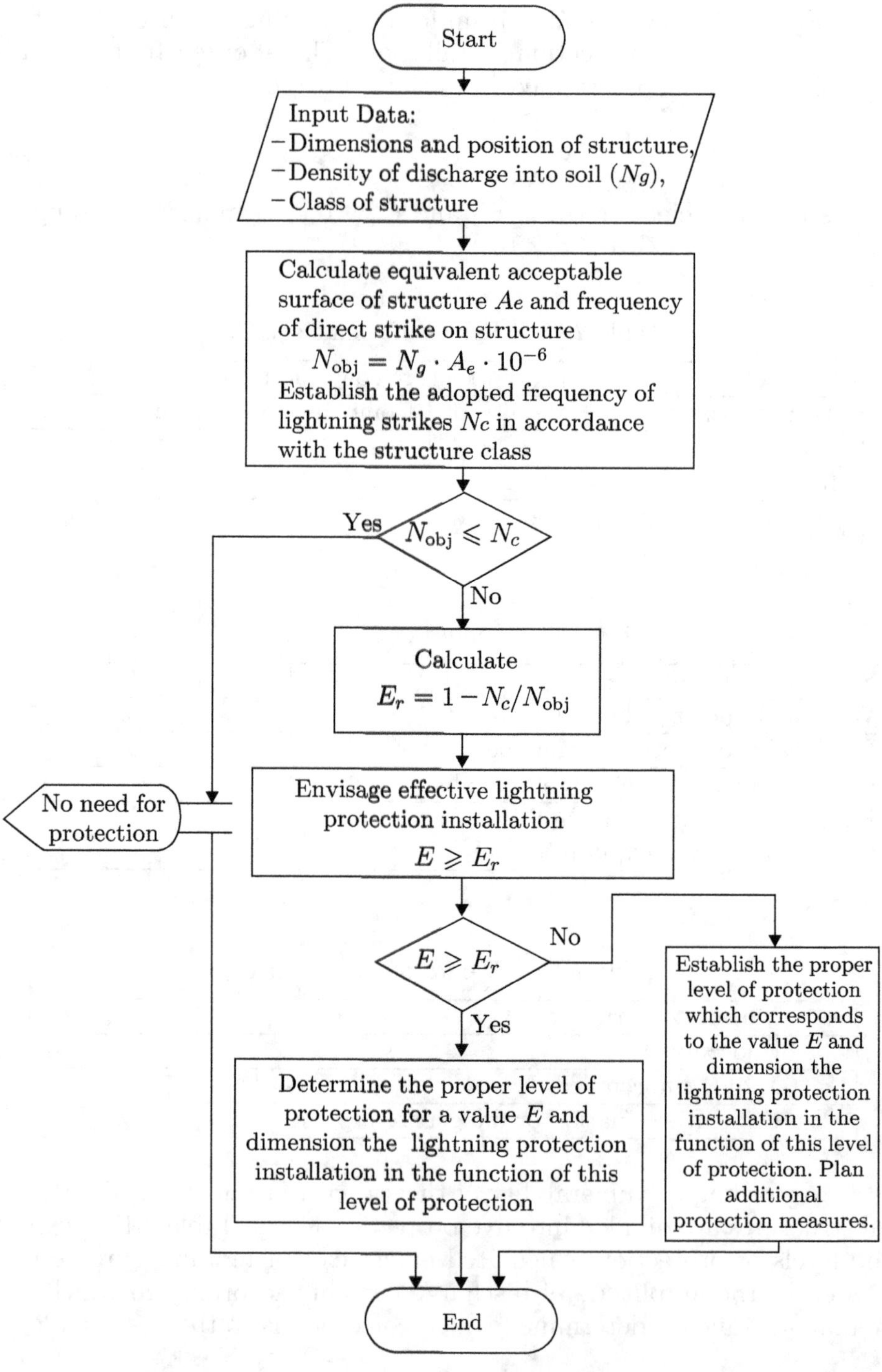


Fig. 2.86 - Algorithm for determining the level of lightning protection installation

Table 2.16 – Values of coefficient $C_4$

| Consequences of lightning strike on structure | $C_4$ |
|---|---|
| Uninterupted operation is not required and no effect on surroundings | 1 |
| Uniterupted operation is required, but no effect on surroundings | 5 |
| Effect (consequences) on surroundings | 10 |

of protection. The striking distance depends on the amplitude of strike current I (kA) and is calculated using the expression:

$$R = K \cdot I^n \tag{2.26}$$

where $K$ and n represent the empirical constants with values defined by several authors [79, 80].

Table 2.17 - Calculated efficiency of lightning protection and selection of protection level

| Amplitude of strike current I (kA) | Striking distance $R_{ud}$ (m) | Calculated efficiency $E_r$ | Level of protection |
|---|---|---|---|
| – | – | $E_r > 0,98$ | Level I with additional measures |
| 2,8 | 20 | $0,98 \geqslant E_r > 0,95$ | Level I |
| 5,2 | 30 | $0,95 \geqslant E_r > 0,90$ | Level II |
| 9,5 | 45 | $0,90 \geqslant E_r > 0,80$ | Level III |
| 14,7 | 60 | $0,80 \geqslant E_r > 0,00$ | Level IV |

The first column in Table 2.17 represents the tolerable amplitude of strike current for the adopted level of protection. With structures requiring protection level I with additional measure for limiting touch voltage and step voltage, measures for limiting the spreading of fire and measures for decreasing the effects of induced overvoltage of lightning origins on sensitive electrical equipment.

The protected zone of the accepted system encompasses the zone in which there is a low probability of a direct lightning discharge occurring. The Protective Angle Method (PAM) and Rolling Sphere Method

(RSM) are two of the most commonly used processes for determining the protected zone of an accepted system. According to the first method, the protective angle is defined as the angle which overlaps the cone generatrix and vertical line placed through the axis of the lightning conductor (Fig. 2.87a). It is believed that a direct lightning discharge will not occur within the cone area if the angle has a value between 30° and 60°. The PAM may be applied in both insulated and non-insulated lightning protection. The RSM is defined by a radius identical to the striking distance for a specific strike current which depends on the level of protection for the structure. The protected zone is figured as the geometric area of points in which the sphere touches the horizontal base when being rotated around the lightning conductor, so the sphere is also in constant contact with conductor.

Fig. 2.87a shows the protected zones of one lightning conductor determined using the PAM and the RSM. Use of the RSM on various structures is illustrated in Fig. 2.87b. The rolling sphere must be "rolled", placed around and above the structure in all possible positions down to the ground, in order to find the touch locations and surfaces on which the elements of the accepted system must be placed because those locations are where a strike can occur. The accepted system should be placed on the thickened lines according to Fig. 2.87b.

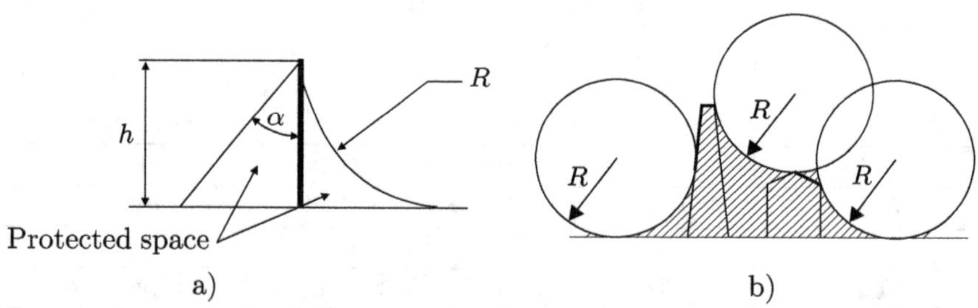


Fig. 2.87 - a) Sketch of protected zone determined by the PAM and the RSM; b) Determining the protected zone of various objects using the RSM

Below the formulas are provided for calculating the protected zone of lightning conductor (Franklin's lightning rod), Early Streamer Emission Lightning Conductor (ESELC) and lightning conductor with a circular ring. The aforementioned lightning conductors are defined by the corresponding standards [75, 76, 78, 81].

*a) Lightning conductor*

The protected zone of one lightning conductor with a height of $h$ is represented by a cone with its outer side in the form of a broken line and the cone base with a radius of $r = 1.5h$ (Fig. 2.88a). The cross section of the protected zone at a height of $h_x$ is represented by a circle with a radius of $r_x$. These values are calculated on the basis of the expression

– for $h \leqslant 60$ m

$$h_x < \frac{2}{3}h, \quad r_x = 1,5(h - 1,25h_x) \tag{2.27}$$

– for $h > 60$ m

$$h_x > \frac{2}{3}h, \quad r_x = 0,75(h - h_x) \tag{2.28}$$

The number and layout of lightning conductors, as well as their height, depends on the size of the protected structure.

*b) Early streamer emission lightning conductor (ESELC)*

The ESELC is basically a lightning conductor equipped with a device at the top which enables early streamer emission. Fig. 2.88b shows the general protected zone of the ESELC. The symbol AC represents the radius of the circle of the protected surface of the structure. For the lightning conductor according to Fig. 2.88a, this radius is determined from the expression:

$$AC = r_{\max} = \sqrt{h \cdot (2R - h)} \quad \text{(m)} \tag{2.29}$$

where: $h$ – vertical distance from the top of the lightning conductor to the level of any other protected point (m), $R$ – radius of rolling sphere, i.e. striking distance in (m) which is determined using expression (2.26).

The radius of the protected zone AC of an ESELC (Fig. 2.88b) is determined using the expression:

$$AC = r'_{\max} = \sqrt{h \cdot \left[2(R + \Delta R) - h\right]} \quad \text{(m)} \tag{2.30}$$

The symbol $R$ and $h$ in expression (2.30) have the same meanings as in expressions (2.26) and (2.29). The value $\Delta R$ represents an

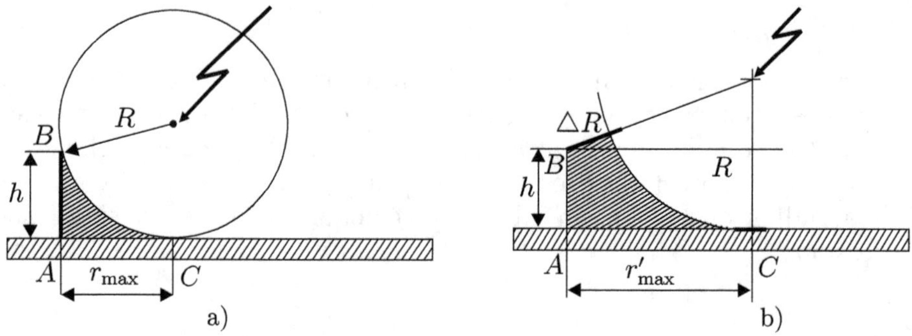


Fig. 2.88 – Protected zone of the lightning conductor (a) and ESELC (b)

increased striking distance which is achieved on the basis of the gain in triggering time and is determined from the expression:

$$\Delta R = v \cdot \Delta t \quad \text{(m)} \tag{2.31}$$

where: $v$ – constant velocity of the upwards progressing discharge with an adopted value of 1 m/$\mu$s, $\Delta t$ – time advantage in $\mu$s; this time is defined by the manufacturer.

By comparing Figs.2.88a and 2.88b, i.e. expressions (2.29) and (2.30), it can be concluded that an ESELC, when compared to a regular lightning conductor, provides an increased protected zone.

In the scientific field an opposite stance has also been taken regarding the effectiveness of the application of this protective device [82].

*c) Lightning conductor with circular ring*

A conductor with a circular ring is basically a lightning conductor which is upgraded with a horizontal ring in order to improve its lightning protection capabilities (Fig. 2.89). A galvanized link is provided between the rod and ring. By using a lightning conductor with a circular ring which has the geometric relations from Fig. 2.89, with the effect of strengthening the field, a gain is achieved as though the height of the conductor is increased by 60% over the actual height of the rod [81].

The protected zone is determined as though a lightning conductor is used which has a height of:

$$h_l = 1.6 \cdot (h + l) \tag{2.32}$$

where: $h$ – vertical distance between the foot of the rod and the level of protected surface, $l$ – height of the conductor.

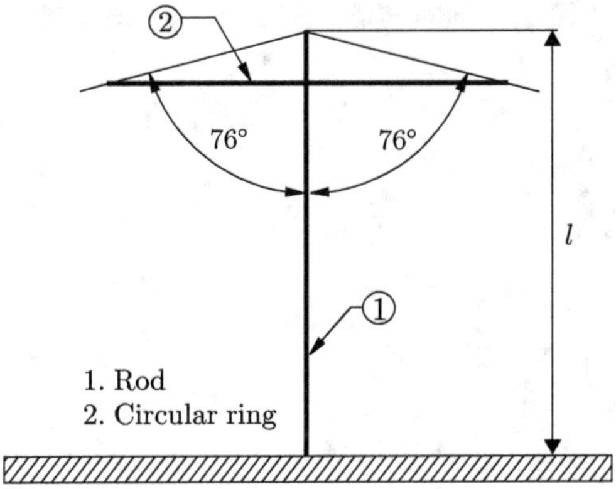


Fig. 2.89 – Geometric relations of a lightning conductor with circular ring

### 2.4.4.3 Examples of calculations of the protected zone for a general purpose structure

Fig. 2.90 shows the dialog box of the program for designing lightning protection for structures of an arbitrary shape and size [83]. The drawing of the structure is achieved in AutoCAD, while the calculation of the protected zone is conducted using the program VB.

The algorithm for calculation of the protected zone consists of the following steps:

- By clicking on the AutoCAD icon, the program is launched and the structure is drawn in a 2D or 3D display,
- for the drawn structure it is possible to select one of the three types of lightning conductors and assign its dimensions,
- the value of impulse current is then adopted depending on the level of protection,
- the striking distance is calculated for the defined impulse current,
- on the basis of the striking distance, the protected zone is determined which is represented by a cone with its external side in the form of a broken line,
- the process is repeated for the next lightning conductor which may be positioned arbitrarily in relation to the structure.

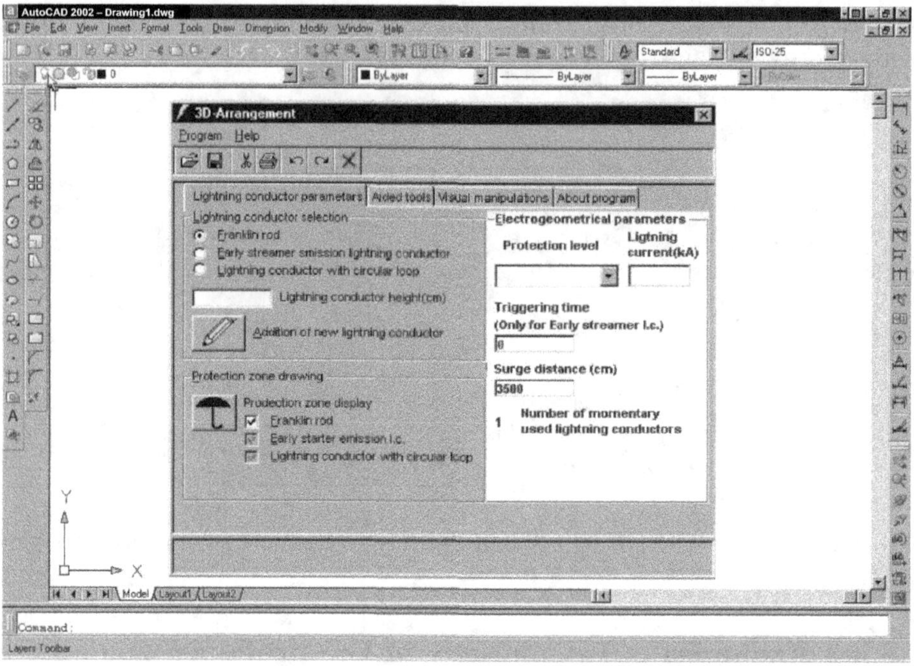

Fig. 2.90 - Dialog box for the program for designing lightning protection for general and special purpose structures

For a general purpose structure with dimensions $8 \times 9 \times 19$ m and an adopted protection level of IV, the following is provided: 3D display of protected zones in the form of wire-frame models, realized using lightning conductors with a height of 4 m and 8 m (Fig.2.91a), 3D display of the protected zones in the form of full bodies for a lightning conductor height of 4 m (Fig. 2.91b) and 8 m (Fig. 2.91c). Based on Fig. 2.91a, it can clearly be seen that the height of the lightning conductor has an effect on the shape and size of the protected zone. Using higher lightning conductors leads to a greater and more effective protected zone. This conclusion is illustrated in Figs. 2.91b and 2.91c. An ineffective protected zone was formed using lightning conductors with a height of 4 m (Fig. 2.91b), while lightning conductors with a height of 8 m provide an effective solution for the analyzed structure (Fig. 2.91c).

For a lightning conductor with a height of 4 m which is placed at the corner of the roof of the considered structure, the protected zone is determined in the form of a wire-frame model (Fig. 2.92a). A protection

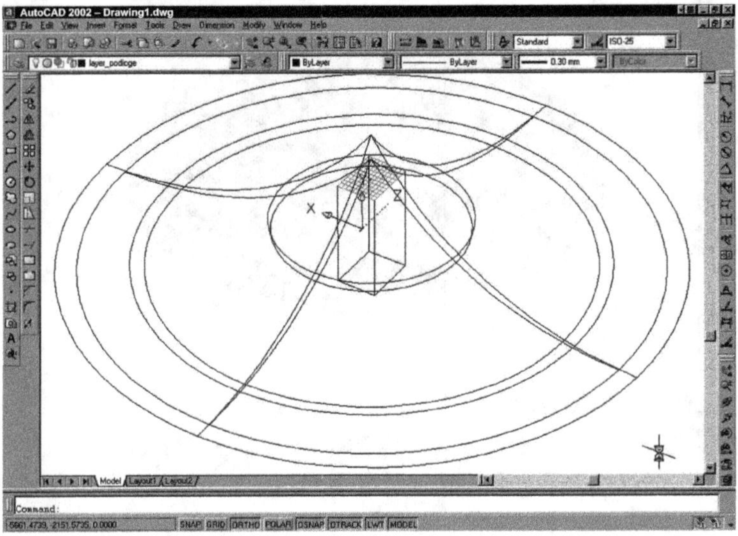
AutoCAD 2002 – Drawing1.dwg
File Edit View Insert Format Tools Draw Dimension Modify Window Help
Standard
ISO-25
layer_podloge
ByLayer
0.30 mm
ByColor
Model
Layout1
Layout2
Command:
SNAP GRID ORTHO POLAR OSNAP OTRACK LWT MODEL

a)

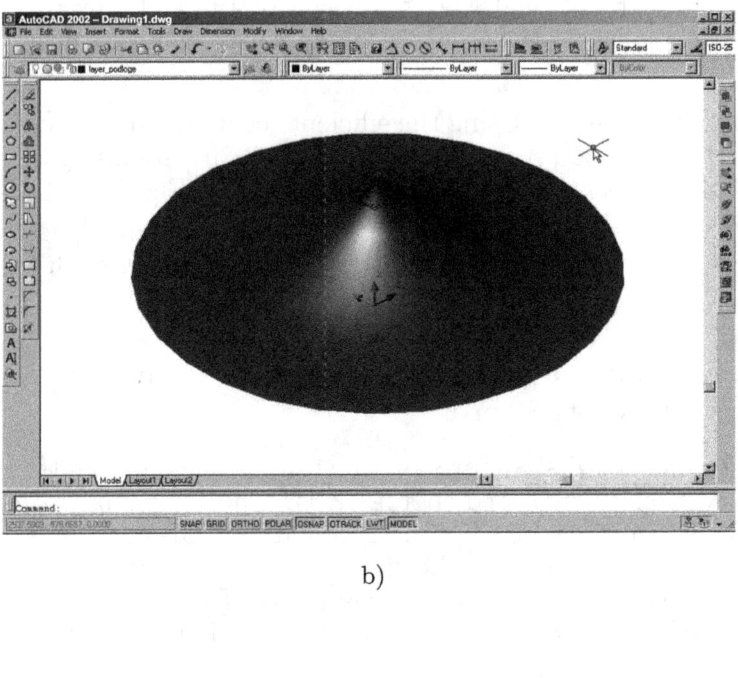
AutoCAD 2002 – Drawing1.dwg
File Edit View Insert Format Tools Draw Dimension Modify Window Help
Standard
ISO-25
layer_podloge
ByLayer
ByColor
Model
Layout1
Layout2
Command:
SNAP GRID ORTHO POLAR OSNAP OTRACK LWT MODEL

b)

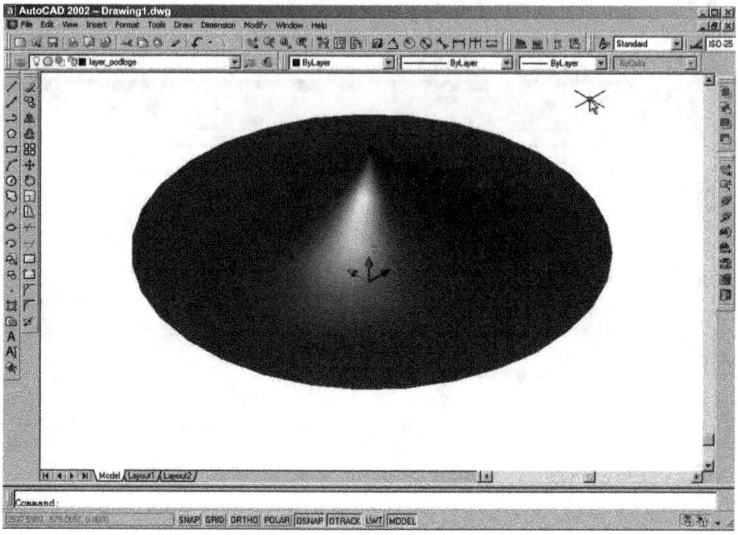

c)

Fig. 2.91 - 3D display of protected zones of a general purpose structure with dimensions 8 × 9 × 19 m realized using lightning conductors; a) Wire-frame models with a height of 4 m and 8 m; b) Full body for a height of 4 m; c) Full body for a height of 8 m

level of III was adopted. Using the efficient techniques of rasterization in AutoCAD [59, 83-85], it is possible to form a corresponding 3D display of the protected zone in the form of a full body (Fig. 2.92b).

For the previous structure and lightning conductor height of 4 m, an analysis was conducted of the effect of the protection level on the size of the protected zone (Figs. 2.93a-d).

#### 2.4.4.4 Example of calculation of protected zone for a special purpose structure

The considered structure for which it is necessary to design lightning protection is displayed in Fig. 2.94. The lightning protection is achieved by placing seven lightning rod towers around the structure. The symbols and height of the lightning rod towers are provided in Table 2.18. The height of the lightning rod towers also includes the dimension of the lightning conductor of 0.5 m, placed at the top of the tower.

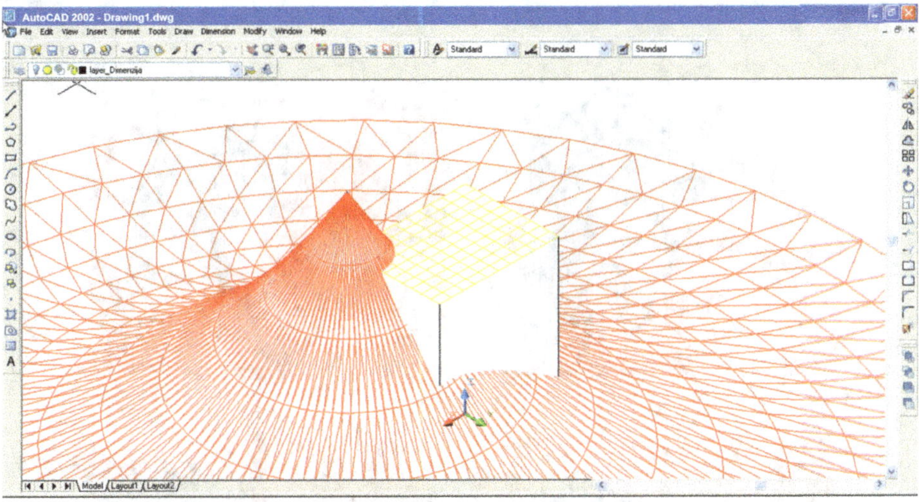


a)

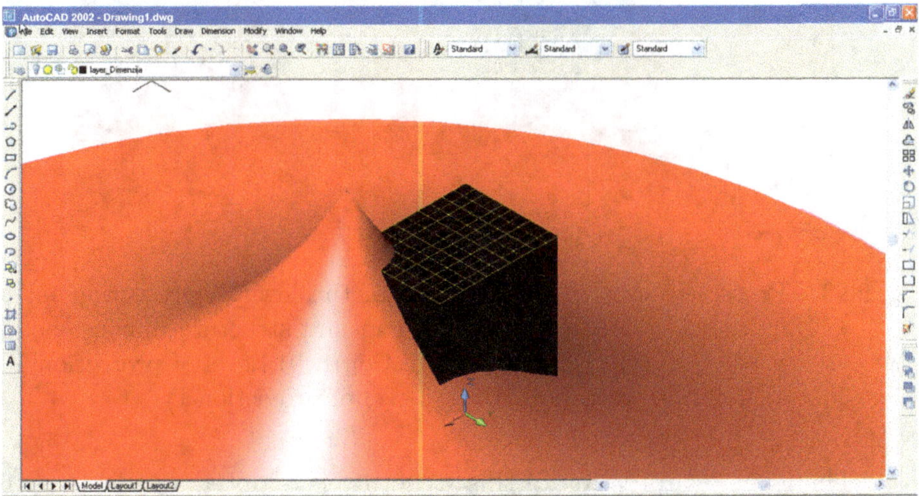


b)

Fig. 2.92 - 3D display of protected zone realized using one lightning conductor with a height of 4 m, placed at the corner of the roof of a structure with dimensions $8 \times 9 \times 19$ m; a) Wire-frame model; b) Full body

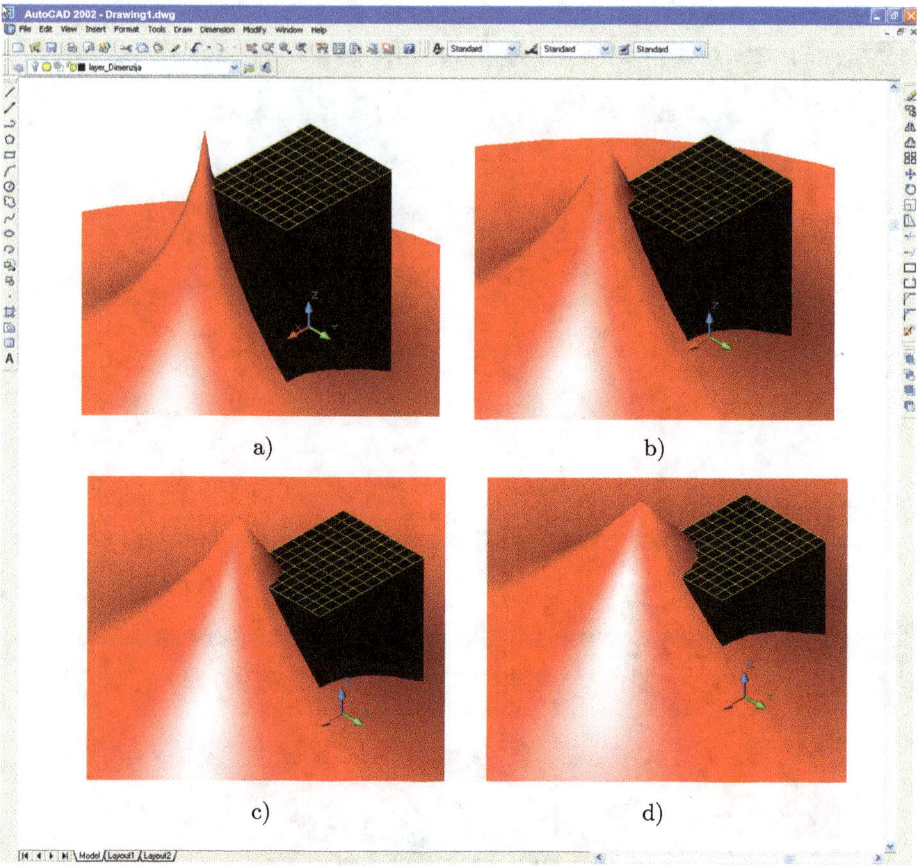


Fig. 2.93 - Protected zone in the function of the level of protection for a lightning conductor height of 4 m and structure dimensions of 8 × 9 × 19 m; a) Level of protection I; b) Level of protection II; c) Level of protection III; d) Level of protection IV

The level of protection of this structure is adopted as a level with combined protection, i.e. insulated (tower) protection and uninsulated lightning protection (implemented on the structure). The protected area of the insulated (tower) lightning protection is determined using the RSM. In accordance with the required level of protection I, a radius of 20 m was adopted for the rolling sphere (Table 2.19). Table 2.20 provides, along with the symbols and height of lightning rod towers, the levels of the bottoms of the towers and tops of conductors on the towers, as well as the maximum radii of the protected area. The maxi-

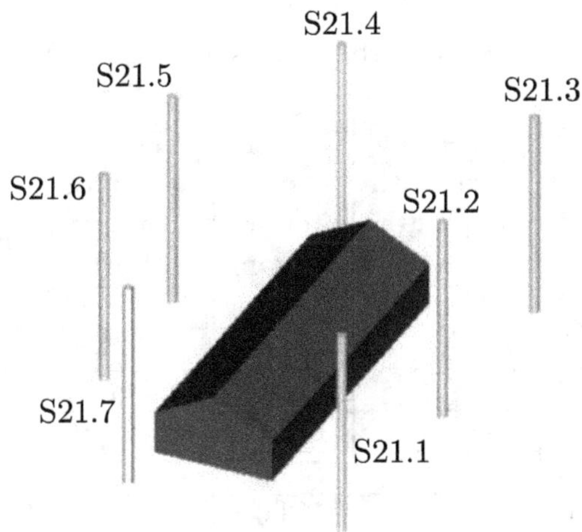


Fig. 2.94 - 3D display of structure with configuration of lightning conductor towers

mum radius of the protected area of an individual lightning conductor is represented by the radius of the same on the level of the terrain and is determined from the expression:

$$r_{\max} = \begin{cases} \sqrt{h \cdot (2R - h)}, & \text{za } h \leqslant R \\ R, & \text{za } h > R \end{cases} \tag{2.33}$$

where: $r_{\max}$ – maximum radius of protected area (m), $h$ – height of lightning conductor (m), $R$ – radius of rolling sphere with a value of 20 m for the adopted level of protection.

Using the graphic method, the protected area of the placed towers is defined, i.e. the area in which there is no penetration of the rolling sphere which "rolls" around and above the structure. The protected area is indicated in the graphical section by cross-hatching. The characteristic of the obtained protected area is that it narrows towards the top of the towers and in between the towers "deepens" due to the partial penetration of the rolling sphere (Figs. 2.95b and 2.95c).

The vertical penetration of the rolling sphere is determined on the basis of the expression:

$$p = R - \sqrt{R^2 - (d/2)^2} \tag{2.34}$$

where: $p$ – depth of vertical penetration of rolling sphere (m), $R$ – radius of rolling sphere ($R = 20$ m), $d$ – diameter of circle around top of considered tower (m).

Table 2.18 - Levels of lightning rod towers and maximum radii of protected area

| Tower | Type of tower | Height of tower with conductor (m) | Level of tower bottom | Level of top of conductor on the tower | Maximum radius of protected area $r_{\max}$ (m) |
|---|---|---|---|---|---|
| S21.1 | SG - 24.0 | 24.5 | 742.90 | 767.40 | 20.00 |
| S21.2 | SG - 24.0 | 24.5 | 742.70 | 767.20 | 20.00 |
| S21.3 | SG - 24.0 | 24.5 | 742.60 | 767.10 | 20.00 |
| S21.4 | SG - 20.5 | 21 | 745.80 | 766.80 | 20.00 |
| S21.5 | SG - 17.0 | 17.5 | 753.60 | 771.10 | 19.84 |
| S21.6 | SG - 17.0 | 17.5 | 753.60 | 771.10 | 19.84 |
| S21.7 | SG - 20.5 | 21 | 746.60 | 767.60 | 20.00 |

Table 2.19 - Vertical penetration of rolling sphere for the adopted configuration of lightning rod towers

| Lightning rod towers | Diameter of described circle around the top of the tower $d$ (m) | Vertical penetration of rolling sphere $p$ (m) | Symbol for apex of rolling sphere | Level of apex of rolling sphere |
|---|---|---|---|---|
| S21.1, S21.2, S21.7 | 32.4 | 8.27 | T1 | 758.93 |
| S21.2, S21.6, S21.7 | 34.0 | 9.46 | T2 | 757.74 |
| S21.2, S21.5, S21.6 | 34.2 | 9.63 | T3 | 757.57 |
| S21.2, S21.3, S21.4 | 34.0 | 9.46 | T4 | 757.34 |

Table 2.19 shows the values of factor $p$ for the adopted configuration of lightning rod towers. As the apexes of vertical penetration of the rolling sphere are at a level which is above the highest level of the structure (752.01), a direct lightning discharge into the structure from the top side cannot be expected. The protected area of uninsulated lightning protection is implemented through a network of conductors around the structure. According to the criteria for protection level I, this network is supplemented by a network of accepted lines so that the meshes have an average width of 5 m, and the descent lines are at a distance of 10 m.

Using the program AutoCAD, the protected zone of the considered structure is determined in a 2D and 3D coordinate system for the

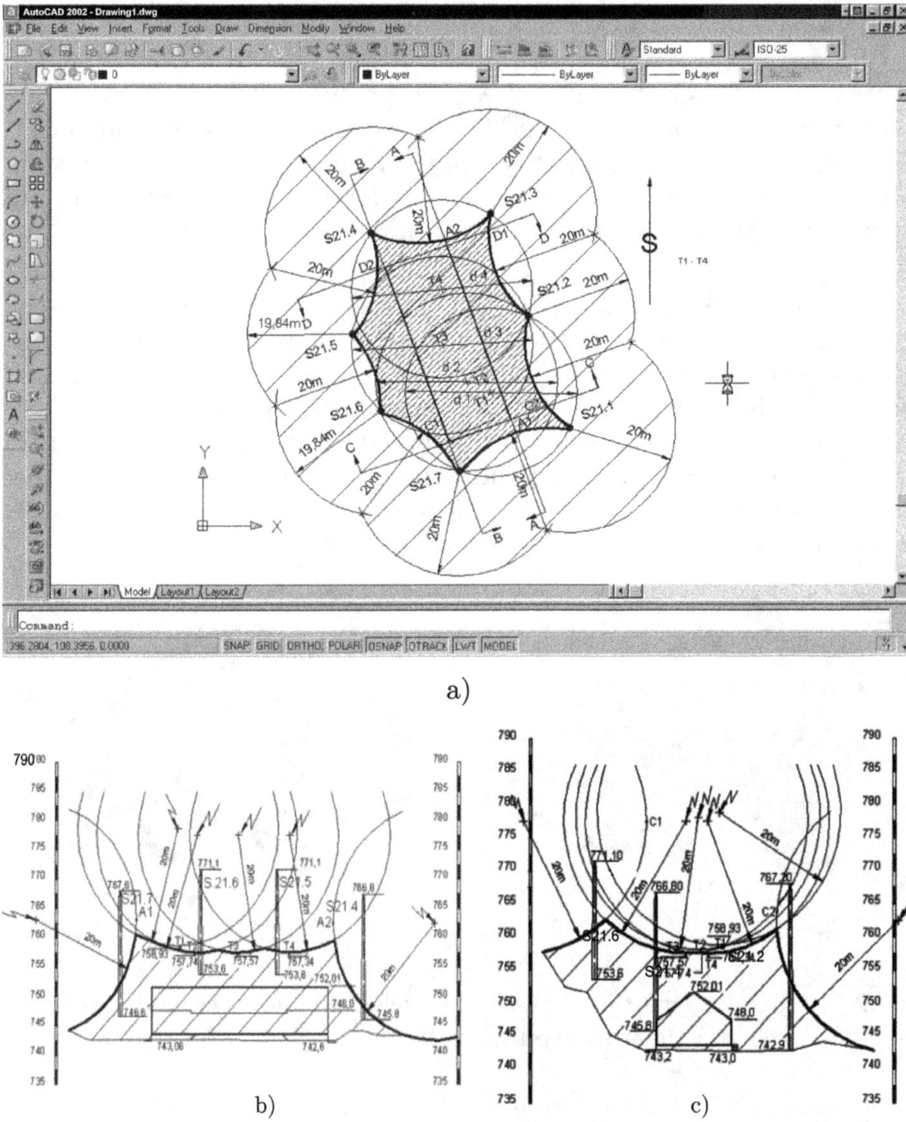


Fig. 2.95 - 2D display of the situation of the protected area: a) Basic; b) Cross section of A-A; c) Cross section of C-C

adopted number and configuration of lightning rod towers [84, 86, 87].

Fig. 2.95a illustrates the 2D display of the situation of the protected area achieved using lightning conductors. The protected area which corresponds to cross sections A-A and C-C from Fig. 2.95a are illustrated in Figs. 2.95b and 2.95c, respectively. On the basis of the displays in Figs. 2.95a, 2.95b and 2.95c, it can be concluded that the application of the adopted number, configuration and height of lightning rod towers provides effective protection for the considered structure, considering that the protected zone is significantly higher than the protected structure.

Using AutoCAD and effective techniques of rasterization, the corresponding 3D displays of the protected zone were created. The protected area of insulated (tower) lightning protection is determined using the PAM. Fig. 2.96a illustrates the protected zone of the realized application of the first two lightning rod towers. The cross section of the shared protected zone realized through the first five lightning rod towers is shown in Fig. 2.96b. Figs. 2.97a and 2.97b illustrate the shared protected zone formed by all lightning rod towers.

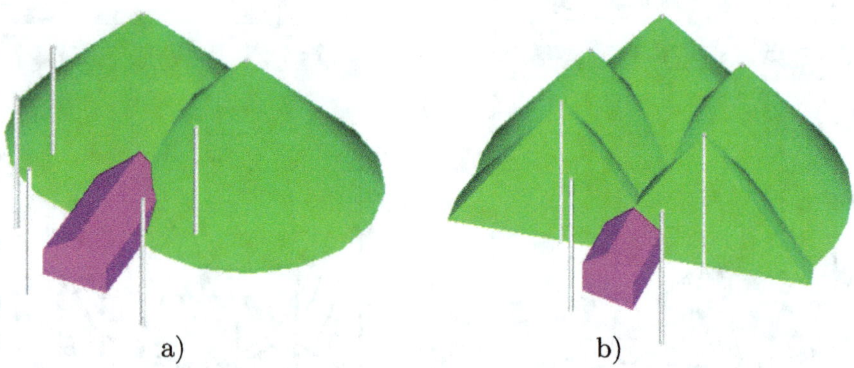


Fig. 2.96 - 3D display of protected zones achieved using the first two towers (a) and the first five towers (b)

Based on the 3D displays of the protected zones provided in Figs. 2.96 and 2.97, one can clearly see the advantages of visualization of the process of designing lightning protection using the proper techniques in AutoCAD. Through the view of the shared protected zone of the considered structure (Fig. 2.97a), it can be concluded that the application of the adopted number, configuration and height of lightning rod towers provides effective lightning protection. The protected structure is located completely within the cone area which provides a great certainty that it is protected from direct discharge. Placing tower

S21.2 further from the considered structure would lead to ineffective lightning protection. In this case a portion of the structure is outside of the shared protected zone, which is clearly illustrated in Fig. 2.97b.

a) b)

Fig. 2.97 - 3D display of shared protected zone achieved using all lightning conductor towers; a) Effective protection; b) Ineffective protection

### 2.4.5. Designing lightning protection for substations

#### 2.4.5.1 General considerations

The following software tool represents the continuation of the previously formed tool for designing lightning protection for general and special purpose structures [83-87]. Some of the capabilities of the software tool for evaluating substation protected zones are presented below. This tool is based on the application of the program AutoCAD and VB [88, 89], and is a user oriented tool for constructing protected zones in a manner which is suitable for the engineering practice. The protected zone is determined using the PAM and the RSM. The application of the tool is illustrated in the example for determining the protected zone of a concrete 110/35/10 kV substation [85, 90].

#### 2.4.5.2 Overview of the method for evaluating the protected zone of substation as a special purpose structure

The term protected zone of the accepted system implies the zone in which there is a low probability of a direct lightning discharge occurring. The PAM and the RSM are the primary methods used in the design practice.

*a) Lightning rod*

The protected zone of a lightning rod with a height $h$ (m) above the ground or $h_a$ (m) above the protected structure, with a height of $h_x$ (m), is represented by the circle at a height of $h_x$ with a radius of $R_x$ (m). When designing lightning protection of high-voltage substations and other structures according to the RSM, the radius of the protected zone is determined using the following expression:

$$R_x = h_a \cdot \frac{1,6}{1 + \frac{h_x}{h}} \cdot p, \tag{2.35}$$

where: $p = 1$ for $h \leqslant 30$ m and $p = 5.5\sqrt{h}$ for $h > 30$ m.

The symbols in expression (2.35) are explained in Fig. 2.98.

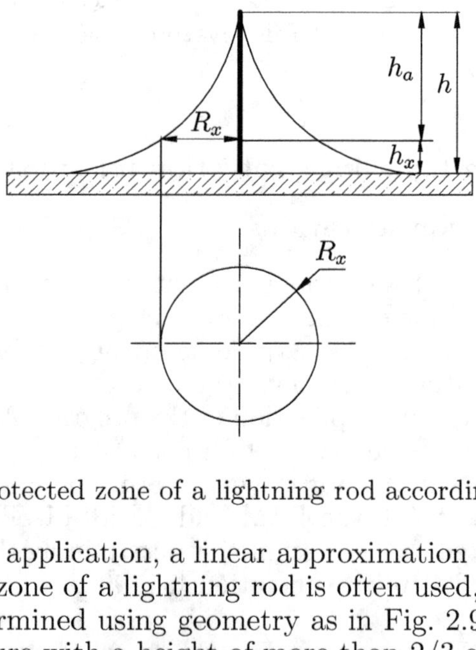


Fig. 2.98 – Protected zone of a lightning rod according to the RSM

In practical application, a linear approximation of the boundaries of the protected zone of a lightning rod is often used, so the protected zone can be determined using geometry as in Fig. 2.99. The protected zone for a structure with a height of more than 2/3 of the total height of the rod from a grounded surface $h$, is determined as the zone within the cone with generatrix at an angle of 36.8° in relation to vertical. The ratio of the height and radius of the base of the cone is 4 : 3.

For structures which are lower, the protected zone is within a cone with a height of $0.8h$, with a generatrix at an angle of $\alpha = 56.3°$. The ratio of the height and radius of the base of the cone is 2 : 3.

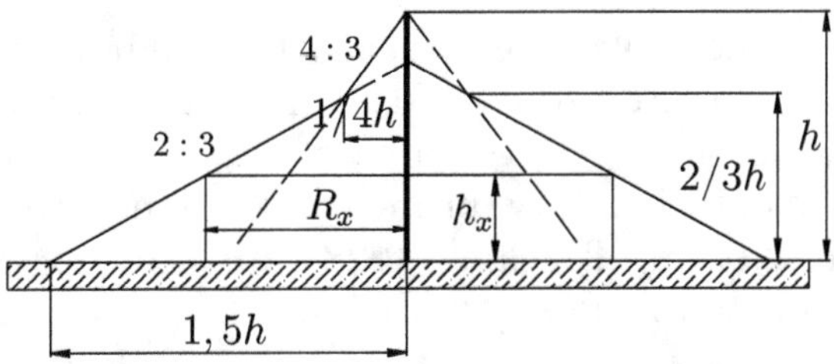


Fig. 2.99 – Protected zone of a lightning rod in linear form

The protected zone of one lightning conductor with a height of $h$, which is represented by a cone with its external side in the form of a broken line and base of the cone with a radius of $r = 1.5h$, is displayed in Fig. 2.100.

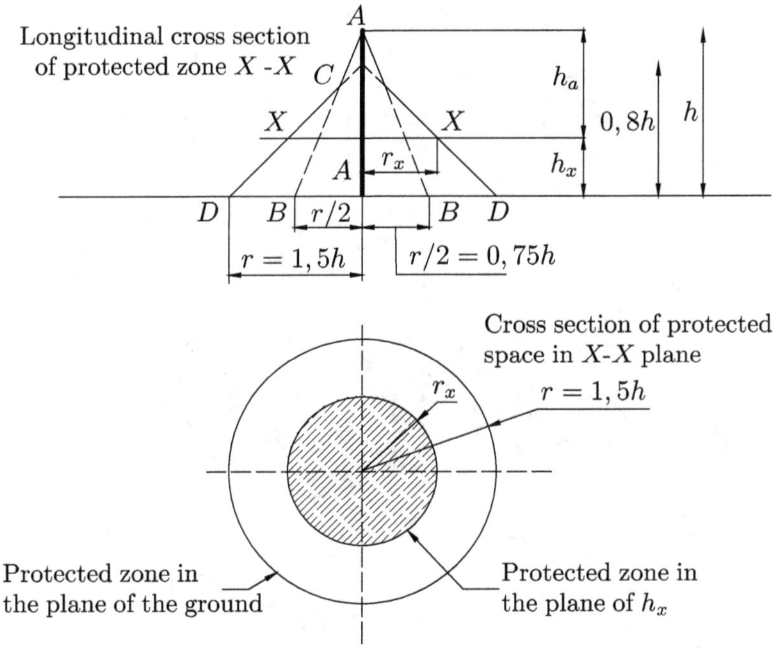


Fig. 2.100 - Lightning conductor and protected zone in the form of a broken cone

The horizontal cross section of the $X - X$ protected zone at the height of the protected structure $h_x$ is represented by a circle with a radius of $r_x$. The symbol $h_a$ in Fig. 2.100 represents the active height of the lightning rod. The broken line which is formed in this way represents the border of the protected zone, and with the rotation around the axis of lightning conductor $A - A$, the area of the broken cone is obtained, which with sufficient certainty protects the structure from lightning discharge. Expressions (2.27) and (2.28) are used for calculating the radius of the protected zone $r_x$ and height of $h_x$.

The protected zone of two lightning conductors which are placed lower than 60 m at a distance of "a" is presented in Fig. 2.101. The border between the equally high conductors is represented by the arch of

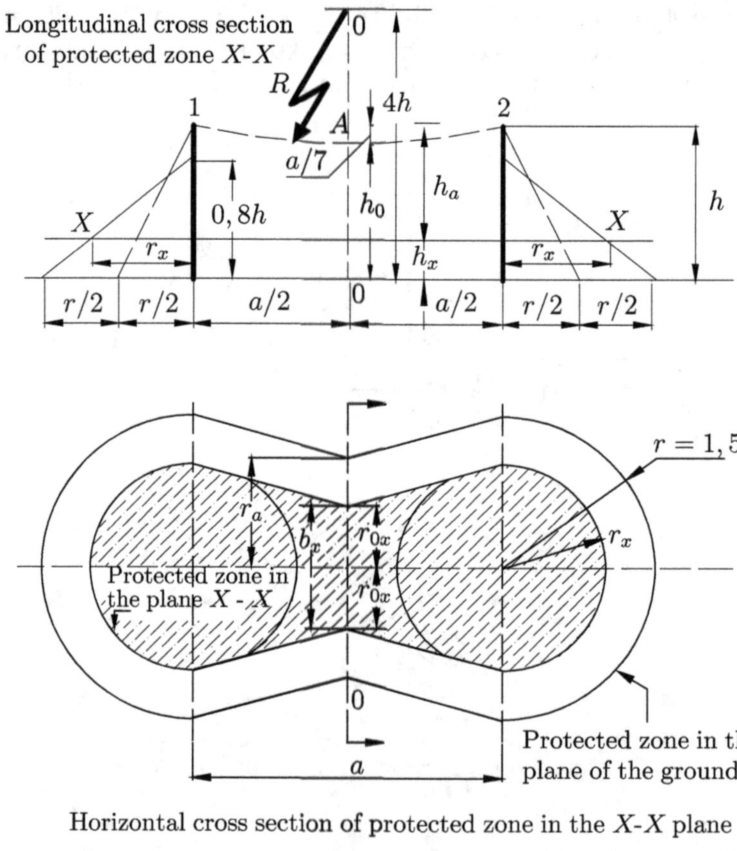


Fig. 2.101 – Protected zone of two lightning conductors of the same height

the circle crossing through the tops of the conductors with the center at vertical O-O, which is placed in the middle at a height of $H = 4h$. The sides of the protected zones are determined as the zones of individual the lightning conductors. The contour of the zone in the cross section of the vertical plane between two conductors is determined according to the rule for forming a protected zone for only one conductor with a height of $h_o$, and that being the lowest height of the zone between the conductors. The values of $r_o$ and $r_{ox}$ are equal to half of the width of the protected zone at the middle point between the conductors, so at the level of the ground it will be $r_o$, and at a height of $h_x$ will be $r_{ox}$.

The lowest height of the protected zone between two lightning conductors is obtained from the following expressions:
if:

$$h \leqslant 30m \Rightarrow h_0 = h - \frac{a}{7} \quad \text{(m)} \tag{2.36}$$

if:

$$h > 30m \Rightarrow h_0 = h - \frac{a}{7} \cdot \frac{5,5}{\sqrt{h}} \quad \text{(m)} \tag{2.37}$$

where: $a$ – distance between two lightning conductors, $h$ – height of lightning conductors.

The protected zone of two lightning conductors of varying heights $h_1$ and $h_2$ are displayed in Fig. 2.102.

The width of the protected zone $r_{ox}$ below a height of $h_o$, for varying heights of the lightning conductors and protected structure is calculated using the following expressions:

for $h \leqslant 30$ m

$$h_x \leqslant \frac{2}{3} \cdot h, \quad r_{ox} = 1.5(h_o - 1.25h_x) \tag{2.38}$$

for $h > 30$ m

$$h_x > \frac{2}{3} \cdot h, \quad r_{ox} = 0.75(h_o - h_x) \tag{2.39}$$

Along with this, when the height of the lower lightning conductor $h_1 \leqslant 30$ m, then:

$$h_0 = h_2 - \frac{a_f}{7} \quad \text{(m)} \tag{2.40}$$

The protected zone of three lightning rods (Fig. 2.103) is determined as the circle with a diameter $D$, which is given in expressions

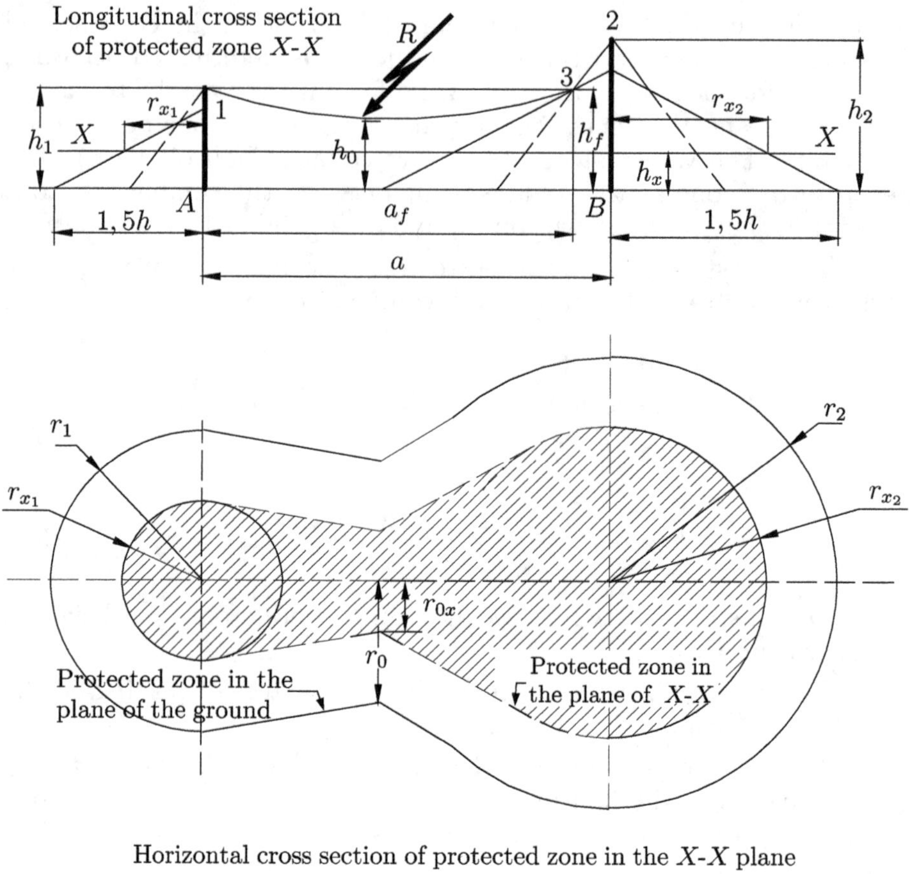


Fig. 2.102 - Protected zone of two lightning conductors of the varying heights

(2.41) and (2.42). The basic condition for protection of a structure with a height $h_x$, or a group of structures with their highest height of $h_x$, is that $r_{ox} > 0$ for all, according to pairs, of the considered conductors.

Additionally, for four conductors (Figs. 2.104 and 2.105) or a larger number of conductors, it is necessary to fulfill the following conditions:

- for lower structures it is necessary for the condition $h_o = h_x$ to be fulfilled for pairs of conductors according to the diagonal of the polygon created from individual conductors,
- for other structures it is necessary for the condition $D \geqslant 5h$

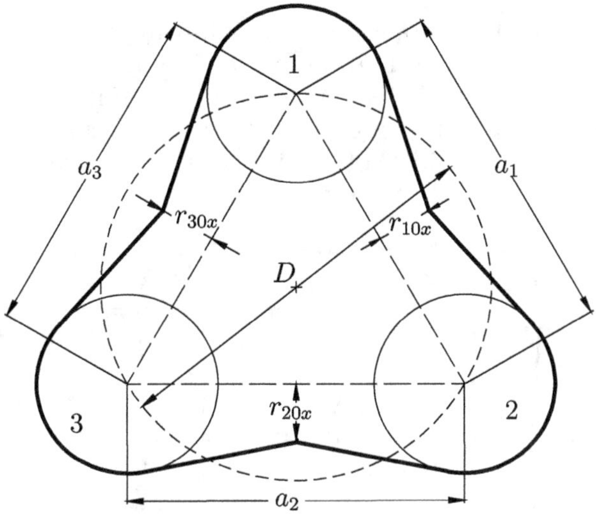


Fig. 2.103 - Protected zone of three lightning conductors placed in an equilateral triangle

to be fulfilled, where $D$ is the length of the diagonal of the polygon of individual conductors.

The configuration of lightning conductors depends on the geometric parameters of the protected structure.

The diagonals of the polygon, created from the lightning conductors, are calculated using the following expressions:

if $h \leqslant 30$ m

$$D \leqslant 8 \cdot (h - h_x) \tag{2.41}$$

if $h > 30$ m

$$D \geqslant 8 \cdot \frac{5,5}{\sqrt{h}} \cdot (h - h_x) \tag{2.42}$$

where: $h$ – height of lightning conductor, $h_x$ – height of protected structure.

Figs. 2.104 and 2.105 present the protected zone at a height of $h_x$ for four lightning conductors configured in a square and the protected zone at a height of $h_x$ for four arbitrarily configured lightning conductors, respectively. All conductors are of the same height and $r_x$ indicates the radius of the protected zone of each lightning conductor.

In Fig. 2.104 the symbols have the following meanings: $D$ – length of diagonal of right-angle quadrangle; $a_1$ – distance between lightning conductors 1 and 4, 2 and 3; $a_2$ – distance between lightning conductors 1 and 2, 3 and 4; $r_{10x}$ – width of protected zone between conductors 2 and 3, 1 and 4; $r_{20x}$ – width of protected zone between conductors 1 and 2, 3 and 4.

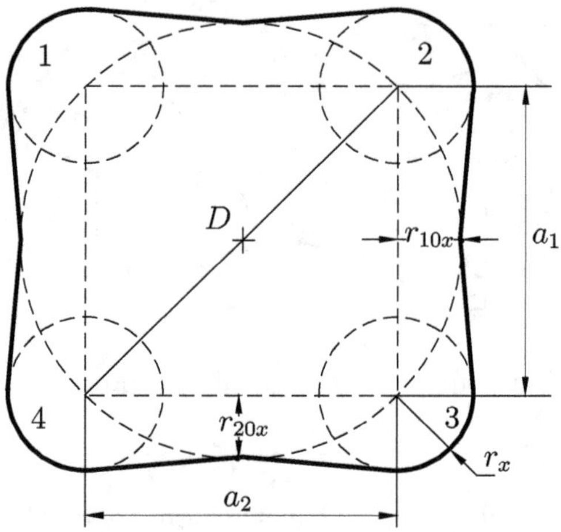


Fig. 2.104 - Protected zone of four lightning conductors configured in a square

In Fig. 2.105 the symbols have the following meanings: $D_1$ – diameter of circle drawn around lightning conductors 1, 2 and 4; $D_2$ – diameter of circle drawn around lightning conductors 2, 3 and 4; $a_1$ – distance between lightning conductors 1 and 2; $a_2$ – distance between lightning conductors 2 and 3; $a_3$ – distance between lightning conductors 3 and 4; $a_4$ – distance between lightning conductors 4 and 1; $r_1$ – width of protected zone between conductors 1 and 2; $r_2$ – width of protected zone between conductors 2 and 3; $r_3$ – width of protected zone between conductors 3 and 4; $r_4$ – width of protected zone between conductors 1 and 4.

*b) Ground wires*

The protected zone is achieved using one ground wire shown in Fig. 2.106 and is determined according to the expression:

$$r_x = \frac{0,8 \cdot h_a}{1 + h_x/H} \quad \text{for } H \leqslant 30 \text{ m} \tag{2.43}$$

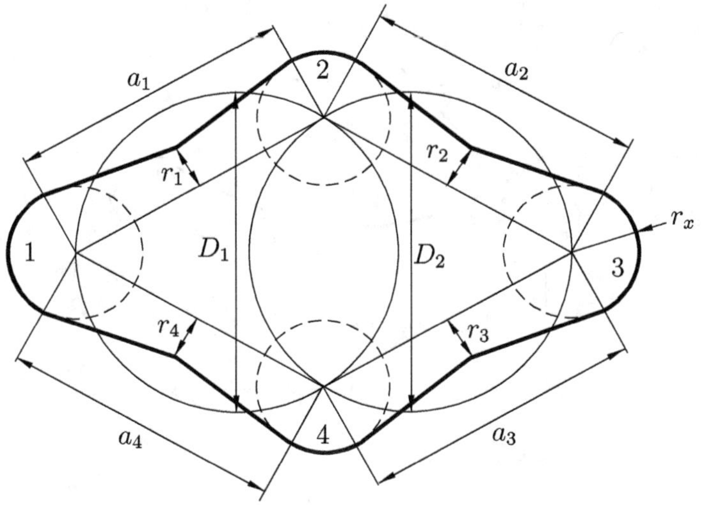


Fig. 2.105 - Protected zone of four arbitrarily configured lightning conductors

The symbols in expression (2.43) have the same meanings as in Fig. 2.106.

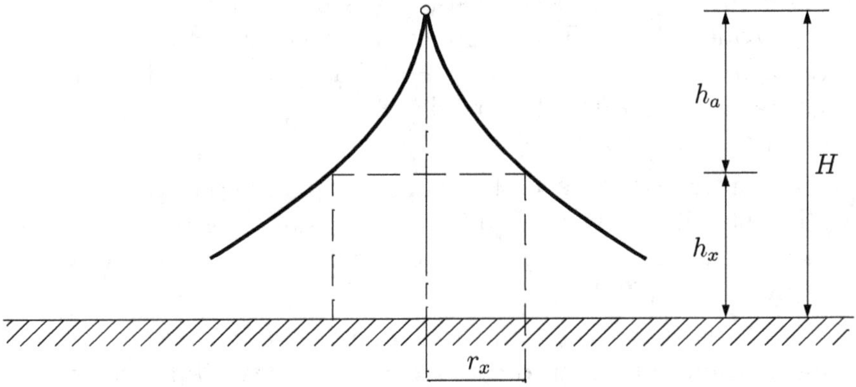


Fig. 2.106 – Protected zone achieved using one ground wire

The protected zone achieved using two ground wires encompasses the area below the arch which touches both ground wires with its center above the wires and an arch radius of $0.58d$, where d is the distance between the ground wires (Fig. 2.107).

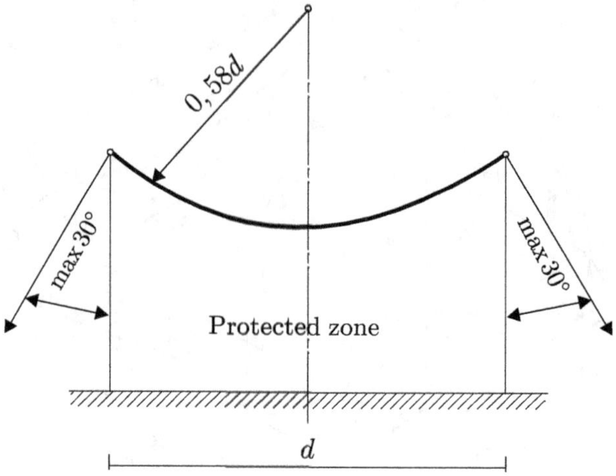


Fig. 2.107 – Protected zone achieved using two ground wires

### 2.4.5.3 Example of calculation of protected zone for substation

In this section the process of visualization of the protected zone of a concrete 110/35/10 kV substation is presented. Figs. 2.108a and 2.108b provide a 2D and 3D display of the considered substation, respectively. Towers 1 - 4 are used both for lighting and for lightning conductors. The height of tower 1 is 28 m, while the heights of towers 2, 3 and 4 are each 15 m.

The protected zone of the substation is determined using the RSM. The striking distance is determined using expression (2.26) for a strike current amplitude of 5 kA and parameters $K = 27$ and $n = 0.42$, defined by Young [79, 80].

Fig. 2.109 shows the protected zone determined for tower 1. Various illustrations of the protected zone for tower 1 are provided in Figs. 2.109a – 2.109c. The entire protected zone of the substation for towers 1-4 is displayed in Fig. 2.110. The front view of the protected zone and 3D display are provided in Figs. 2.110a and 2.110b, respectively. Based on that which is displayed in Figs. 2.110a and 2.110b, it can be concluded that the 3D display is important for recognizing the effects of lightning conductors on the form and size of the protected zone.

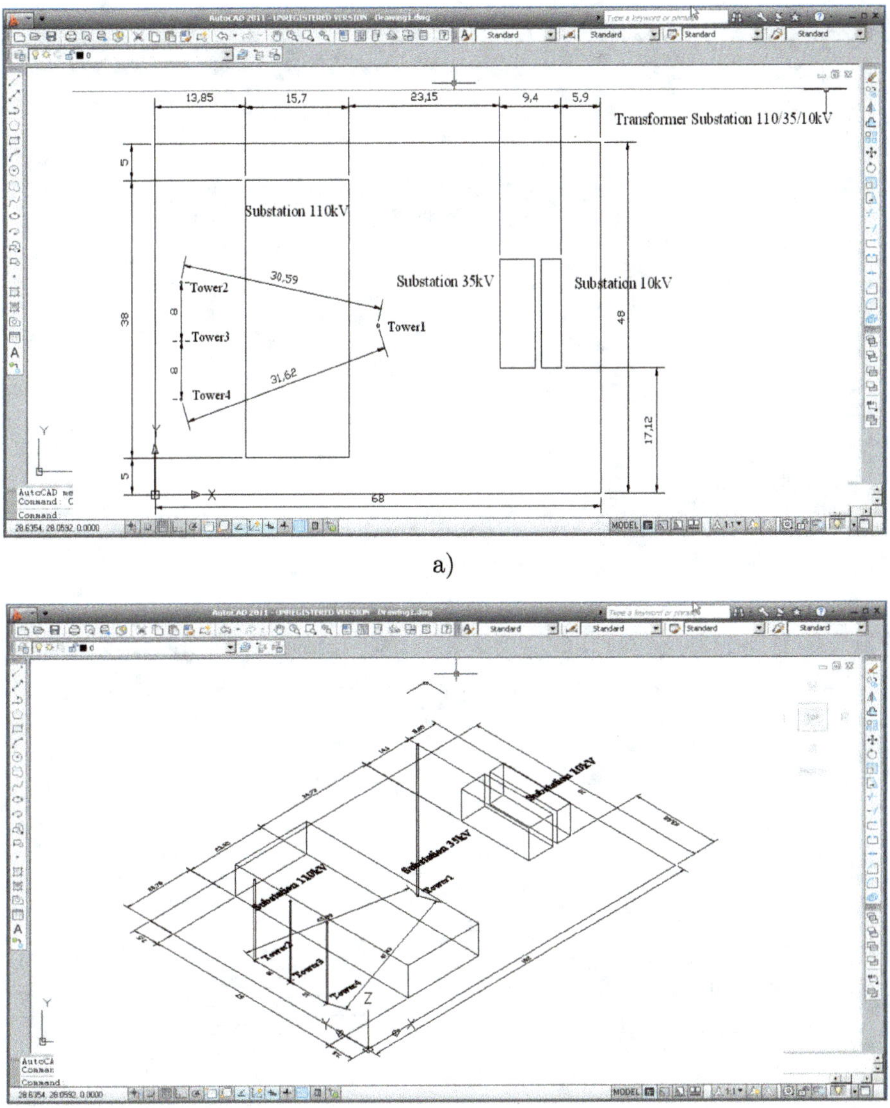


a)

b)

Fig. 2.108 – 110/35/10 kV substation; a) 2D display; b) 3D display

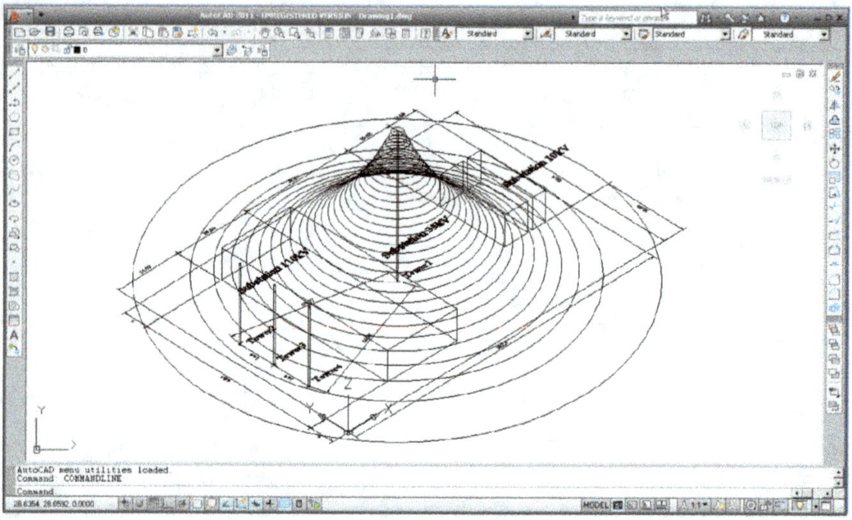
AutoCAD menu utilities loaded.
Command: COMMANDLINE
Command:
MODEL

a)

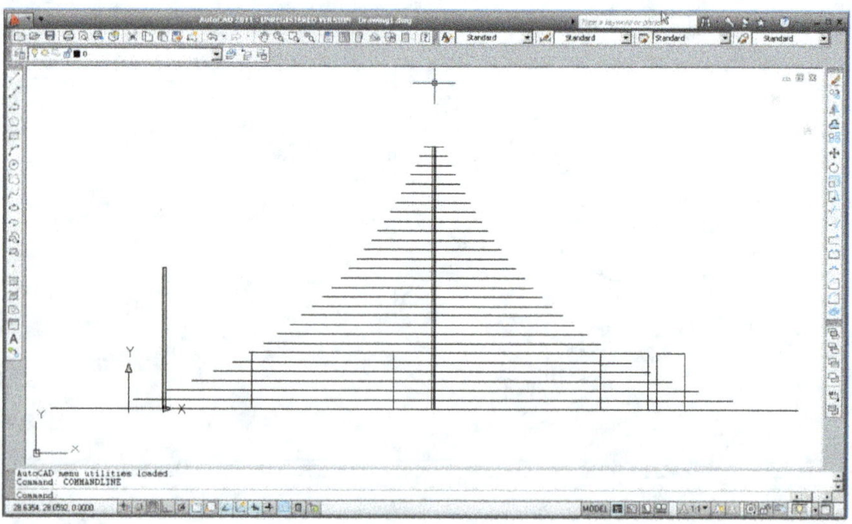
AutoCAD menu utilities loaded.
Command: COMMANDLINE
Command:
MODEL

b)

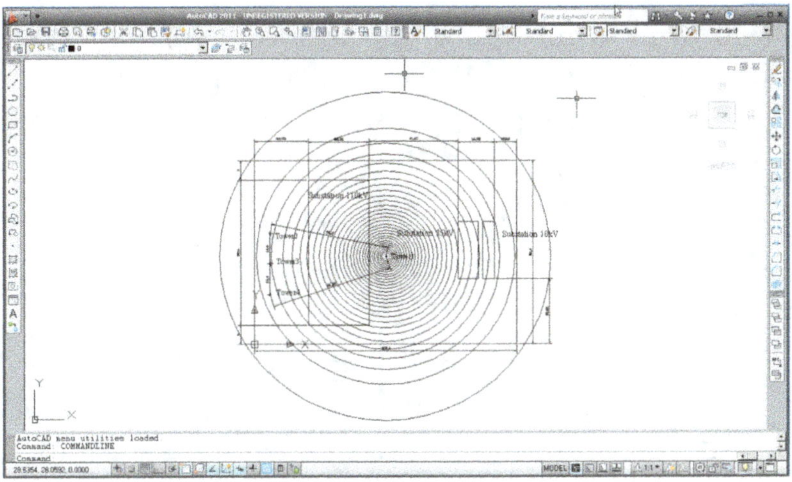

c)

Fig. 2.109 - Protected zone of the substation achieved using tower 1 with a strike current amplitude of 5 kA and Young model; a) 3D display; b) 2D front view display; c) 2D top view display

## 2.4.6. Designing lightning protection for overhead lines

### 2.4.6.1 General considerations

Lightning discharges through their effects can cause a break in the transfer of electrical energy, which is especially pronounced in lines with a lower rated voltage. From the aspect of behavior during lightning discharges, overhead lines can be divided into three groups [91]:

- lines with steel-lattice or reinforced concrete towers without ground wires,
- lines with steel-lattice or reinforced concrete towers with ground wires,
- lines on wooden poles without ground wire.

When dealing with lines with steel-lattice or reinforced concrete towers without ground wires, a lightning discharge can damage the insulation in three ways:

- by direct strike into the phase conductor,
- by strike into the top of the tower, which causes a flashover across the insulation towards the phase conductor (back flashover),

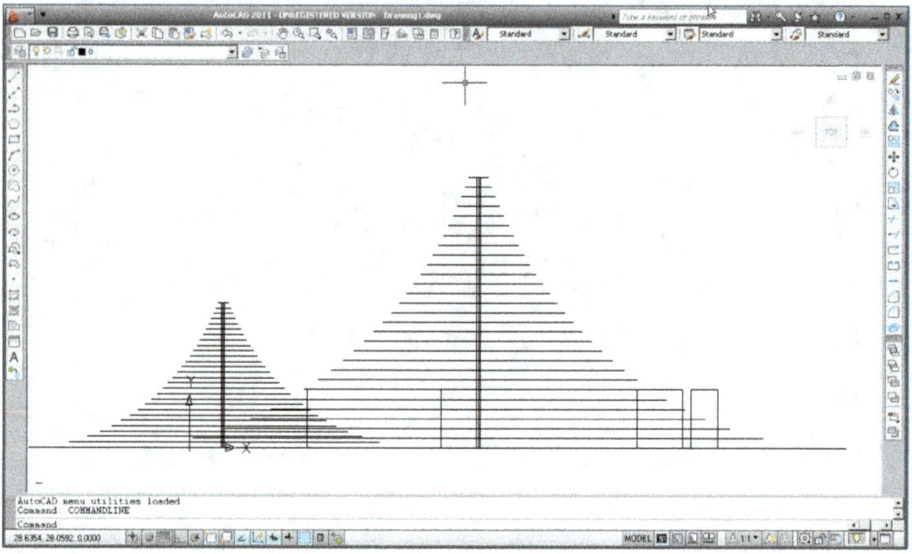


a)

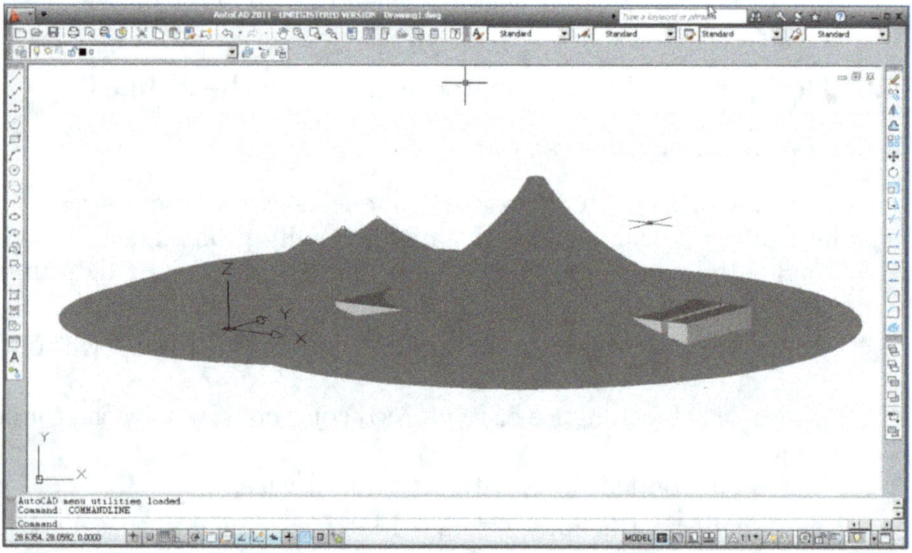


b)

Fig. 2.110 - Entire protected zone for towers 1 - 4, strike current amplitude of 5 kA and Young model; a) 2D front view display; b) 3D display

– by strike in the area around the overhead line, which causes induced overvoltage on phase conductors.

A direct strike into the phase conductor and a strike in the area around the overhead line, which causes induced overvoltage in phase conductors, are ways in which insulation is damaged on lines held up by wooden poles without ground wire.

Finally, when dealing with lines with ground wires, three cases stand out:

– strike into the ground wire or top of the tower, which causes a flashover across the insulation towards the phase conductor (back flashover),
– direct strike to the phase conductor (shielding failure),
– strike in the area around the overhead line, which causes induced overvoltage on phase conductors.

Further along in this section a review is presented of the method for calculating the protected zone of overhead lines as structures with limited danger. The calculation of the protected zone was done using the PAM and the RSM. Calculations of the striking distance were done using the geometric method, electrogeometric method and generic method [80, 84]. The application of software tools is illustrated in the examples for determining the protected zones of 220 kV overhead line tower with horizontal arrangement of the phase conductors and two ground wires and a 400 kV double overhead line, respectively.

#### 2.4.6.2 Review of the method for calculating the protected zone of overhead lines as structures with limited danger

The calculation of the protected zone was conducted using two different methods. The PAM represents the first method and is based on the calculation of the protective angle. That is the angle which covers the vertical line, placed through the center of the ground wire, with the line which connects the centers of the phase conductor and ground wire. According to valid recommendations, the maximum value for this angle is 30° regardless of the rated voltage of the line.

The ground wire does not provide absolute protection from penetration of direct strike to the phase conductor (shielding failure). The probability of direct lightning strikes to the phase conductor of overhead line, equipped with ground wires P can be estimated using the following expression [91]:

$$\log P = \frac{\alpha \cdot \sqrt{h}}{A} - B \tag{2.44}$$

where: $\alpha$ – protective angle, $h$ – effective height of ground wire, $A$, $B$ – the empirical constants which values are taken to be 90 and 4, respectively.

Based on expression (2.44), it can be concluded that the probability $P$ is independent on the strike current magnitude. For this reason, this model is labeled in the literature as a current-independent model.

The RSM represents the second method and is based on the theory of striking distance. For the calculation of striking distance the following three models are used:

- geometric model,
- electrogeometric model (EM),
- generic model (GM).

The first model is the geometric method according to which it is assumed that the striking distance is independent from the strike current amplitude and local geometry of the considered structure. The basic assumption of the EM refers to the correlation between the length of the last step of the downward leader and stroke current magnitude. Generally, the striking distance is defined by expression (2.26). Table 2.20 shows the values of empirical constants $K$ and n [79, 80]. The symbol $\beta$ represents a constant with a value in the range of 0.64 - 1 depending on the voltage level of the line [92].

Table 2.20 – Relations for striking distance $R = K \cdot I^n$

| Literature [79]<br>Author | $R$ to phase conductor or ground wire ($R_1$) | | $R$ to earth or ground ($R_2$) | |
|---|---|---|---|---|
| | $K$ | $n$ | $K$ | $n$ |
| Wagner | 14.2 | 0.42 | 14.2 | 0.42 |
| Armstrong and Whitehead | 6.7 | 0.80 | 6.0 | 0.80 |
| Love | 10.0 | 0.65 | 10.0 | 0.65 |
| Young | 27.0 | 0.32 | 27.0 | 0.32 |
| Mousa and IEEE 1993 | 8.0 | 0.65 | 8.0 | 0.65 |
| Anderson and IEEE 1985 | 8.0 | 0.65 | $\beta$. R1 | 0.65 |

Based on Table 2.20 it can be concluded that there are generally two striking distances, one for the phase conductor or ground wire $R_1$

and one for the earth or ground $R_2$. That is because the breakdown gradient for a rod – plane gap (core of downward leader to ground) differs from the breakdown gradient for a rod – rod (downward leader to top of tower). To simplify the use of these striking distance equations, both IEEE and CIGRE only vertical strokes have considered and recommended for general use.

The generic model represents the third type. This model is based on significant expansion of knowledge on the physical model of a long arc gap, and assumes that the striking distance is a function of the current of lightning discharge and the local composition of the structure. The Eriksson [79, 93] and Petrov-Waters [80] models are covered by a developed software tool.

Based on the review of experimental results, Eriksson developed a quantitative model for calculation of the attractive radius $r_a$ (m) for a vertical rod on flat terrain as a function of return current of an lightning discharge $I_o$ (kA) and the height of the rod $h$ (m):

$$r_a = 0,84 \cdot I_0^{0,74} h^{0,6} \tag{2.45}$$

Petrov and Waters showed that for a negative discharge and strike on the rod, the striking distance amounts to:

$$r_a = 0,8\Big[(h+15)\, I_0\Big]^{2/3} \quad [\text{m}, \ \text{kA}] \tag{2.46}$$

The EM and GM represent current-dependent models considering that they take into consideration the amplitude of strike current.

### 2.4.6.3 Program organization

The algorithm for overhead line shielding zone construction using AutoCAD / AutoLISP programs for both presented methods (PAM and RSM) is displayed in Fig. 2.111.

AutoCAD offers several possibilities for automation of calculation of the protected zone. A corresponding program can be formed using Visual LISP (VLISP), VBA and AutoLISP. VLISP enables the import and editing of any AutoCAD object using AutoCAD ActiveX interface. VLISP has a user-friendly graphic interface for writing code, removing errors and testing the program.

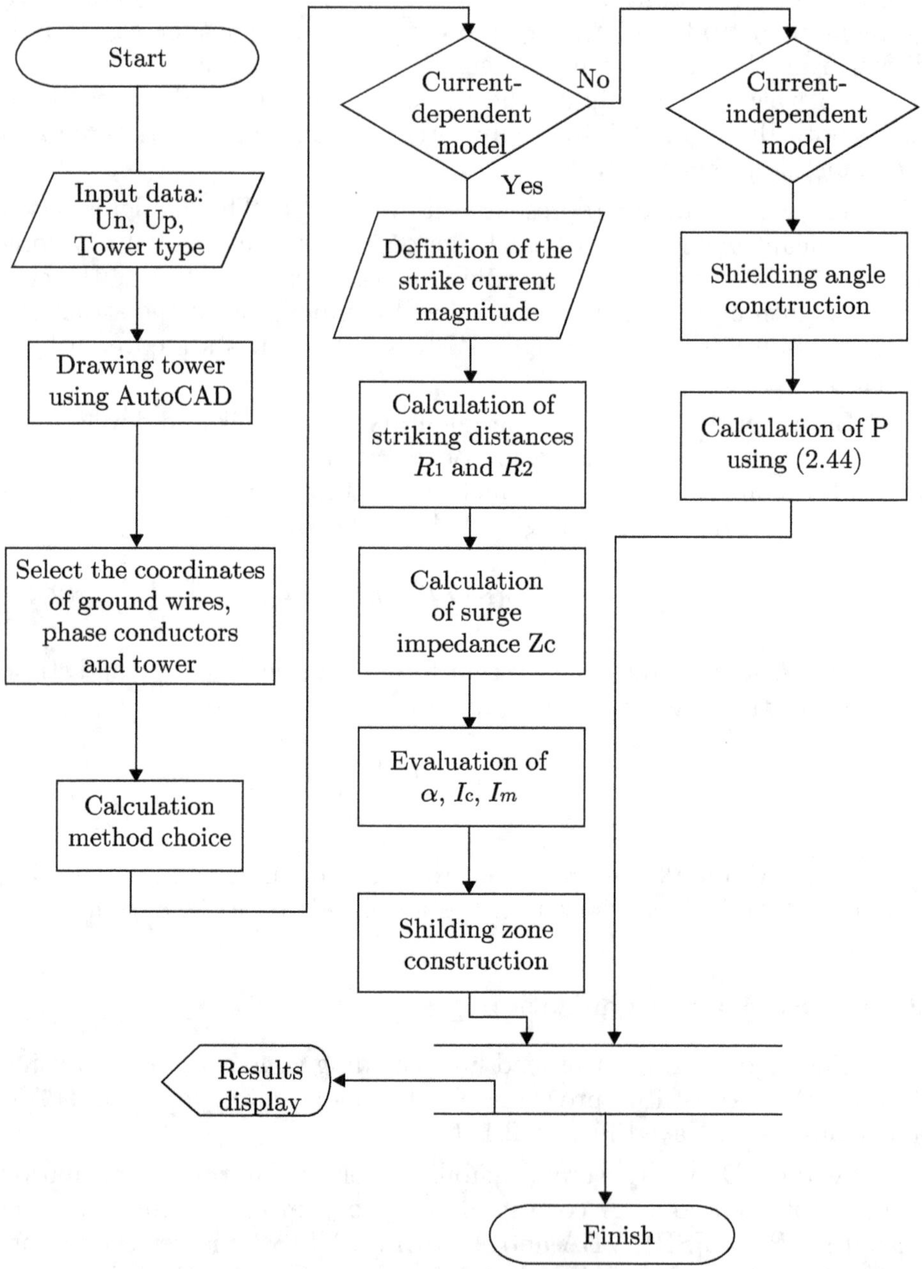


Fig. 2.111 – Algorithm for calculation of protected zone of overhead line

VBA is a software tool and software environment which is contained in several Microsoft programs. VBA enables simple importing, editing and working with objects which are organized in the library. Using VBA it is also possible to create dialog boxes in AutoCAD.

AutoLISP is a simplified version of the LISP artificial intelligence language. Lists represent the basic structure of programming in AutoLISP. AutoCAD places all data on objects in a list and the list consists of several smaller lists. The lists are used for modifying an object (entity) in AutoCAD's database. The principles of forming an AutoLISP program are displayed in the selected literature [89].

The calculation of overhead line shielding zone was conducted using the AutoLISP program which is shown in Fig. 2.112. It is important to mention that in AutoLISP, data can be entered through interactive work, by defining numerical values or through direct reading of coordinates from a drawing. The second method was used in the considered example.

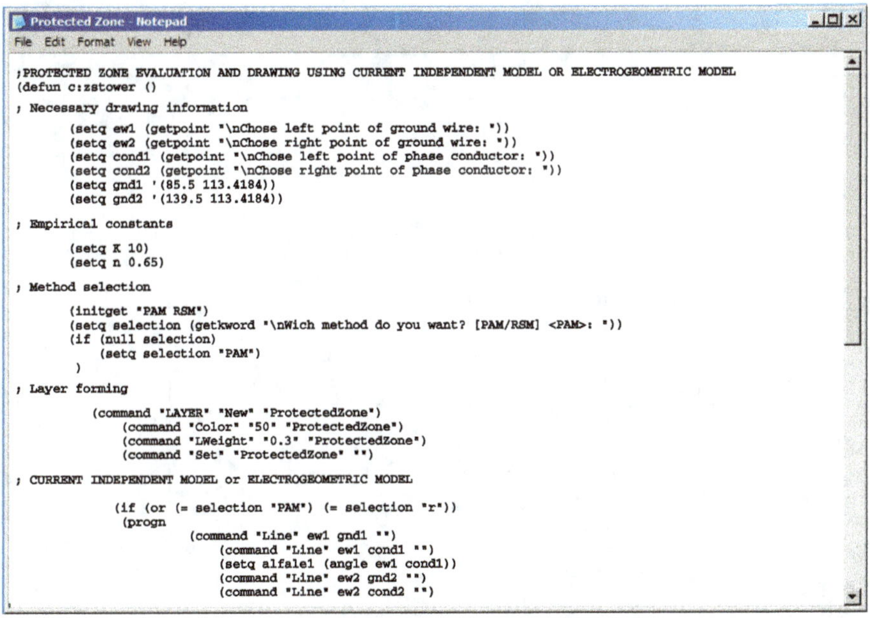

```
Protected Zone - Notepad
File Edit Format View Help

;PROTECTED ZONE EVALUATION AND DRAWING USING CURRENT INDEPENDENT MODEL OR ELECTROGEOMETRIC MODEL
(defun c:zstower ()
; Necessary drawing information
       (setq ew1 (getpoint "\nChose left point of ground wire: "))
       (setq ew2 (getpoint "\nChose right point of ground wire: "))
       (setq cond1 (getpoint "\nChose left point of phase conductor: "))
       (setq cond2 (getpoint "\nChose right point of phase conductor: "))
       (setq gnd1 '(85.5 113.4184))
       (setq gnd2 '(139.5 113.4184))
; Empirical constants
       (setq K 10)
       (setq n 0.65)
; Method selection
       (initget "PAM RSM")
       (setq selection (getkword "\nWich method do you want? [PAM/RSM] <PAM>: "))
       (if (null selection)
           (setq selection "PAM")
        )
; Layer forming
           (command "LAYER" "New" "ProtectedZone")
               (command "Color" "50" "ProtectedZone")
               (command "LWeight" "0.3" "ProtectedZone")
               (command "Set" "ProtectedZone" "")
; CURRENT INDEPENDENT MODEL or ELECTROGEOMETRIC MODEL
              (if (or (= selection "PAM") (= selection "r"))
               (progn
                       (command "Line" ew1 gnd1 "")
                           (command "Line" ew1 cond1 "")
                           (setq alfale1 (angle ew1 cond1))
                           (command "Line" ew2 gnd2 "")
                           (command "Line" ew2 cond2 "")
```

Fig. 2.112 - Portion of AutoLISP program for calculation of overhead line shielding zone

After activating the program, the user is asked for the coordinates of the attachment points for the ground wire and phase conductor. Next comes the selection of the method for determining the protected zone and then finally the calculation of the protected zone. In the case that the PAM is used, the user is provided with the information on whether the protective angle has a value which is within the maximum allowed values.

### 2.4.6.4 Examples of calculation of overhead line shielding zone

*a) 220 kV overhead line tower with horizontal arrangement of the phase conductors and two ground wires*

An illustration of the application of the program is provided in the example of the calculation of the protected zones of 220 kV overhead line tower with horizontal arrangement of the phase conductors and two ground wires (Fig. 2.113). The same drawing shows the results of calculation of the protected zone according to both methods. The

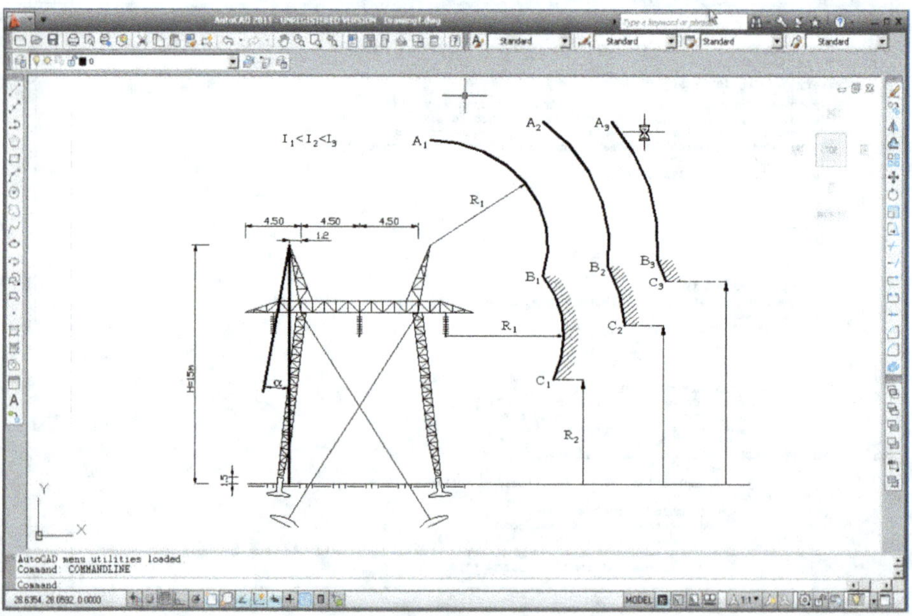


Fig. 2.113 - Illustration of the protective angle (left side) and the EM application (right side) for the presented 220 kV overhead line tower with horizontal arrangement of the phase conductors and two ground wires

protective angle is determined for the left ground wire. The EM concept is applied for the right side of the ground wire – outside phase conductor diagram in Fig. 2.113. A family of shielding zones for different strike current magnitudes is constructed.

Consider the general concept as depicted in Fig. 2.113. For a specific value of strike current, arcs of radii $R_1$ are drawn from the outside phase conductor and from the ground wire. In addition, a horizontal line a distance $R_2$ from the earth's surfee is constructed. The intersections of these arcs and the intersection of the arch with the horizontal line for the corresponding strike current $I_i$ are marked $A_i$, $B_i$ and $C_i$ $(i = 1, 2, 3)$. Downward leaders that reach the arch between $B_i$ and $C_i$ will terminate on the phase conductor. Those that reach the arch between $A_i$ and $B_i$ will terminate on the ground wire, and those that terminate beyond $C_i$ will terminate to earth. Finally, the whole shielding zone for the presented overhead line tower consists two symmetrical parts, the construction of which is in accordance with described procedure.

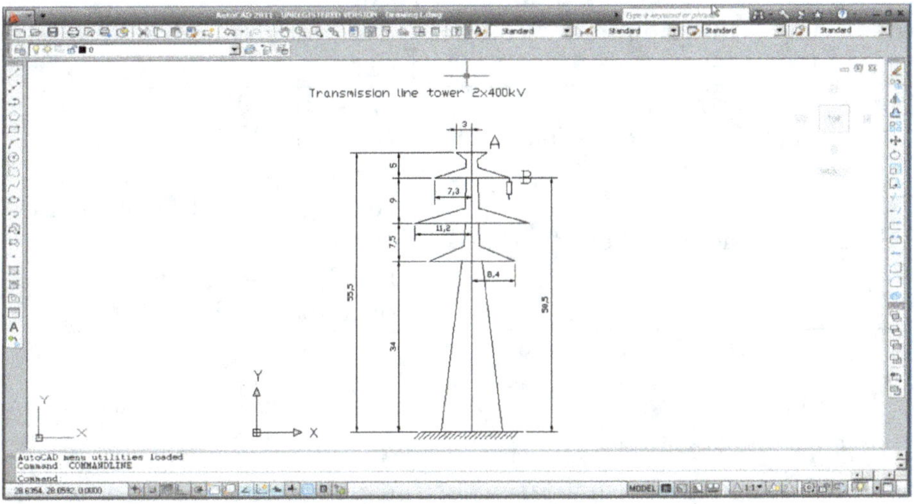


Fig. 2.114 – Analyzed 2x400 kV overhead line tower

Table 2.21 provides the input data and calculation results for denoted 220 kV overhead line. The symbol $Z_c$ represents the surge impedance of the phase conductor which is determined according to the calculations provided in literature [91]. The symbol $I_m$ represents the maximum shielding failure current, while $I_c$ indicates the critical value of strike current which causes a flashover on the line. These values are determined using the methodology provided in [79, 84].

The heights of the ground wire and the phase conductors equal 15 m and 10 m, respectively.

Table 2.21 - Input data and calculation results for the 220 kV overhead line tower according to Fig. 2.113

| **Input data** | |
|---|---|
| Nominal line voltage $U_n$ (kV) | 220 |
| Basic insulation level $U_p$ (kV) | 950 |
| Ground wires and phase conductors arrangement | Horizontal |
| Strike current magnitude $I$(kA) (only for current – dependent method) | $I_1 = 0.5$kA $I_2 = 1.0$ kA $I_3 = 1.5$ kA |
| Constants $K$ and n (only for current dependent model) | $K = 10$, $n = 0.65$ |
| **Calculation results** | |
| **Current independent model (PAM)** | **Current dependent model (EM)** |
| $\alpha = 22°$ ($< 30°$)<br>$P = 0.000891$ from expression (2.44)<br><br><br>Shielding zone on the left side of Fig. 2.113 | $Z_c = 313\Omega$ [91]<br>$I_m = 2.1$ kA [84]<br>$I_c = 6.07$ kA [84]<br>For $I_m < I_c$, Risk of failure $R = 0$<br>Family of shielding zones on right side of Fig. 2.113 |

*b) 400 kV double overhead line tower with vertical arrangement of the phase conductors and two ground wires*

The process of visualization of the protected zone of a 400 kV double overhead line is based on the use of various methods. Fig. 2.114 shows the analyzed 2 × 400 kV overhead line tower. In accordance with the generic model, the attractive surface for terminating the phase conductor of denoted overhead line was evaluated (Fig. 2.115). The striking distance is calculated according to the Eriksson model. The impulse current amplitudes are taken to be 10 kA and 25 kA, respectively.

Attractive surface for terminating the phase conductor of the analyzed 2×400 kV overhead line tower for the impulse current amplitude of 10 kA according to the Eriksson model (arch $A_1B_1$) and the Petrov-Waters model (arch $A_2B_2$) was illustrated in Fig. 2.116.

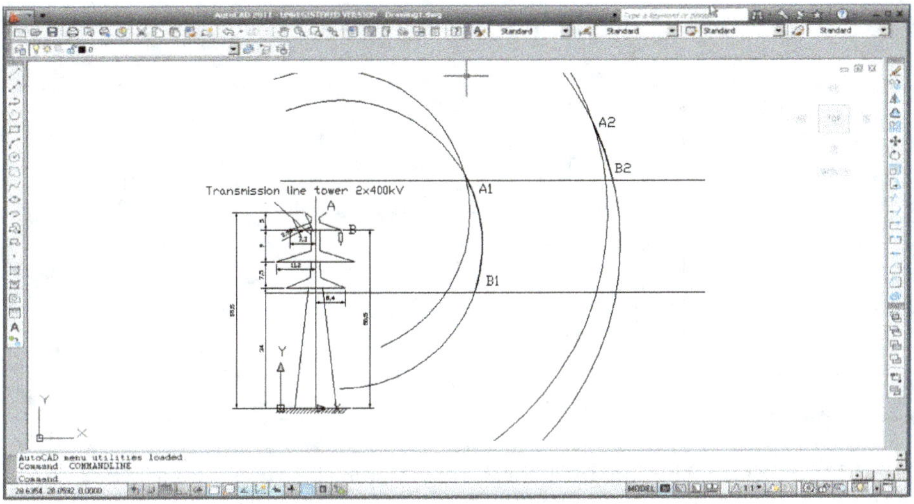


Fig. 2.115 - Attractive surface for terminating the phase conductor of the analyzed $2 \times 400$ kV overhead line tower in accordance with the generic Eriksson model, impulse current amplitudes 10 kA (arch $A_1B_1$) and 25 kA (arch $A_2B_2$)

Determination of the attractive surface enables the evaluation of the lightning performance of the overhead lines using various processes defined in the selected literature [79, 80, 91, 92, 94].

### 2.4.7. Conclusions

Some aspects of the lightning protection design of structures, substations and overhead lines are described. The protection zone evaluation and its visualization represent important and difficult tasks in this procedure.

Different methods for the protection zone evaluation are used. The structure protection zone evaluation using the PAM and the RSM is performed. In general, the PAM is recommended for structures not exceeding 20 m height. The magnitude of protective angle is based on the protection level adopted. The use of the RSM enables all possible positions for the leader approach and, therefore, it is recommended for higher buildings and special structures.

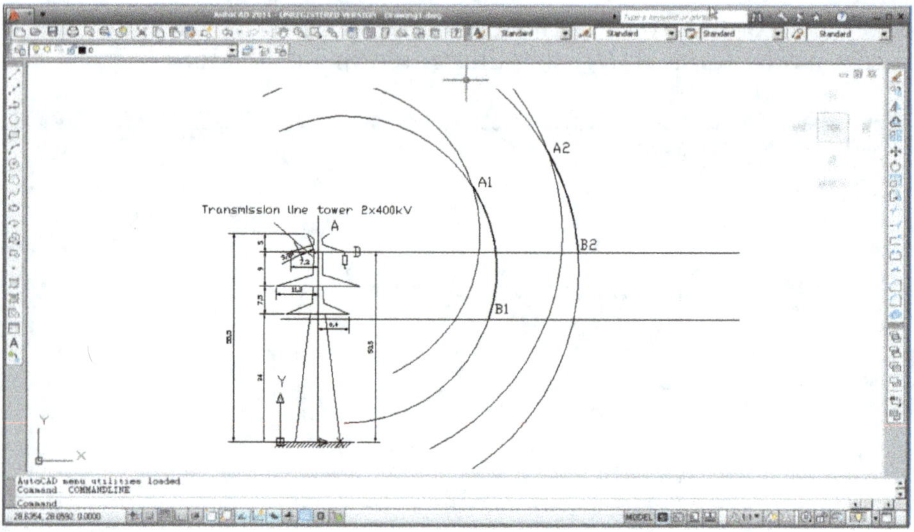


Fig. 2.116 - Attractive surface for terminating the phase conductor of the analyzed $2 \times 400$ kV overhead line tower for the impulse current amplitude of 10 kA according to the Eriksson model (arch $A_1B_1$) and the Petrov-Waters model (arch $A_2B_2$)

The evaluation of the substation protection zone is based on the EM and GM. The overhead line shielding zone evaluation based on the classical method of shielding angle is independent on the strike current magnitude. On the other hand, the EM is based on the striking distance theory. As result of this, the EM application in the designer's practice enables to estimate the shielding angle involving the most important parameters, such as the strike current magnitude, tower height and ground flash density.

The software tool developed enables easy and quick visualization of lightning protection zones for structures, substations and overhead lines. This tool is a user-oriented system formed through the application of powerful software AutoCAD and the programming languages AutoLISP and VB. The purpose of this user-oriented tool is to construct the lightning protection zone in a way suitable for the designer's practice.

# LITERATURE


[1] MATLAB® – The Language of Technical Computing, Version 7.10.0 (R2010a), The MathWorks Inc, 2010.

[2] J.H. Mathews, K.D. Fink: Numerical Methods using MATLAB, Fourth Edition, Pearson Education International, 2004.

[3] S. T. Karris: Numerical Analysis Using MATLAB and Spreadsheets, Second Edition, Orchard Publication, ISBN 0-9744239-1-2.

[4] A. Gilat: Introduction to MATLAB 7 with examples, Translation of second edition, Mikro Knjiga, 2005.

[5] Z. Stojković: Computer-Aided Design in Power Engineering – Software Tools, Monography, II edition, Faculty of Electrical Engineering, Belgrade, Academic Mind, Belgrade, March 2003.

[6] Z. Stojković, J. Mikulović, Z. Stojanović: Workshop for Software Tools in Power Engineering, Faculty of Electrical Engineering, Belgrade, Academic Mind, Belgrade, June 2006.

[7] A. M. Gole, A. Daneshpooy: Towards Open Systems : A PSCAD / EMTDC to MATLAB Interface, IPST'97 – International Conference on Power Systems Transients, Sietle, June 1997, pp.145 – 149.

[8] Simulink®, User's Guide, Version 7.5 (R2010a), The MathWorks Inc, 2010.

[9] H. Bode: Matlab-Simulink, Analyse und Simulation dynamischer Systeme, 2. Auflage, Teubner Verlag, Wiesbaden, 2006.

[10] J. Nahman, V. Mijailović: High-Voltage Substations, Monography, Beopres, Belgrade, 2000.

[11] R. Natarajan: Computer-Aided Power System Analysis, Marcel Dekker Inc, New York, NY, USA, 2002.

[12] ATP – EMTP Rule Book, Canadian-American EMTP Users Group, 1997.

[13] Power Tools for Windows, SKM Systems Analysis, Inc, Manhattan Beach, California.

[14] Power System Simulator – Program Application Guide, PSS documentation, PT1.

[15] J. Nahman, D. Salamon, V. Mijailović: High-Voltage Substations – Collections of Solved Tasks with Supplements, Faculty of Electrical Engineering, Academic Mind, Belgrade, 2002.

[16] Z. Stojković, P. Vučetić, M. Bubnjević, A. Mijalčić, B. Bjelanović: Application of Databases in the Design of High-Voltage Substations, Elektroprivreda, No. 1, 2005, pp. 3-19.

[17] M. Marinković, Z. Stojković: Software Tool SPLCAD for Designing Medium-Voltage Overhead Lines, Faculty of Electrical Engineering, Belgrade, 2009.

[18] Development of a Magnetic Field Spatial Scanner for Diagnostic Equipment in Power Systems and Environmental Protection, Technological Development Project TR-17031, Ministry of Science and Technological Development of the Republic of Serbia, 2008-2011, (Manager Z. Stojković).

[19] G. Dotlić: Excerpts from Technical Standards, SMEITS, Belgrade, 2007, pp. 51–65.

[20] M. Đurić: Elements of Power Systems, Beopres, Belgrade, 2005.

[21] G. Dotlić: Power Systems Rulebooks, SMEITS, Belgrade, 2007, pp. 75–138.

[22] Modeling and Simulation of the Power Plants and Power Systems in the Balkan Region under a new Technological and Market Environment, Faculty of Electrical Engineering, Rostock, FR Germany, Faculty of Electrical Engineering, Belgrade, Serbia, 2006-2007, (Manager Z. Stojković).

[23] M. Lovrić, Z. Stojković: Modeling the Turbine Regulator of the "Kokin Brod" Hydroelectric Power Plant Using Matlab® / Simulink®, Faculty of Electrical Engineering, Belgrade, 2007.

[24] D. Stojanović, Z. Stojković: Modeling the Hydraulic and Mechanical Parts of the "Kokin Brod" Hydroelectric Power Plant Using Matlab® / Simulink®, Faculty of Electrical Engineering, Belgrade, 2007.

[25] M. Đurić: Regulation of Power Systems, Faculty of Electrical Engineering, Belgrade, BEOPRES, Belgrade, 2004.

[26] M.S. Ćalović: Regulation of Power Systems, Volume 1: Regulation of Frequency and Active Powers and Volume 2: Regulation of Voltage and Reactive Powers, Faculty of Electrical Engineering, University of Belgrade, Belgrade, 1997.

[27] Electronic Governor MIPREG DGC 600c – Service Manual.

[28] L. Dube: MODELS in ATP, Language Manual, Feb 1996.

[29] L. Prikler, H. Kr. Hoidalen: ATPDraw™ for Windows 5.6, Users' Manual, Nov 2009, p. 270.

[30] C.M. Wiggins, S.E. Wright: Switching Transient Fields in Substations, IEEE Transactions on Power Delivery, Vol.6, No.2, April 1991, pp. 591-599.

[31] D.E. Thomas, C.M. Wiggins, T.M. Salas, F.S. Nickel, S.E. Wright: Induced Transients in Substation Cables: Measurements and Models, IEEE Transactions on Power Delivery, Vol.9, No.4, Oct 1994, pp. 1861-1867.

[32] C.M. Wiggins, D.E. Thomas, F.S.Nickel, S.E. Wright: Transient Electromagnetic Interference in Substations, IEEE Transactions on Power Delivery, Vol.9, No.4, Oct 1994, pp. 1869-1881.

[33] H. Bauer, M. Claus, J. Gorablenkov, G. Hentschel, B. Suermann: Zur Storbelastung und EMV der Leit- und Schutztechnik von $SF_6$ – isolierten Schaltanlagen, Elektrizitatswirschaft, Jg.94, 1995, Heft 9, S. 498-503.

[34] W. Kohler, T. Dischinger, U. Scharli: Measurement of Fast Transient in HV Substations and their Effects on Secondary Equipment, Proceedings EMC Symposium Zurich, 1993, pp. 365-370.

[35] Elaboration No. 89/IU-003: Calculation of Switsching Overvoltages and Growth of Potential of the Metal Enclosure of the 400 kV $SF_6$ Gas Insulated Switchgear (Višegrad), Institute for Power Systems "Energoinvest" Sarajevo, Sarajevo, 1989.

[36] Study No. 319903: Analysis of Effectiveness of Overvoltage Protection in "Višegrad" HPP, Institute of Electrical Engineering "Nikola Tesla" Belgrade, Belgrade, 1999.

[37] M. D'Amore, M. S. Sarto : Modelling of Lossy Ground Parameters in the EMTP for Very-Fast Transient Analysis, IPST'97 – International Conference on Power Systems Transients, Sietle, June 1997, pp. 49-54.

[38] A. Eriksson, K.G. Petterson, A. Krenicky, R. Baker, J.R. Ochoa, A. Leibold: Experience with Gas Insulated Substations in the USA, IEEE Transactions on Power Delivery, Vol.10, No.1, Jan 1995, pp. 210-216.

[39] V.V. Kumar, J. Thomas, M.S. Naidu: Influence of Switching Conditions on the VFTO Magnitudes in a GIS, IEEE Transactions on Power Delivery, Vol.16, No.4, Oct 2001, pp. 539 – 544.

[40] A. Ametani, N. Nagaoka, N. Mori, K. Shimizu: Switching Overvoltages on a Pipe in a Gas - insulated Substation, IPST'97 – International Conference on Power Systems Transients, Seattle, June 1997, pp. 286-291.

[41] T. Irwin, J. Lopez-Roldan: Substation Earthing: Special Considerations for GIS Substations; Substation Earthing, IEE Seminar on, 2000, Ref. No. 2000/033, pp. 5/1-5/5.

[42] A.M. Miri, Z. Stojković: Transient Electromagnetic Phenomena in the Secondary Circuits of Voltage- and Current Transformers in GIS (Measurements and Calculations), IEEE Transactions on Power Delivery,

Vol.16, No.4, Oct 2001, pp. 571 – 575.

[43] A. M. Miri, Z. Stojković: Transient Electromagnetic Phenomena in the Secondary Circuits of Measuring Transformers in GIS, Elektroprivreda, No. 4, 1999, pp. 61-66.

[44] Z. Stojković, A. Miri, G. Mitrić: Estimation of Metal Clad Gas $SF_6$ Insulated Substation Potential Pickup Caused by Switching Disconnector Operations, Elektroprivreda, No.2, 2003, pp. 54-59.

[45] Z. Stojković, A. M. Miri, G. Mitrić: Estimation of Metal Clad Gas $SF_6$ Insulated Substation Potential Pickup Caused by Switching Disconnector Operations, 26th JUKO CIGRE Conference, R33-06, Teslić, May 2003.

[46] CIGRE Working Group 33.02: Guidelines for Representation of Network Elements when Calculating Transients (Internal Overvoltages), Paris, 1991.

[47] D. Povh, H. Schmitt, O. Volcker, R. Witzmann: Modeling and Analysis Guidelines for Very Fast Transients, IEEE Transactions on Power Delivery, Vol.11, No. 4, Oct 1996, pp. 2028-2035.

[48] Katalog Siemens Badenwerk, Umspannanlage Oberwald.

[49] L. Mahserdijian, M. Landry, B. Khodabakhchian: The new EMTP Breaker Arc Model, IPST '97 – International Conference on Power Systems Transients, Sietle, June 1997, pp. 245-249.

[50] 420 kV $SF_6$ Switchgear Catalog, Energoinvest, Sarajevo.

[51] Z. Stojković: Tempus Project "Power System Platforms for Steady State and Transients Calculations", thema "Alternative Transients Program (ATP) Features", Faculty of Electrical Engineering, Belgrade, May 11th to May 14th, 2009.

[52] P. Blattner, L. Urlich, K. Cook, T. Dysk: Microsoft Excel 2000 Guide (Special Edition), CET, Belgrade, 2000.

[53] J. Walkenbach: Microsoft Office Excel 2007, Mikro Knjiga, Belgrade.

[54] P. Litwin, K. Getz, M. Gilbert: Access 2000 Handbook for Programmers, Mikro Knjiga, Belgrade, 2001.

[55] S. Roman: Access Database Design & Programming, 3rd Edition, O'Reilly, Cambridge, 2002

[56] S. Roman: Writing Excel Macros with VBA, Second Edition, O'Reilly, Cambridge, 2002

[57] P.G. Mc Keown, C.A. Piercy: Learning to Program with Visual Basic, Second Edition, John Wiley & Sons Inc., New York, 2002

[58] Estimation of the Value of Telecommunications Equipment for Telekom Srbija a.d., Faculty of Electrical Engineering, Belgrade, customer Telekom "Srbija" a.d., 2005. (Z. Stojković - participant in Study).

[59] Z. Stojković, D. Medan, M. Nikolić, Ž. Stankić, S. Oparnica: Some Aspects of AutoCAD Program Application in Computer-Aided Design, Elektroprivreda No. 4, 2003, pp. 18-26.

[60] C. Fleischhaner: Excel in Naturwissenschaft und Technik, Grundlagen und Anwendung, 2. uberarbeitete und erweiterte Auflage, Addison-Wesley, Munich, 2000.

[61] K. Redžić, Z. Stojković: Designing Power Substations Using Macros in Excel, Faculty of Electrical Engineering, Belgrade, 2008.

[62] Instructions for Designing Power Systems and Installations for Supplying Power to Base Stations of Mobile and Wireless Systems, customer Republic Agency for Telecommunications - RATEL, Belgrade, 2007. (Group Manager Z. Stojković).

[63] I. Jovanov, Z. Stojković: Application of Excel in Designing Power Supply of Telecommunications Equipment, Faculty of Electrical Engineering, Belgrade, 2007.

[64] Main Project for Direct Voltage Power Supply of Devices for Expansion 2 of the Control-Commutation Center "Belgrade" GSM Network of Mobile Telecommunications of Serbia Company "Telekom Srbija" a.d., Faculty of Electrical Engineering, Belgrade, June 2007, (Chief of Project Z. Stojković).

[65] Main Project for Power Supply of the Control-Commutation Center "Belgrade" GSM/UMTS Network "VIP MOBILE" (a part of the Mobilkom Austria Group) – phase 1, Faculty of Electrical Engineering, Belgrade, June 2007, (Chief of Project Z. Stojković).

[66] Main Project for Construction of the "Podgorica" Control-Commutation Center for the GSM/UMTS Network for the Telecommunications Company "MTEL" D.O.O. (Power Supply of Devices), Faculty of Electrical Engineering, Belgrade, June 2007, (Chief of Project Z. Stojković).

[67] Z. Stojković, A. Jovanović: Software Tool for Automation in the Design of Power Supply for Telecommunications Equipment, Faculty of Electrical Engineering, Belgrade, 2009.

[68] User's Guide AutoCAD 2011, Autodesk, 2010.

[69] E. Finkelstein: AutoCAD 2002 Bible, Mikro knjiga, Belgrade, 2002.

[70] H. J. Engelke: 3D – Konstruktion mit AutoCAD 2002, Volumen-Modellieren fur Einsteiger, Hanser Verlag Munchen Wien, 2002.

[71] ANSYS 12.1 – Technical Overwiew, ANSYS Inc, USA, 2010.

[72] D. Medan, Z. Stojković: Three-Dimensional Modeling of Power System Structures Using AutoCAD, Faculty of Electrical Engineering, University of Belgrade, 2004.

[73] Z. Stojković: Check of High-Voltage Equipment According to Power of Short Circuit in 2000 Bajina Bašta HPP, Work for Expert Exam in the Field of Electrical Engineering, Belgrade, 1991.

[74] P. Hasse: Overvoltage Protection of Low Voltage Systems, IEE Power Series 12, Peter Peregrinus Ltd., London, 1992.

[75] IEC 1024-1-1 Protection of Structures Against Lightning; Part 1: General Principles, Section 1: Guide A – Selection of Protection Levels for Lightning Protection Systems, 1993.

[76] BS 6651 Code of Practice for Protection of Structures Against Lightning, 1999.

[77] IEEE Standard 998: Guide for Direct Stroke Shielding of Substations, 1996.

[78] French National Standard NF C 17-102: Protection of Structures and Open Areas Against Lightning Using Early Streamer Emission Air Terminals, July 1995.

[79] A.R. Hileman: Insulation Coordination for Power Systems, Marcel Decker, Inc, New York-Basel, 1999.

[80] A. Haddad, D.F. Warne: Advances in High Voltage Engineering, IEE Publishing London, 2004.

[81] Lightning Protection Installations. Lightning Conductors with Circular Ring, SRPS N.B4.811, 1996.

[82] A. M. Mousa: Scientists Oppose Early Streamer Air Terminals, 1999, National Lightning Safety Institute, 1999, http://www.lightningsafety.com/nlsi_lhm/ charge_transfer_opp.html

[83] Z. Stojković, D. Medan: Software Tool for Lightning Protection Design of General and Special Buildings, International Scientific-Trade Symposium INFOTEH-JAHORINA 2005, Vol. 4, Ref. D-11, March 2005, pp. 186-190.

[84] Z. Stojković, Ž. Stankić: AutoCAD-Based Concept for Estimating Lightning Protection Zone of Transmission Lines and Structures, International Journal of Electrical Engineering Education (IJEEE), Vol. 43, No. 4, pp.299-317, Oct 2006.

[85] Z. Stojković: Evaluation of Lightning Protection Zone using AutoCAD-Based Software Tool, Institute of Power Transmission and High Voltage Technology, University of Stuttgart, FR Germany, Annual Report 2006, pp. 64-67.

[86] Z. Stojković, Ž. Stankić: Lightning Protection Design of General and Special Buildings, Elektroprivreda, No.3, 2005, pp. 84-91.

[87] Z. Stojković, Ž. Stankić: Lightning Protection Design of General and

Special Buildings, 27th JUKO-CIGRE Conference, RC4-01, Zlatibor, May 29th – June 3rd, 2005.

[88] G. Omura: AutoCAD 2002, SYBEX, Inc, San Francisco, 2003.

[89] G. Omura: The ABC's of AutoLISP, SYBEX, Inc, San Francisco, 2001.

[90] Z. Stojković, A.Grujić, S. Tenbohlen: Lightning Protection Design of Substations and Transmission Lines, 28th JUKO-CIGRE Conference, RC4-01, Vrnjačka Banja, September 30th – October 5th, 2007.

[91] M. Savić, Z. Stojković: High-Voltage Technique – Lightning Overvoltages, Monography, II corrected and amended edition, Faculty of Electrical Engineering, Belgrade, 2001.

[92] IEEE Working Group on Lightning Performance of Transmission Lines: A Simplified Method for Estimating the Lightning Performance of Transmission Lines, IEEE Trans. on PA&S, Apr. 1985, pp.919-932.

[93] A.J. Eriksson: An Improved Electrogeometric Model for Transmission Line Shielding Analysis, IEEE Transactions on Power Delivery, July 1987, pp. 871-886.

[94] G. Baldo: Lightning Protection and the Physics of Discharge, High Voltage Engineering Symposium, London, No. 467, Vol. 2, pp. 2.169.S0-2.176.S0, 1999.

# 3. APPLICATION OF SOFTWARE TOOLS IN DRAFTING GRAPHICAL DOCUMENTATION

## 3.1. Introduction

The basic content of electrotechnical documentation is displayed in Fig. 3.1 [1].

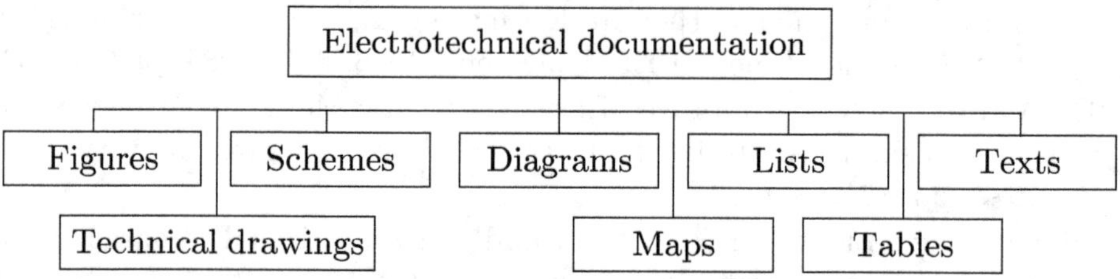


Fig. 3.1 – Basic content of electrotechnical documentation

When drafting electrotechnical documentation, it is necessary to observe the following requirements:

- contents and form must be adapted to practical application,
- displays, descriptions and pictures must be clear and understandable,
- texts must be concisely and adequately formulated,
- markings must enable quick identification of operating devices,
- revision and expansion of documentation must be made possible.

With computer supported documentation it is necessary to:

- provide a transition between all documents and supplements,
- apply standardized information formats and collections of symbols,
- use graphical symbols, markings and information formats according to industry standards,

Z. Stojković, *Computer-Aided Design in Power Engineering*,
DOI: 10.1007/978-3-642-30206-0_3, © Springer-Verlag Berlin Heidelberg 2012

– use standardized formats for databases and corresponding graphics,
– enable fast addition and modification of data.

The graphical documentation is a part of the electrotechnical documentation and represents an integral component of the design. This documentation consists of [2]:

– the situation,
– installation plans,
– dispositions,
– single-pole diagrams,
– wiring diagrams,
– connection diagrams,
– details for carrying out electrical installation works.

In this chapter the elements of graphical documentation are considered. Descriptions are provided for the most commonly used types of electrical diagrams along with examples of their formation. Electrotechnical graphic symbols and markings for devices, and connections within diagrams are displayed. The elements of the program EPLAN, which enables the efficient formation of wiring diagrams and connection diagrams, are also considered. Application of the new version (R.6.1) of the program SIMARIS SIVACON for design and installation of 8PT SIEMENS SIVACON low-voltage switchgear is displayed in the example of a transformer station with a single-pole diagram created using the program SIMARIS DESIGN.

## 3.2. Graphical documentation

### 3.2.1. Electrotechnical graphic symbols

Electrotechnical graphic symbols facilitate the efficient display of all parts of an electrical design, beginning from global power plans to designing individual functional structures. The use of electrotechnical graphic symbols has been established within all national and international standards [2-4]. It is necessary to note that standards give several variants for individual symbols, so there is no unique and generally accepted list of graphic symbols. Table 3.1 provides commonly used electrotechnical graphic symbols and their definitions [2].

Table 3.1 – Electrotechnical graphic symbols

| Symbol | Definition |
|---|---|
| | One conductor or a group of conductors |
| | Separator |
| | Three conductors or three elements of an electrical circuit |
| | Grounding |
| | Transformer |
| | AC generator |
| | Asynchronous motor with caged rotor. General symbol |
| | Sliding-ring asynchronous motor. General symbol |
| | Capacitor |
| | Coil (restrictor) |
| | Resistor |
| | Fuse (D type) |
| | Fuse with switches (N type) |
| | Switch disconnector |
| | Current transformer |
| | Voltage transformer |
| | Surge arrester |
| | Arc gap |

| Symbol | Definition |
|---|---|
| | Coil relay. General symbol |
| | Coil relay with slowed release |
| | Coil relay with slowed excitement |
| | Closing contact with opening delay |
| | Normally open (NO) contact. General symbol |
| | Normally closed (NC) contact. General symbol |
| | Turn switch contact (contact opens before closing) |
| | Measuring converters x-size |
| | Thermal element |
| | Key |
| | Signaling lamp |
| | Display instrument (e.g. ammeter) |
| | Rotary switch |
| | Bimetal relay |
| | Installation automatic circuit breaker |
| | Centrifugal switch |
| | Electrical heating device. General symbol |
| | Variability. General symbol |

### 3.2.2. Types of electrical diagrams

Electrical diagrams display the way in which various parts of a grid, substation, group of devices or device are interconnected, as well as what type of functional interconnected relation they have. It is a basic rule that electrical diagrams are displayed in a no-voltage state for relay and contactor coils, and switches and circuit breakers in a shut-off state.

During design, the following types of electrical diagrams are used [2]:

– dispositional diagram (abbreviated disposition),
– principal (block) diagram,
– single-pole diagram,
– multiple-pole (e.g. three-pole) diagram,
– wiring diagram,
– connection plan (or diagram) of linear terminals (installation diagram or connection diagram),
– equivalent diagram,
– logical diagram.

Short descriptions of the individual types of electrical diagrams are provided below [2]. The disposition represents a display of the placement of the parts of a device at the installation location, such as connectors, plug-in contacts, groups of devices, etc. The disposition does not have to be drawn to scale. Fig. 3.2 shows the disposition of the 110 kV substation of the 110 kV / 35 kV "Vreoci" transformer station [5]. Disposition diagrams are most commonly formed using the program AutoCAD [6-8].

A principal (block) diagram is a simplified display of a substation for a simple understanding of operation. This diagram, through symbols, provides a display of the most important parts of the substation or one part of the substation with functional interconnections. This diagram also displays the division of the substation according to function and space.

A single-pole diagram is displayed through one line (one, two or several conductors). A single-pole diagram is especially useful for displaying the conductors of a multiple-phase system, conductors with similar electrical functions, parallel conductors and conductors with the same physical route or identical spatial location within the diagram. In a single-pole display, several identical elements can be shown by one symbol.

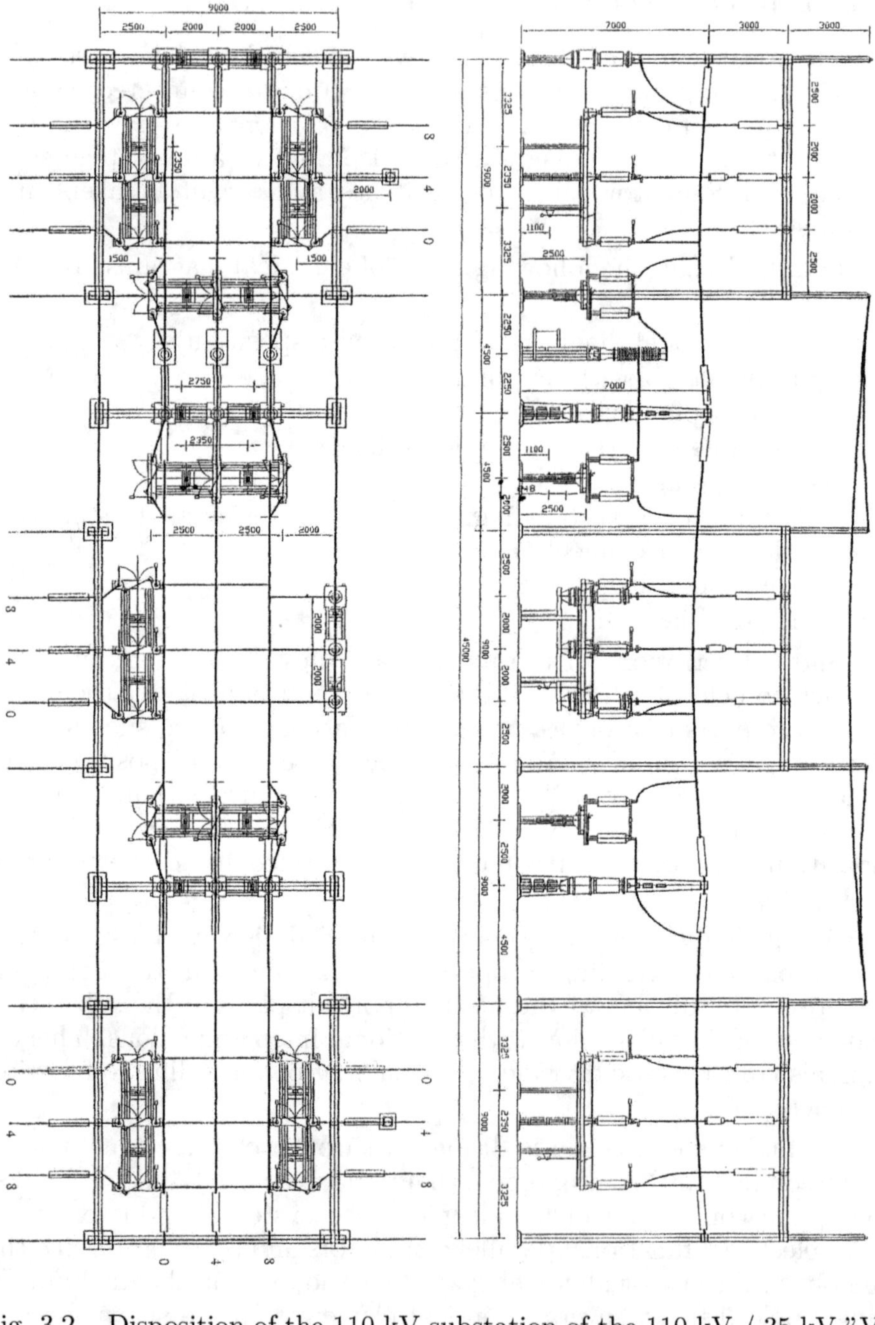

Fig. 3.2 - Disposition of the 110 kV substation of the 110 kV / 35 kV "Vreoci" transformer station

The single-pole diagram is an important document which is required in several rulebooks [2, 9]. The content of a single-pole diagram has not been established (prescribed) by a single valid document, but rather its viewability is based on the requirements of the investor. Generally speaking, the content of a single-pole diagram depends on the voltage level of the power system structure.

Single-pole diagrams, by rule, must contain the following data [2]:

a) Content of a single-pole diagram in low-voltage electrical installations
   - nominal voltage, frequency and power (installed and peak) of the main distribution cabinet and other distribution cabinets,
   - cross section and material for busbars (or type of linear terminal if a bus is not used),
   - symbol (number) of all current circuits- symbol of all installed devices of all electrical circuits (circuit breakers, switches and fuses) with nominal values,
   - cross section and material of conductor or type of cable for all current circuits,
   - power of each current circuit (motor, transformer, rectifier, connector, light source, etc.) with basic data (e.g. number of phases, etc.),
   - applied measures of protection from excessive touch voltage.

b) Content of a single-pole diagram in high-voltage electrical installations
   - nominal voltage, frequency and power (current) of short circuit on busbars,
   - type of busbar, material and cross section,
   - symbols of all cells and bays,
   - basic rated values of high-voltage devices (circuit breakers, disconnectors, measurement transformers, etc.),
   - number and rated power of generators, transformers, etc. with basic data,
   - symbol of the type of protective device (with basic technical data),
   - elements of signalization and control (local and/or remote),
   - type, cross section and material of cables (conductors) in cells (bays).

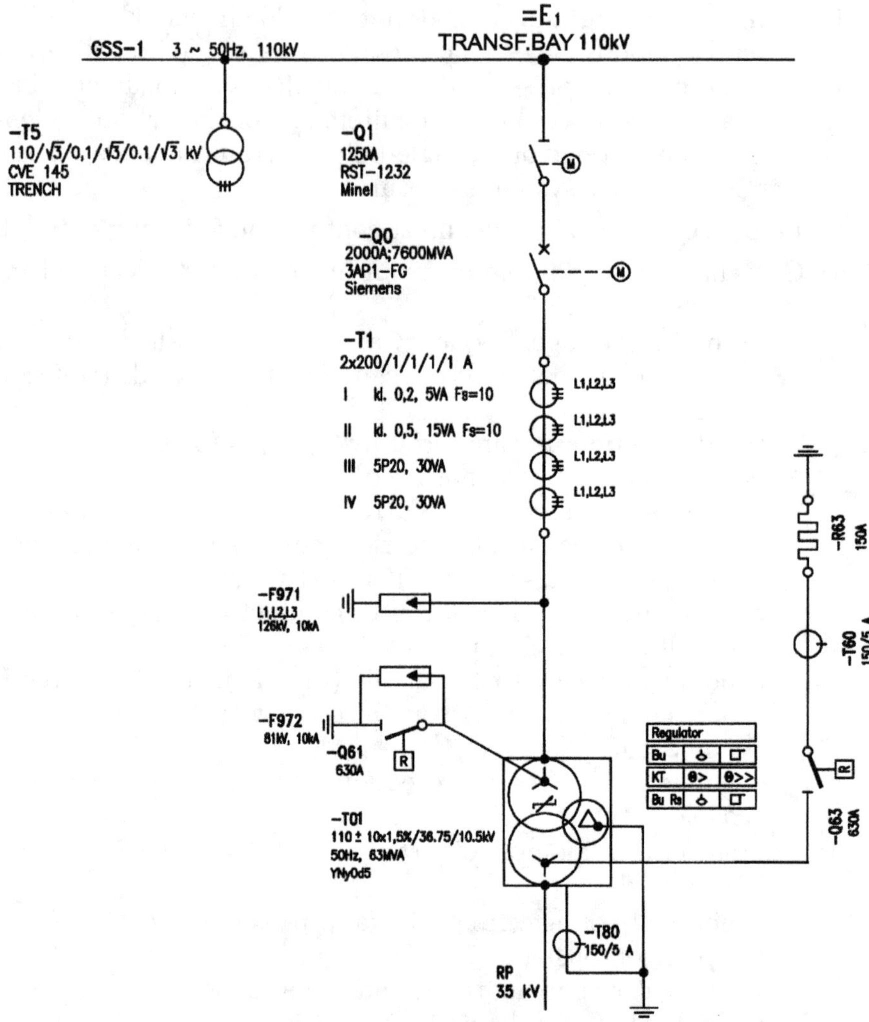


Fig. 3.3 - Single-pole diagram of the 110 kV transformer bay, No. 1 of the 110 kV / 35 kV "Vreoci" transformer station

An example of a single-pole diagram of the 110 kV transformer bay of the 110 kV /35 kV "Vreoci" transformer station is displayed in Fig. 3.3 [5].

Along with single-pole diagrams, so-called functional diagrams are also commonly added. These diagrams must contain all details about the functional operation of the system, installation of devices, software, etc., without data on the construction.

The following rules must be applied for functional diagrams:

- relays, contactors, equipment, brakes and junctions must be displayed in an inactive state. If not, active state must be indicated in the diagram,
- the function of circuit breakers must be explained from several positions, for example by diagram,
- semiconductor switches must be shown in an operational-ready state,
- dependencies of non-electric and electric functions must be displayed in such a way that they are clearly recognizable,
- basic connections must be displayed uniformly, for example, connections of bridges, star-delta starters, bistable switch stages, etc.,
- necessary symbols for functions, signalization and control must be drawn,
- if necessary, provide waveshapes, formulas and algorithms.

Multiple-pole diagrams display each conductor through a separate line and each element through one symbol. Multiple-pole diagrams are used less often because they make the display of complicated functions confusing.

A wiring diagram is a detailed display of the behavior of a substation or part of a substation including all details. Using symbols, all electrical connections are displayed along with other connections with other parts of the substation which are important for functioning. When forming a wiring diagram it is necessary to abide by the following rules:

- graphic symbols for displaying operating devices and functions must be used,
- all connections between devices and functions must be drawn,
- devices and connectors must be marked,
- logical signals and the level of signals must be coordinated,
- provide necessary data for monitoring current circuits and paths,
- provide information (e.g. about contactors) for the purpose of understanding functions.

A connection plan (or diagram) of linear terminals (connector terminals) is a display of linear terminals of an electrical device or structure which are coupled to internal and external connections. In practice these diagrams are referred to as installation diagrams or connection diagrams and contain the following [1]:

- link between elements, devices, structures and substations,
- markings for connections,
- used cables and conductors,
- method of placement,
- route of conductors,
- used connections,
- method of tightening,
- method of wrapping,
- length of cables or conductors,
- markings for signals,
- classification,
- special data.

The rules for connection diagrams are:

- provide a topographical display (layout), but not to scale,
- display operating devices using simple contours (squares, rectangles or circles) or symbols with connection markings,
- display conductors using a solid line,
- draw groups of conductors, cables or bundled cables with one line,
- mark conductors with markings of operating devices and terminals,
- when possible make the display in the form of a matrix.

The rules for connection tables and lists are:

- display reference code may be connectors or couplings,
- devices are marked with operating device signs,
- mark connectors with numbers, letters, symbols and colors,
- mark conductors with operating device signs, markings, color or terminal markings.

An equivalent diagram is a very simplified display of a current cycle or its pàrt and is used for the purpose of analysis or calculation (e.g. an equivalent diagram of a network for calculating short circuit current). The term equivalent circuit is also commonly used.

A logical diagram is a display of logical functions which must be provided for the operation of individual components of the substation. Logical functions contain at least two conditions connected to the logical relation (e.g. mutual condition of two contacts), except negation for which just one condition is sufficient. Logical diagrams in power engineering are most often used to display blockade conditions. In place of logical diagrams one can use a system of logical conclusion using the so-called Boolean algebra. In one diagram it is allowed to use several of the aforementioned types of diagrams and a behavioral diagram and connection diagram can be displayed on the same drawing.

### 3.2.3. Marking of devices and connections in diagrams

According to [2, 9] the system for marking devices and connections (functionally oriented diagrams) contains:

- a higher degree of upgrades,
- location of installation,
- type, number and function,
- terminal (socket).

The general nomenclature for marking the place of installation and devices has also been recommended and is displayed in Tables 3.2 and 3.3.

The rules for marking electrical equipment are [1]:

- provide only collections of symbols which are necessary for marking,
- do not include unique sign marks,
- the order of collections is arbitrary but the aforementioned order is recommended,
- horizontal and vertical layout of collections is standard,
- a collection of symbols which is mostly valid for the equipment may be provided in a table,
- collections of markings must be entered in directly to the left or below the symbol,
- the marking for a connector is entered in directly outside of the symbol, with the horizontal writing on the right and vertical writing on the left along the jointing line (vertical flow of current circuit) or above the jointing line (horizontal coupling of the current circuit).

Each group of devices has its own system for marking the functions of the devices. For example, some symbols and numbers of relays (general markings of device K) according to standard [3] are provided in Table 3.4. With the function number of the device, letter suffixes can also be used for marking various devices, individual or specific parts of

Table 3.2 – Symbols of location of installation (DIN-40719, part 2)

| Code | Installation location (related to voltage level) | Code | Installation location (is not related to voltage level) |
|---|---|---|---|
| A | - | P | Desk |
| B | Installations for > 400 kV | Q | Measurement panels and cabinets |
| C | Installations for 380 kV – 400 kV | R | Protection panels and cabinets |
| D | Installations for 220 kV – < 380 kV | S | Distribution system panels and cabinets |
| E | Installations for 110 kV – 220 kV | T | Transformer panels and cabinets |
| F | Installations for 60 kV – < 110 kV | U | Management, signalization and auxiliary system panels and cabinets |
| G | Installations for 45 kV – < 60 kV | | |
| H | Installations for 30 kV – < 45 kV | V | Cubicles |
| J | Installations for 20 kV – < 30 kV | W | Command panels |
| K | Installations for 10 kV – < 20 kV | X | Central devices (processing computers and alarm systems) panels and cabinets |
| L | Installations for 6 kV – < 10 kV | | |
| M | Installations for 1 kV – < 6 kV | Y | Telecommunications panels and cabinets |
| N | Installations for < 1 kV | Z | – |

a device, auxiliary contacts of a device, feeders, characteristics or device status, or its contacts in the equipment.

For example, for terminal (-X) number 10 (-X:10) which is located on the fourth relay for verification of condition 4K3, and this relay is located on relay stand W3, according to the IEC, the following combinations of markings are also possible [2]:

– horizontal arrangement:

Table 3.3 – General markings of devices

| Code | Device |
|---|---|
| C | Capacitor |
| E | Various (devices for lighting, heating, etc.) |
| F | Protective device |
| G | Generator (rotating), rotary frequency converters, accumulator batteries, inverters, oscillators |
| H | Optical and acoustic signal devices and relays |
| K | Contactors, relays, etc. |
| L | Inductances |
| M | Machines and motors |
| P | Measuring and testing devices (indicative and registrative), devices (instruments), counters, impulse encoders (converters), clocks |
| Q | Power switch equipment (circuit breakers, disconnectors, etc.) |
| R | Pre-insertion resistors, resistors, potentiometers, starters, adjustors, regulators |
| S | Buttons, command-check switches, multiple switches, packaged switches, etc. |
| T | Power, measuring and auxiliary transformers |
| U | Discriminators, frequency converters with collector, electronic frequency converter, coding devices, converters, inverters |
| V | Diodes, electronic duct, transistors, thyristors |
| X | Terminals, plugs and sockets, linear terminals, socket |
| Y | Unconventional devices driven by electrical supply (brakes, clutches, valves, etc.) |

=W3+4K3-X:10; or
-X:10+4K3=W3;

– vertical arrangement:

=W3
+4K3
-X:10
or
-X:10
+4K3
=W3

Table 3.4 – Numbers of relays (general markings of device K)

| Symbol | | Definition |
|---|---|---|
| IEC[1] | ANSI[2] | |
| 1 | 2 | Pickup – time delay relay |
| $Z <$ | 21 | Distance protection |
| V/Hz | 24 | Overexcitation protection relay, Volts/Hertz relay |
| $\Theta >$ | 26 | Transformer protection (thermal relay, Buchholz relay) |
| $U <$ | 27 | Undervoltage protection |
| $P \leftarrow (I_2)$ | 32 | Reverse power relay |
| $\Phi <$ | 40 | Loss of field protection, field relay |
| $I_{nsc}$ | 46 | Negative (inverse) sequence time overcurrent relay |
| $I$ | 49 | Thermal relay for overload protection |
| $I >, I \gg$ | 50, 51 | Overcurrent or short-circuit relay |
| $U >$ | 59 | Overvoltage relay |
| $U_X >$ | 59N | Zero sequence overvoltage protection |
| $I$ | 64 | Ground detector relay |
| $R_c <$ | 64F | Rotor earth fault protection |
| $I >$ | 67 | Directional overcurrent protection |
| $I$ | 67N | Directional ground overcurrent protection |
| | 74 | Alarm relay |
| O I | 79 | Auto recloser relay, AC reclosing relay |
| $f <, f >$ | 81 | Underfrequency or overfrequency relay |
| $I_d$ | 87 | Differential protective relay |
| $I_d$ | 87N | Differential earth fault protection |
| $U ><$ | 90 | Regulating device |
| I O | 94 | Tripping relay |
| | 95-99 | Relay for specific application in individual installations when no other function number from 1 to 94 is suitable |

Note: [1] IEC - International Electrotechnical Commission, [2] ANSI - American National Standards Institute.

In the lower part of the diagram there is a list of devices which directs the reader of the diagram to quickly find the column and list where the selected contact is located. Depending on whether the contact is on the same list as the coil or on a different list within the same attachment, or within some other attachment, the following combinations are possible.

- (a) contact being considered is located on the same list as the coil, in column (a),
- (b/a) contact being considered is located within the same attachment on list (b) in column (a),
- (c/b/a) contact being considered is located within attachment (c) on list (b) in column (a).

Based on this it can be concluded that the general shape has a form (attachment/list/column), and the following shapes (list/column) are also possible within the same attachment which carries a unique registration number and (column) within the same list and attachment.

The wiring layout makes maintenance of electrical and other installations of devices easier using the wiring diagram only if the order number of the conductor is marked with numbers written at a 90° along the connection. That is a partial-installation diagram according to IEC standard 113-4:1975, which is still often used today because it enables the automatic computerized elaboration of a diagram (for example, automatically generated installation diagram or connection diagram).

A portion of the diagrams which represent an integral part of the project [10] are presented below:

- functional diagram of 110 kV transformer bay, No.1 110 kV / 35 kV "Vreoci" transformer station (Fig. 3.4),
- current circuits of high-voltage equipment of 110 kV transformer bay, No.1 110 kV / 35 kV "Vreoci" transformer station (Fig. 3.5),
- high voltage equipment of transformer No. 1 of 110 kV / 35 kV "Vreoci" transformer station (Fig. 3.6),
- voltage circuits of high-voltage equipment of 110 kV transformer bay, No.1 110 kV / 35 kV "Vreoci" transformer station (Fig. 3.7),
- wiring diagram for circuit breaker Q0 of 110 kV transformer bay, No.1 110 kV / 35 kV "Vreoci" transformer station (Fig. 3.8),
- connection diagram of current transformers (Fig. 3.9),
- connection diagram of voltage transformers (Fig. 3.10),
- cable placement plan (Fig. 3.11).

The diagrams were formed using the program AutoCAD. They can also be drafted using specialized software tools [11-13].

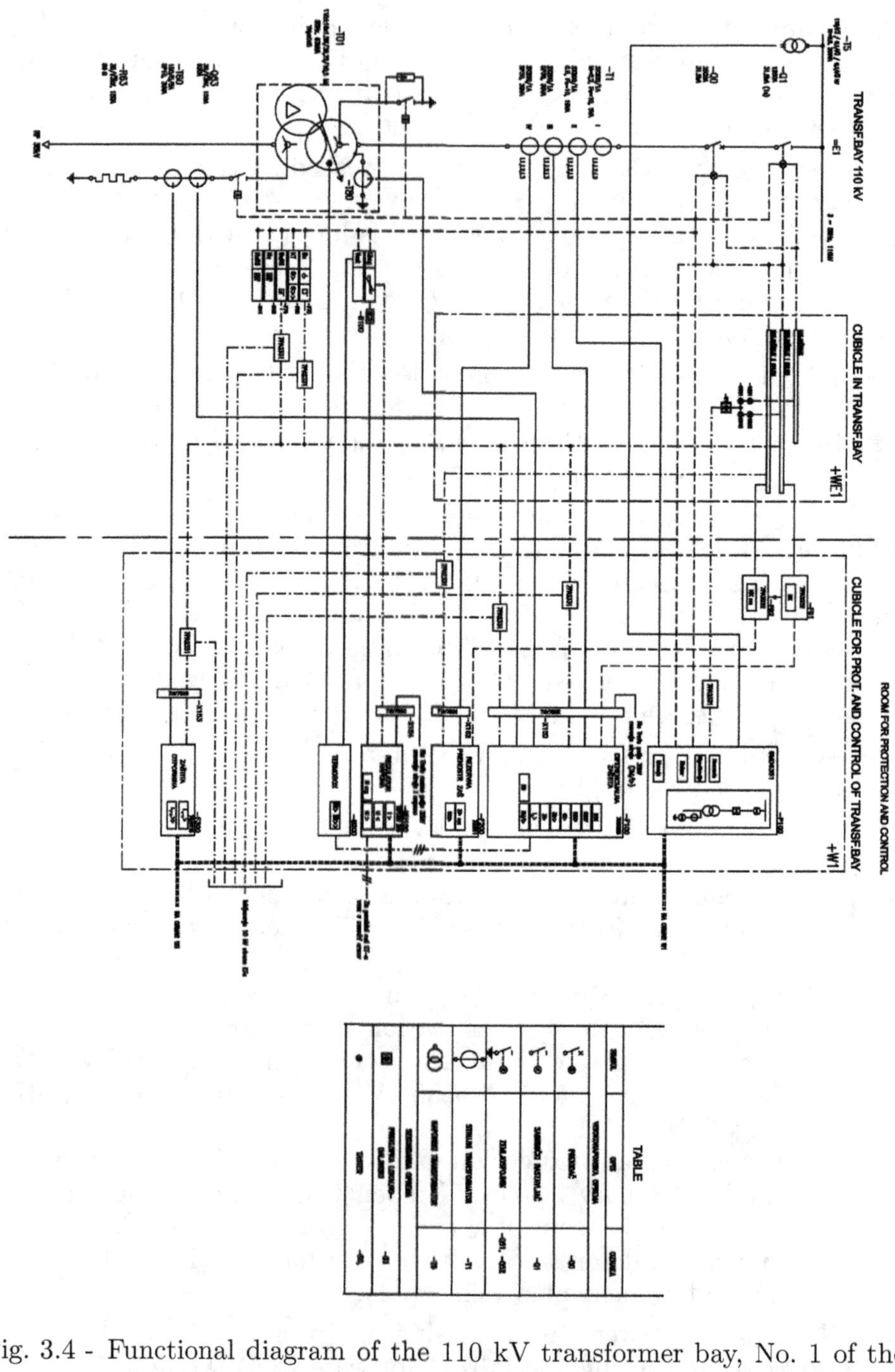


Fig. 3.4 - Functional diagram of the 110 kV transformer bay, No. 1 of the 110 kV / 35 kV "Vreoci" transformer station

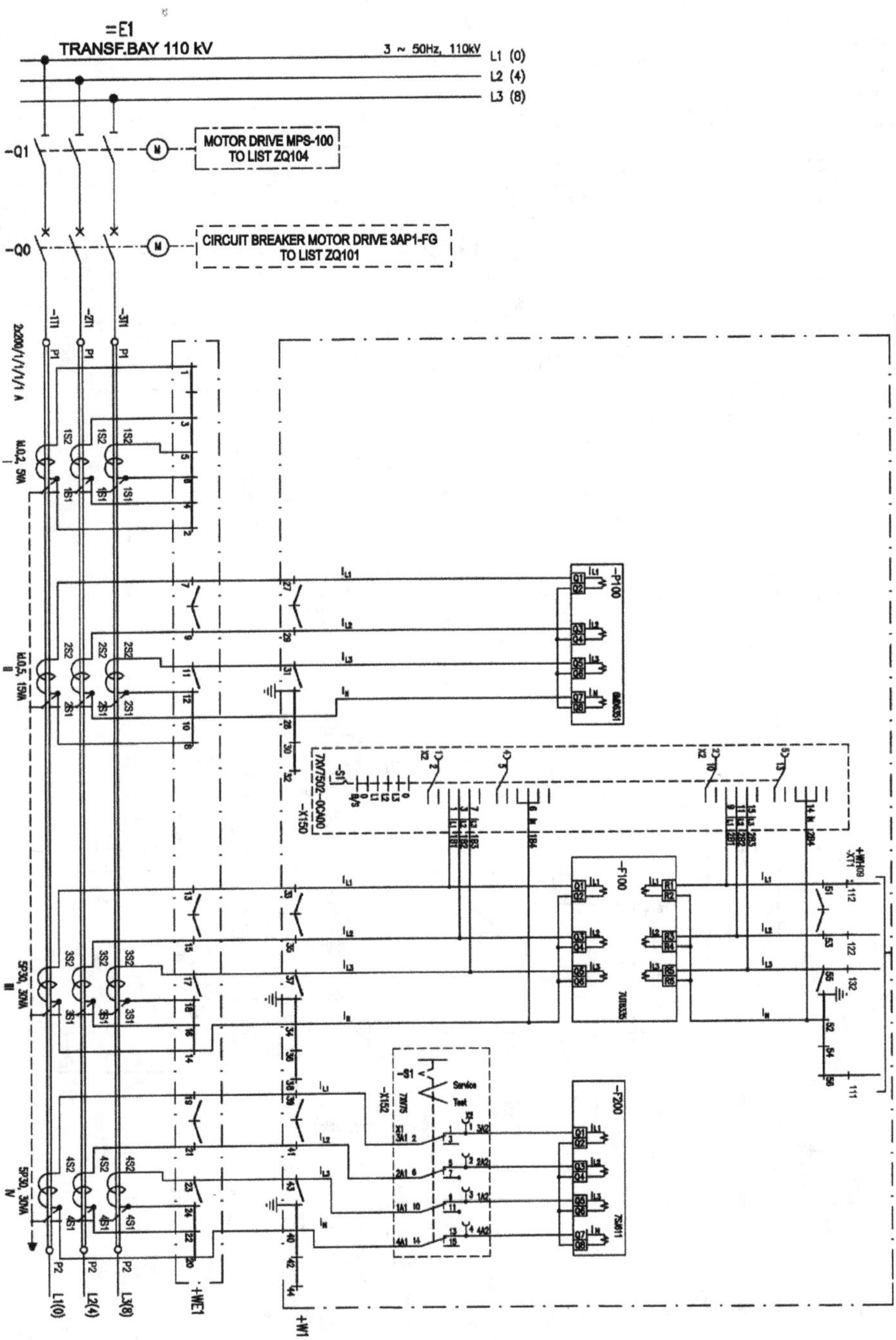


Fig. 3.5 - Current circuits of high-voltage equipment of 110 kV transformer bay, No.1 110 kV / 35 kV "Vreoci" transformer station

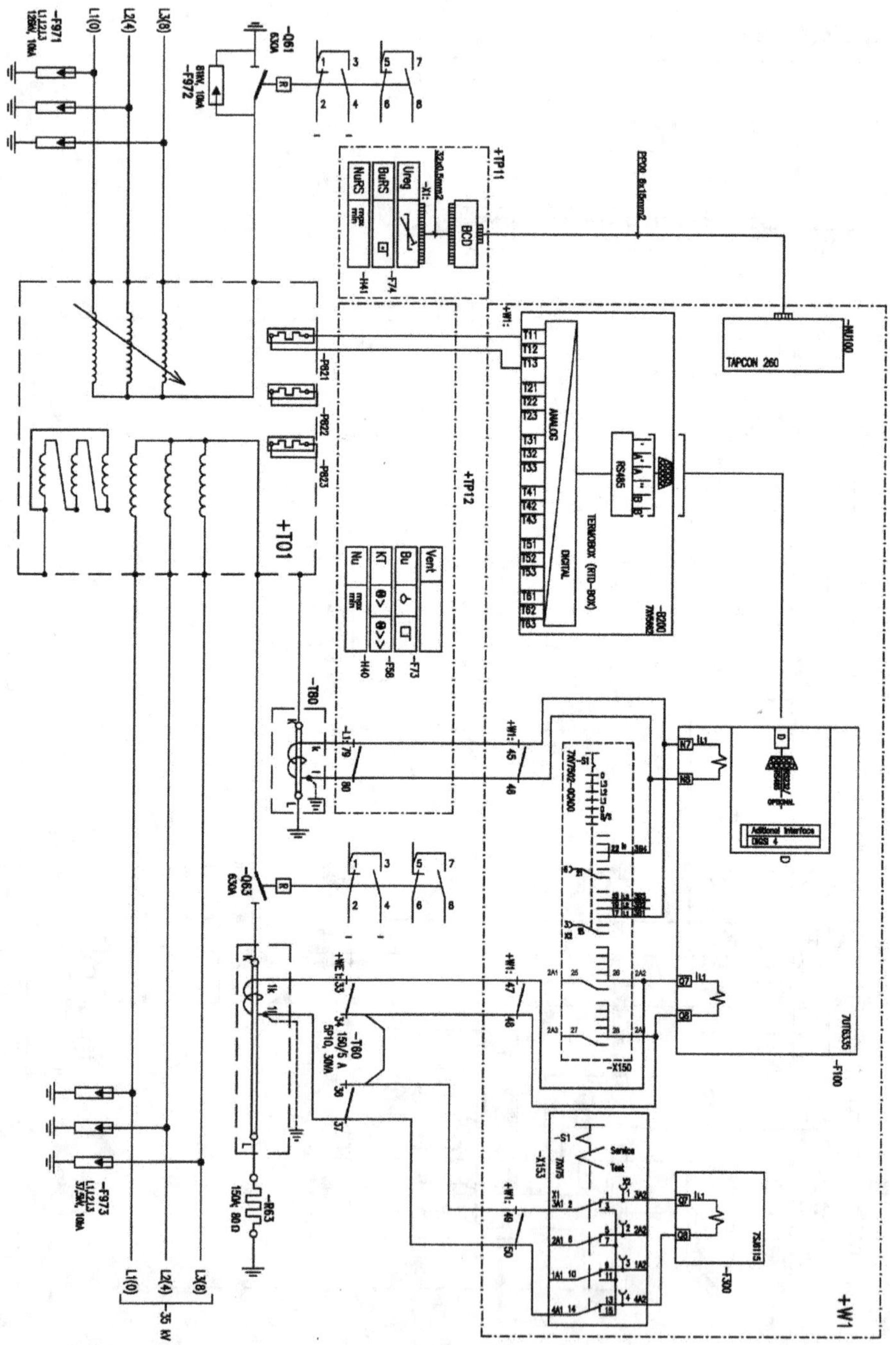


Fig. 3.6 - High voltage equipment of transformer No. 1 of 110 kV / 35 kV "Vreoci" transformer station

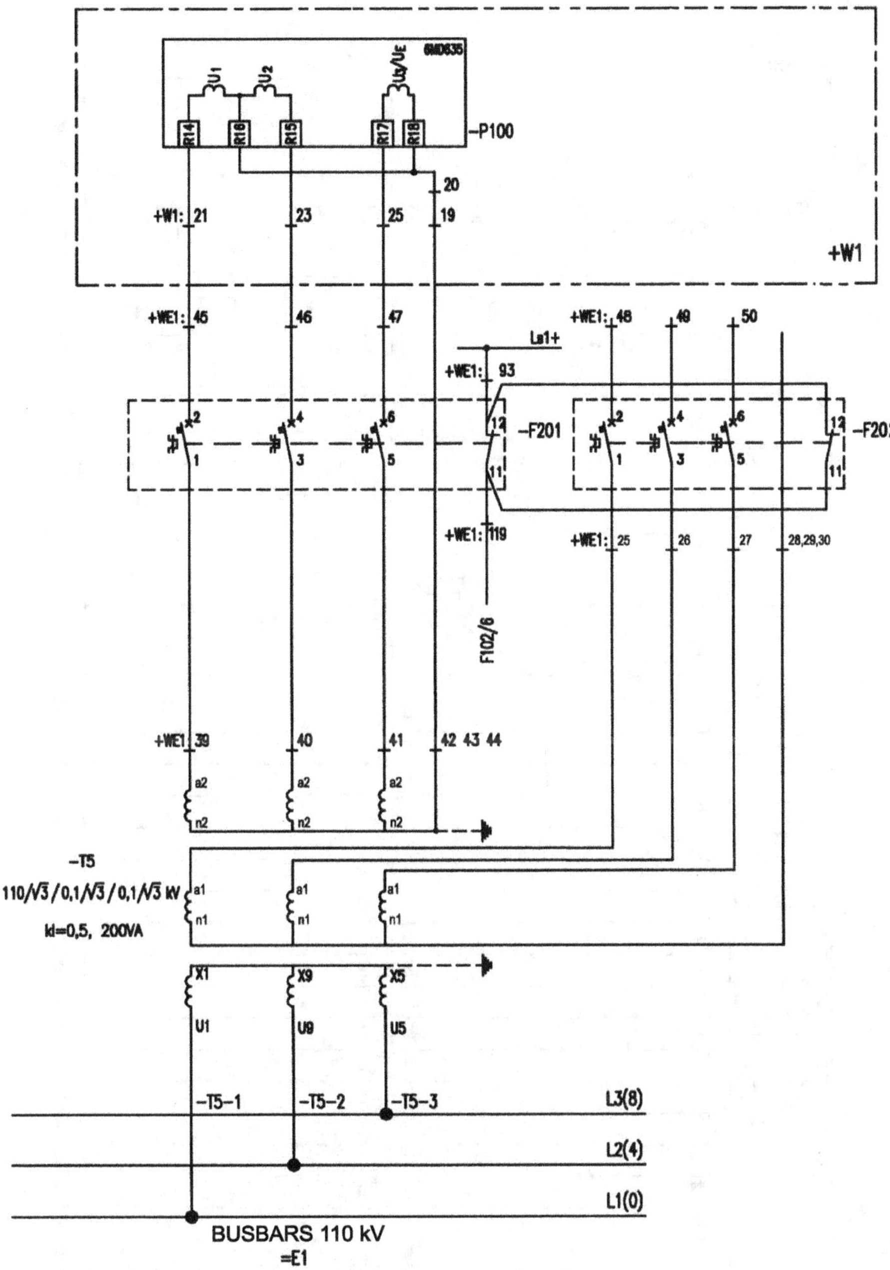


Fig. 3.7 - Voltage circuits of high-voltage equipment of 110 kV transformer bay, No.1 110 kV / 35 kV "Vreoci" transformer station

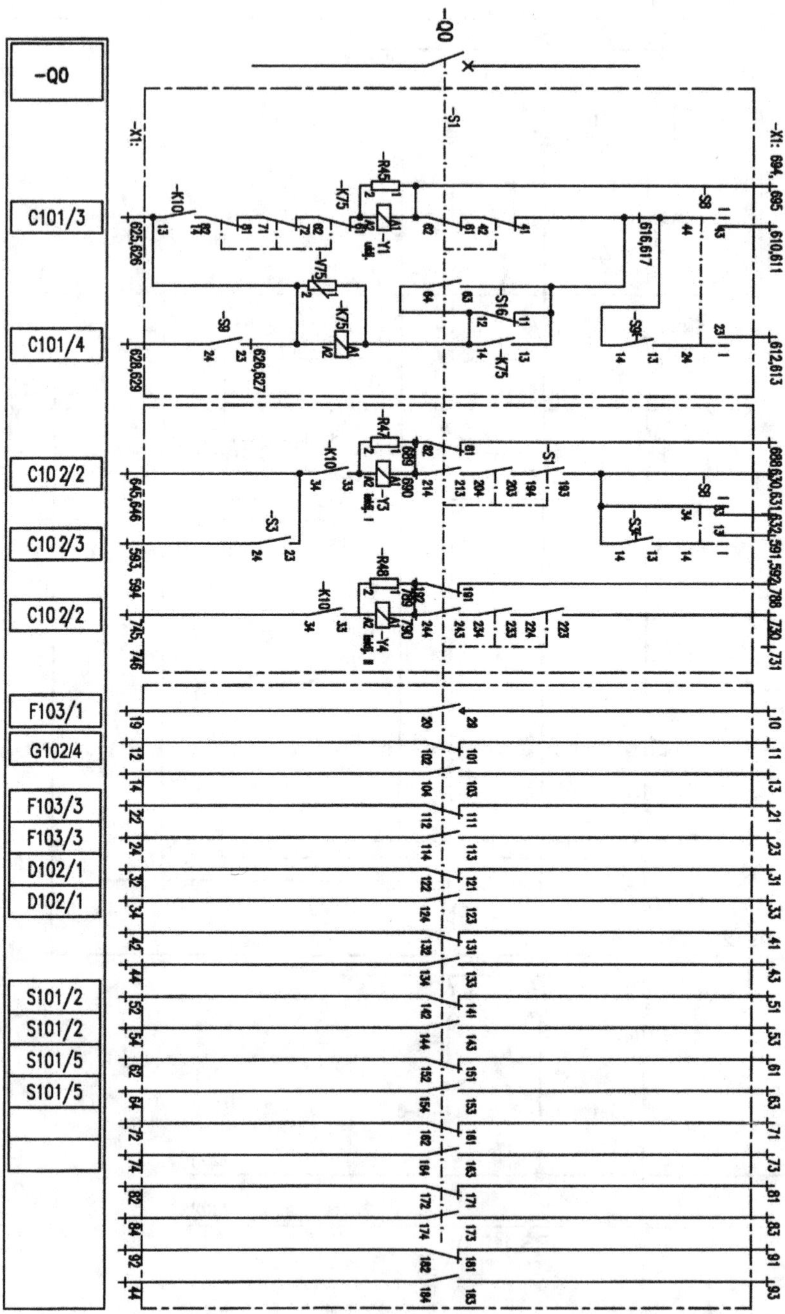


Fig. 3.8 - Wiring diagram for circuit breaker Q0 of 110 kV transformer bay, No.1 110 kV / 35 kV "Vreoci" transformer station

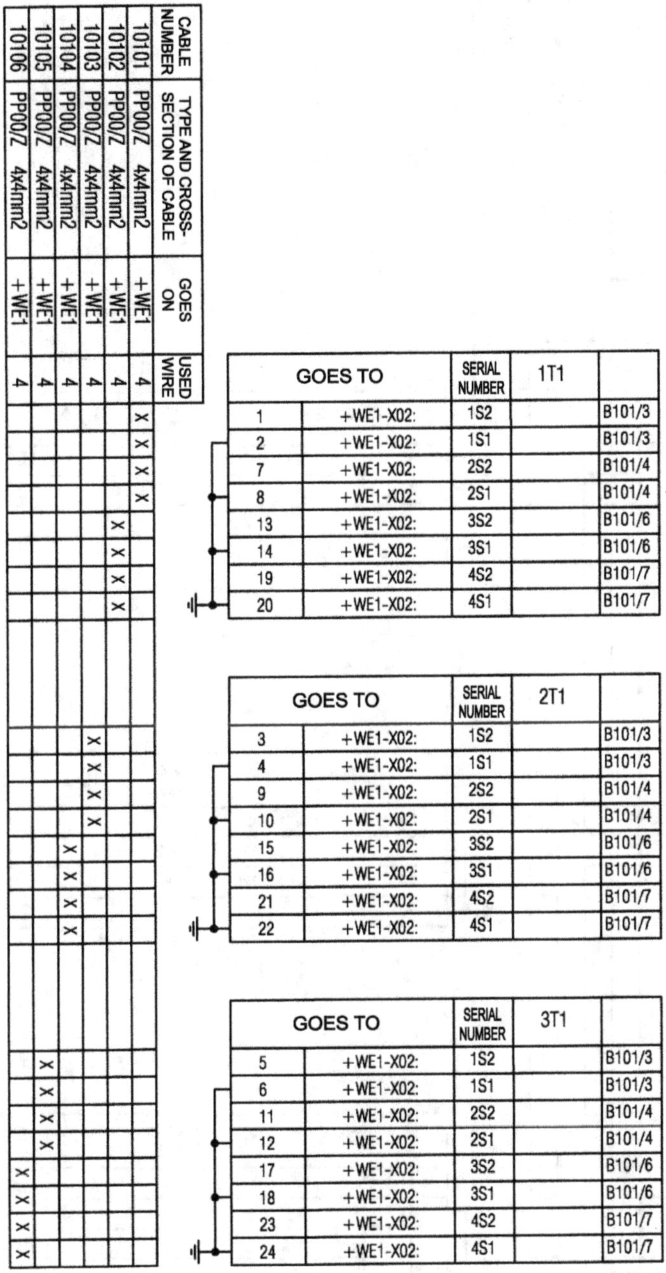


Fig. 3.9 – Connection diagram of current transformers

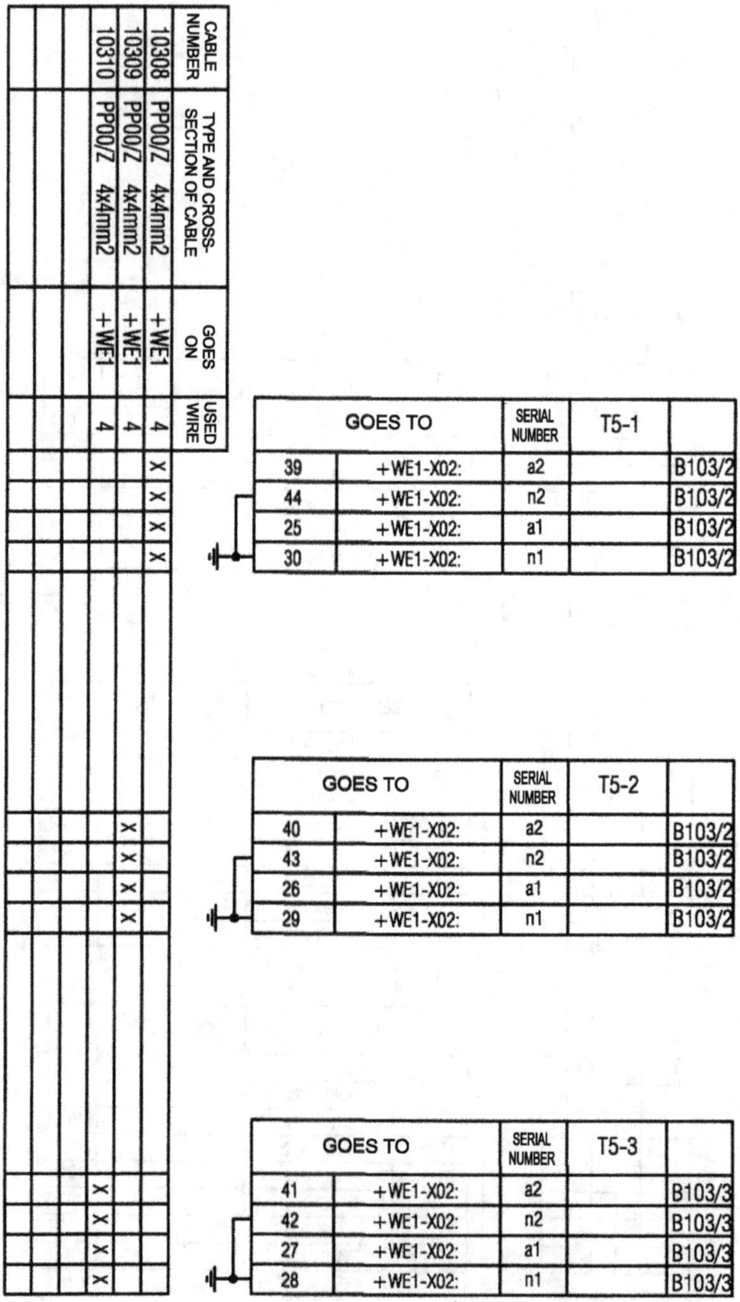


Fig. 3.10 – Connection diagram of voltage transformers

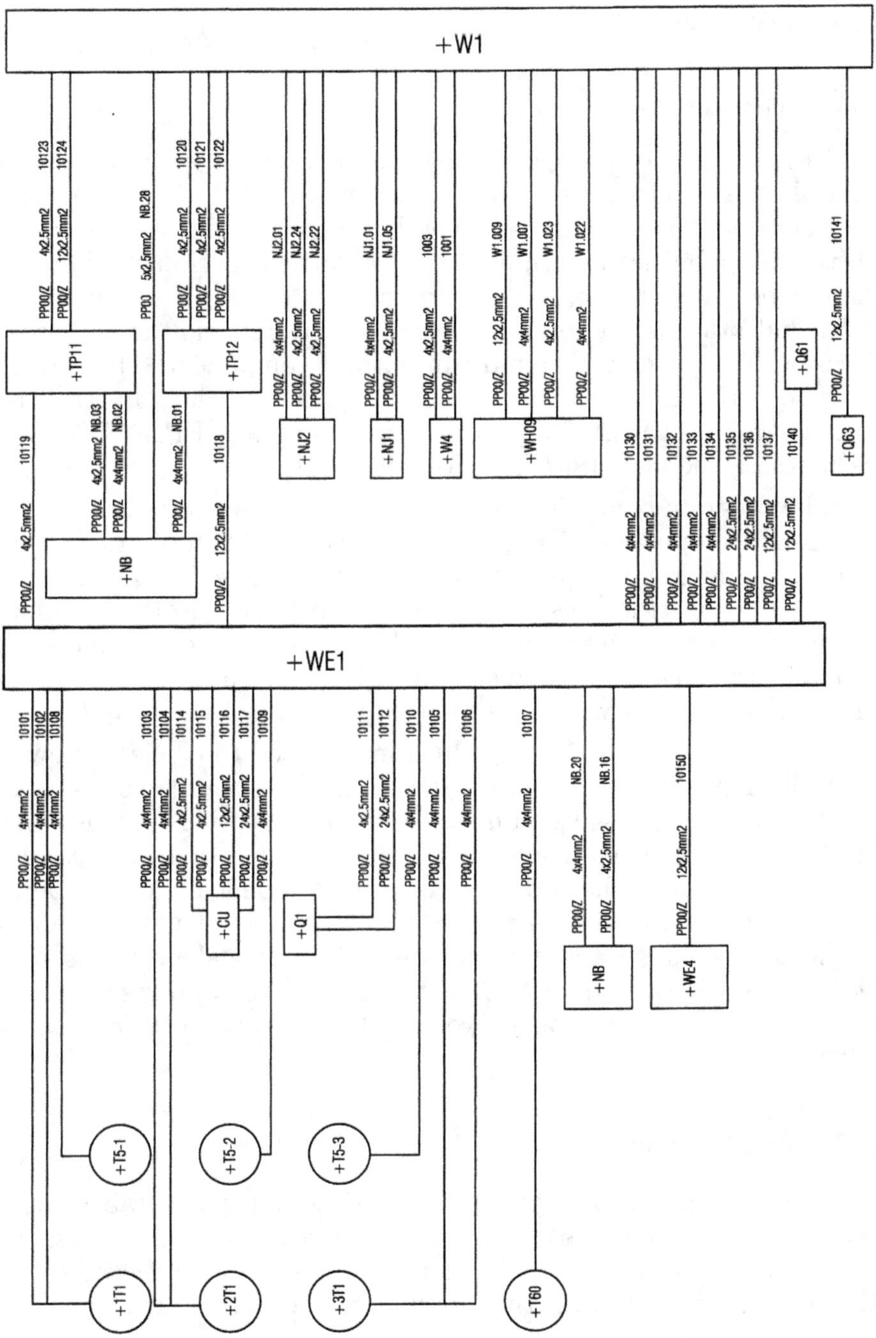


Fig. 3.11 – Cable placement plan

# 3.3. EPLAN

## 3.3.1. Introduction

Development of projects today requires serious effort and use of technical aides which enable one to stay in step with the competition. Since their inception, computers have been used in different ways as a tool, their capabilities determined by the current level of development of hardware and software. Today the power of hardware is greater and greater and prices keep going lower and lower. Selection of the proper software tool is of course the most important component. The software must enable the most efficient use of equipment and human resources. The software tool EPLAN, a product of the company EPLAN Software & Services, is the global leader in this field [11].

The program belongs to the group of software referred to as intelligent CAE (computer-aided engineering) tools. CAD tools are oriented towards specific design requirements where the physical appearance and characteristics of structures are most important. On the other hand, the elements which are used in electrical diagrams have several more hidden parameters and contents which must be systematized in the proper form after completing work on the graphical part of the project.

As opposed to CAD tools which are limited to efficient drawing, CAE tools, for design in power engineering, offer a more expansive and fuller approach in the sense of engineering and creating designs which, aside from the graphical aspect, offer all other information necessary for the realization of the project. Years of experience and capabilities required in the realization of projects have been built into these tools and the built-in logic and functions enable practical problems to be solved in a sophisticated way. CAE tools save time and resources, enable more productive and faster work and ensure competiveness on the market.

## 3.3.2. EPLAN concept

EPLAN is a typical CAE tool. It is, in actuality, a database which holds all physical elements of a project, such as: contactors, relays, terminals, cables, etc. Additionally, the database contains the functional and electrical links between the aforementioned design elements.

The graphical editor is just a tool, an interface through which the database is accessed. All elements drawn on the diagram are intelligent and contain all necessary technical information. The project pages can be divided into two types: logical and graphical. Graphical pages are simple drawings, pictures without a logical background. These pages are used for displaying the physical arrangement of equipment and its appearance.

Logical pages are divided into interactive (diagrams which are drawn) and generated. The idea consists of the diagram drawing with intelligent elements which contain all necessary technical characteristics and their electrical and functional interconnections. In this way the project database is filled in. Already in this section, work with CAE and CAD tools differs greatly because graphical and all other functions of the tool, the approach and the logic of work are adapted to designing a concrete project. The use of CAE tools makes the work of the designer easier in the control of available contacts, online references, direct access to databases of equipment manufacturers, etc. This eliminates a significant source of errors which occur when carrying out routine tasks which are constantly repeated.

The second, less creative and tiring part of the work is the drafting of the corresponding documentation on the basis of the drawn diagram. With CAD tools, the work would now become manual, extracting information from the drawings, systematization of that information and transforming it into the form necessary for drafting a project solution. A team of engineers and technicians, based on the diagrams, would formulate and draft terminal strips, cable plans, internal connections, layout of elements, which represents a routine job where, even with the greatest caution, errors are still generated.

In this phase CAE tools use the information which already exists, contained within the interactive logical pages of the project. Based on the data which has been stored in the database, all other logical units are generated automatically. The information about the elements and their interconnections is analyzed and automatically systematized. As a result, automatically generated corresponding documentation is obtained.

The advantage of CAE tools is reflected in their ability to automatically verify solutions, through which all hidden errors are revealed and provided to the designer with the goal of ensuring the proper corrections are made. Harmonizing the technical solution with the characteristics of the selected equipment, verification of potential, improperly marked elements and cross-reference are just a few of the elements which are subject to automated verification.

This type of approach enables the division of work into the creative portion, which is conducted by the engineer, and the routine portion, which is conducted by the CAE tool. In this way, savings of time, reduction of errors to a minimum and maximum increase of production are all achieved.

The modern concept of the EPLAN platform contains a greater number of tools:

- EPLAN Electric P8,
- EPLAN Fluid,
- EPLAN PPE,
- EPLAN Cabinet,
- M-CAD Data Management.

An integral component of the EPLAN platform is the tool EPLAN Electric P8 which is used for design in power engineering. The main advantages of this tool are displayed below.

### 3.3.3. Main functions of the CAE tool EPLAN Electric P8

The CAE tool EPLAN Electric P8 contains the following:

- basic functions:
  - creating drawings and management diagrams on the basis of the library of symbols,
  - display in accordance with DIN EN 61082,
  - autoconnecting of components,
  - standard text (comments) and text which can be subsequently evaluated,
  - macro support (symbol-macros, window-macros, macro verticals, macro pages),
  - free drawing on schematic and graphical lists,
  - zooming (options "window" – window zoom and square),
  - High Level Architecture (HLA) and location editor according to the DIN,
- special functions:
  - symbol editor,
  - Support for Programmable Logic Controller – PLC,
  - Master Data Management – MDM,
  - specification of materials,
  - marking of components,
  - Hewlett-Packard Graphics Language (HPGL) interface,

    - data storage,
  - automatic functions:
    - numbering of pages and components,
    - cross-reference for contactors / contacts and devices displayed in various positions,
    - cross-reference of potentials,
    - generating cables,
    - selection of contactors from the contactor specification file,
    - marking of terminals / contactors and analysis of couplings,
    - graphical display of terminals and interconnections,
    - project contents.

### 3.3.4. Example of the application of EPLAN

An example of a project done using the tool EPLAN is displayed below. Fig. 3.12 shows the title page of the project. Fig. 3.13 shows the

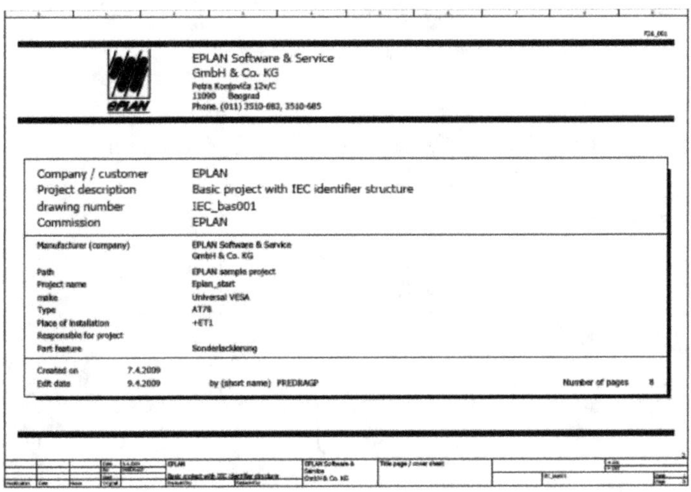


Fig. 3.12 – Project title page

project contents. The first and second pages of the wiring diagram are shown in Figs. 3.14 and 3.15, respectively. An overview of the terminal strip is provided in Fig. 3.16, while Fig. 3.17 provides a display of the terminal strip. Overviews of the cable list and display of the cables plan are provided in Figs. 3.18 and 3.19, respectively.

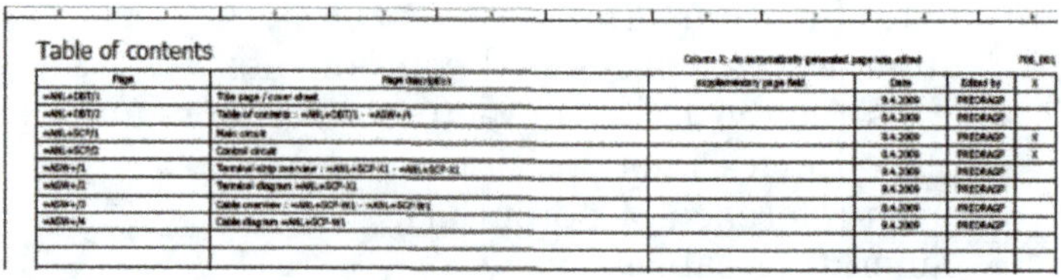

Table of contents

Column X: An automatically generated page was edited

| Page | Page description | supplementary page field | Date | Edited by | X |
|---|---|---|---|---|---|
| =ANL+DBT/1 | Title page / cover sheet | | 9.4.2009 | PREDRAGP | |
| =ANL+DBT/2 | Table of contents : =ANL+DBT/1 - =ASW+/4 | | 9.4.2009 | PREDRAGP | |
| =ANL+SCP/1 | Main circuit | | 9.4.2009 | PREDRAGP | X |
| =ANL+SCP/2 | Control circuit | | 9.4.2009 | PREDRAGP | X |
| =ASW+/1 | Terminal strip overview : =ANL+SCP-X1 - =ANL+SCP-X1 | | 9.4.2009 | PREDRAGP | |
| =ASW+/2 | Terminal diagram =ANL+SCP-X1 | | 9.4.2009 | PREDRAGP | |
| =ASW+/3 | Cable overview : =ANL+SCP-W1 - =ANL+SCP-W1 | | 9.4.2009 | PREDRAGP | |
| =ASW+/4 | Cable diagram =ANL+SCP-W1 | | 9.4.2009 | PREDRAGP | |
| | | | | | |
| | | | | | |

Fig. 3.13 – Project contents

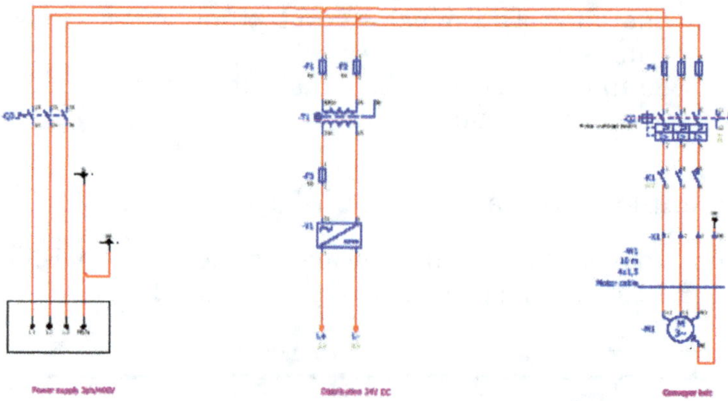



Fig. 3.14 – First page of wiring diagram

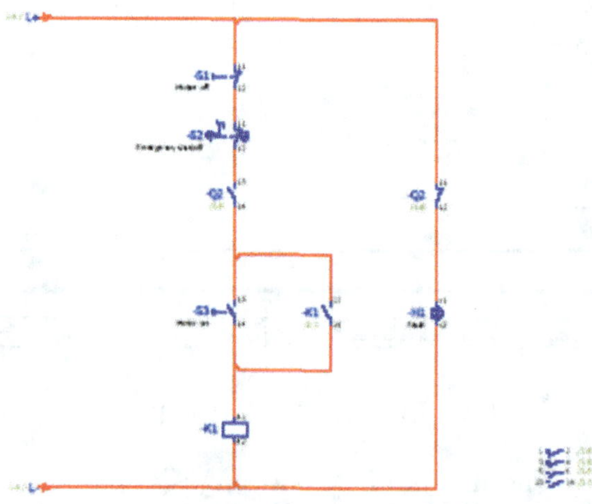



Fig. 3.15 – Second page of wiring diagram

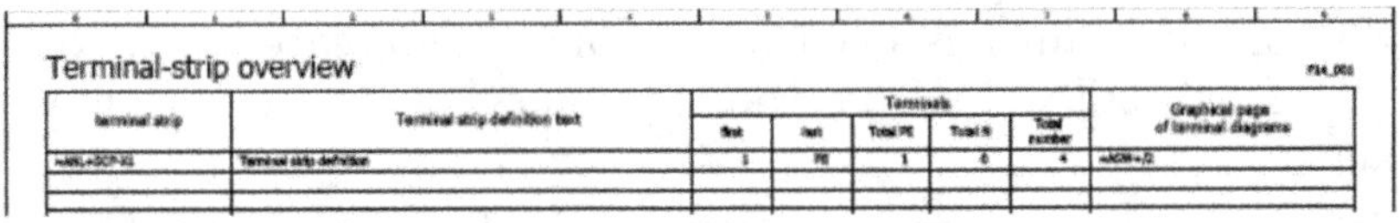


Fig. 3.16 – Terminal-strip overview

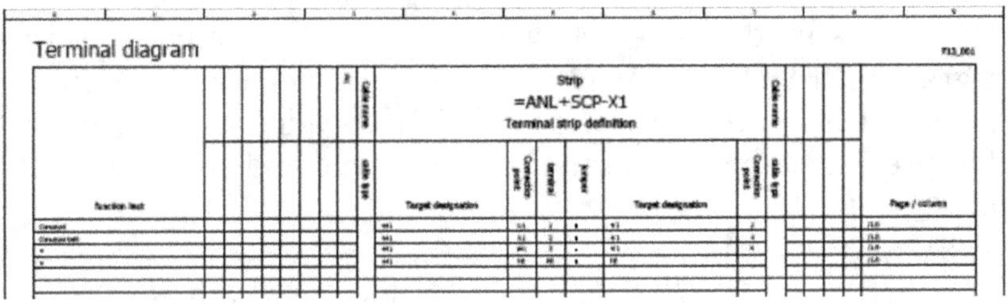


Fig. 3.17 – Terminal diagram display

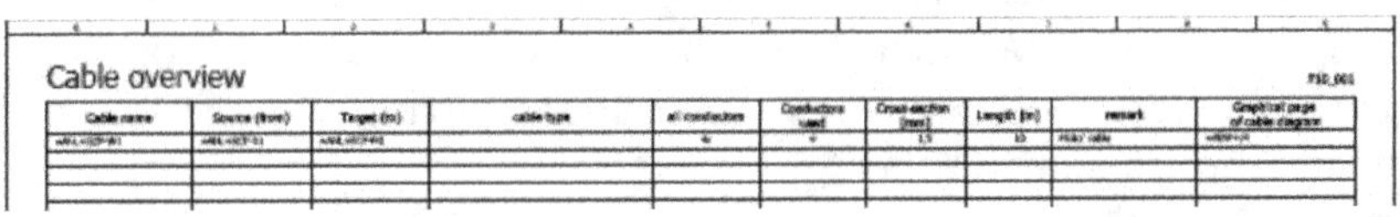


Fig. 3.18 – Cable list overview

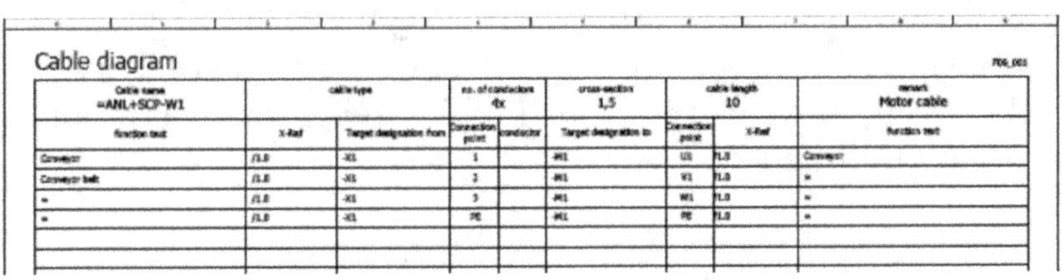


Fig. 3.19 – Cable diagram display

## 3.4. Application of the program SIMARIS DESIGN and SIMARIS SIVACON for design and installation of low-voltage switchgear

### 3.4.1. Introduction

This section presents the application of the new version (R.6.1) of the program SIMARIS SIVACON for the design and installation of 8PT SIEMENS SIVACON low-voltage switchgear. As the basic task, a single-pole diagram has been used which was created with the program SIMARIS DESIGN from the bid request sent to the designer and contractor.

Influenced by economic reasons in the field of energy, SIEMENS developed the 8PT SIVACON energy switchgear. The basic advantage of this switchgear is the local price and global quality of the product. In order for this union to be achieved, a licensed manufacturer receives the program SIMARIS SIVACON. The program enables the simple and easy communication between the SIEMENS headquarters and the local manufacturer, expedient ordering of equipment and complete technical support.

Within the documentation there are standard certifications, descriptions, technical characteristics, drawings, 3D models and instructions for installation of equipment and devices necessary for the 8PT SIVACON switchgear. Without this program, maintaining quality in designing and manufacturing switchgear would not be possible, and with that the entire concept of local manufacturing of SIEMENS products would be unsustainable.

Installation of the program is simple and logical with clearly defined procedures for selection of available options. The installation also includes auxiliary programs for support when viewing documentation.

In the actual example, the application of the program is displayed in detail starting from the task which is placed before the user and going up to the solution and results.

### 3.4.2. Application of the program SIMARIS DESIGN

As an example of the application of the program, the switchgear of one transformer station was considered. Using the program SIMARIS DESIGN, a single-pole diagram will be formed in accordance with the bid request sent to the designer and contractor [14]. According to that request, the transformer station equipment must be placed within the facilities of the existing 10/0.4 kV, 1600kVA transformer station. The transformer station comprises of space for holding the 10 kV and 0.4 kV switchgear, as well as space in which it is necessary to place two 1600 kVA epoxy power transformers.

For the 0.4 kV low-voltage switchgear, the following has been envisaged:

- withdrawable type,
- transformer busbar trunking system,
- low-voltage circuit breakers on the feeders for a nominal current of 3200 A with electro-motor drive and built-in microprocessor protection, possibility of remote command and control, as well as possibility of connecting to computer network,
- automatic coupling bay between transformers,
- measurement of consumption of electrical energy at the feeders with possibility of remote reading and connecting to computer network.

When creating a single-pole diagram using the program SIMARIS DESIGN, loads are divided into two groups ensuring that they are, in accordance with the technological process, evenly distributed in the sense of power. Feeders have been envisaged for several loads (pumps, fans, lighting, etc.) and connections must be made to both transformers.

The 200 kW electro-motor pumps are driven by star-delta starters, while the 160 kW electro-motor valves are driven by soft starters. The remaining electro-motors are driven directly.

Considering that the bill of quantities requests the division of loads into two groups, in this example the formation of individual elements will be elaborated for just one group, which is sufficient to clearly illustrate the capabilities of the practical application of the considered program.

After opening the program SIMARIS DESIGN, three basic options appear:

- project description,
- planning the network,
- output data of the project.

The first option enables the user to open a new project or an existing project. After selecting the option to open a new project, the project name must be defined along with the basic technical data for medium and low voltage. For medium voltage the following is required: rated voltage, maximum and minimum short circuit power, maximum and minimum cross section. The data which is required for low voltage is displayed in Fig. 3.20.

After defining the required parameters, the SIMARIS DESIGN workspace appears (Fig. 3.21), through which the planning of the network is made possible. The symbols library is located on the left side of Fig. 3.21 while the workspace is located on the right side.

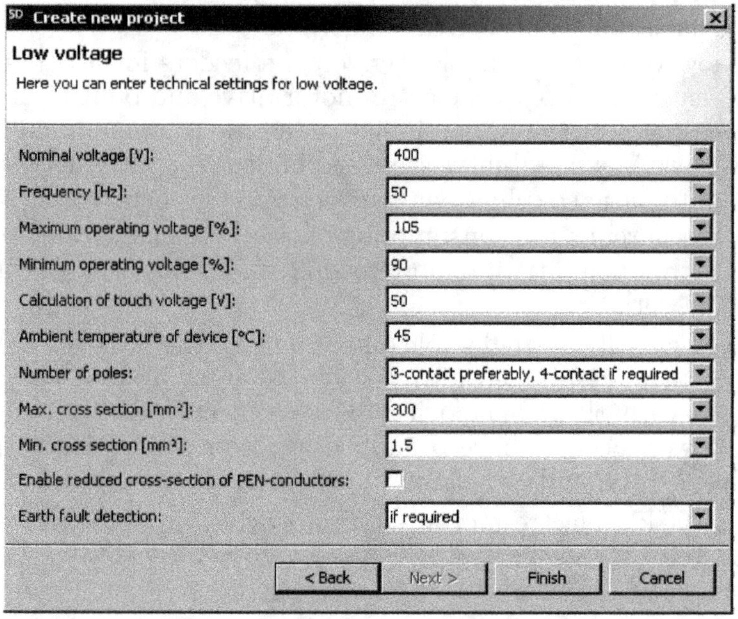


Fig. 3.20 – Required data for low voltage

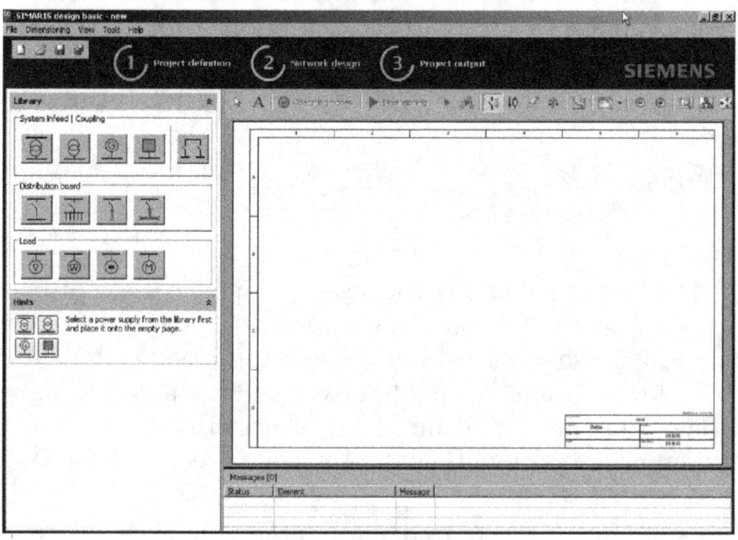


Fig. 3.21 – SIMARIS DESIGN workspace

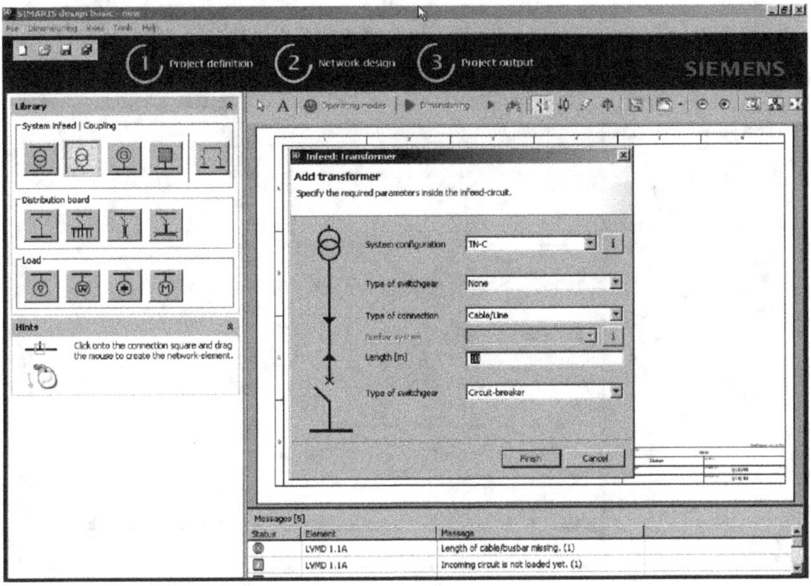

Fig. 3.22 – Procedure for creating a transformer bay

Fig. 3.22 shows the procedure for creating a transformer bay. It is necessary to provide the following:

- system configuration,
- types of circuit breakers,
- types of couplings,
- length.

The following system configurations are available: TN-C, TN-S, IT and TT. Information about the system configurations is obtained by clicking on the *i* icon (Fig. 3.23).

The following options may be selected for the switch: no circuit breaker, circuit breaker, circuit breaker-disconnector with a fuse, disconnector with a fuse or fuse with a stand.

A cable/lead or a busbar trunking system are the possible types of couplings. This is followed by providing the length. Fig. 3.24 shows an example of a transformer bay which is defined using the aforementioned procedure.

The motor feeder and corresponding load distributor are formed in a similar way according to the project task. A complete single-pole diagram is displayed in Fig. 3.25.

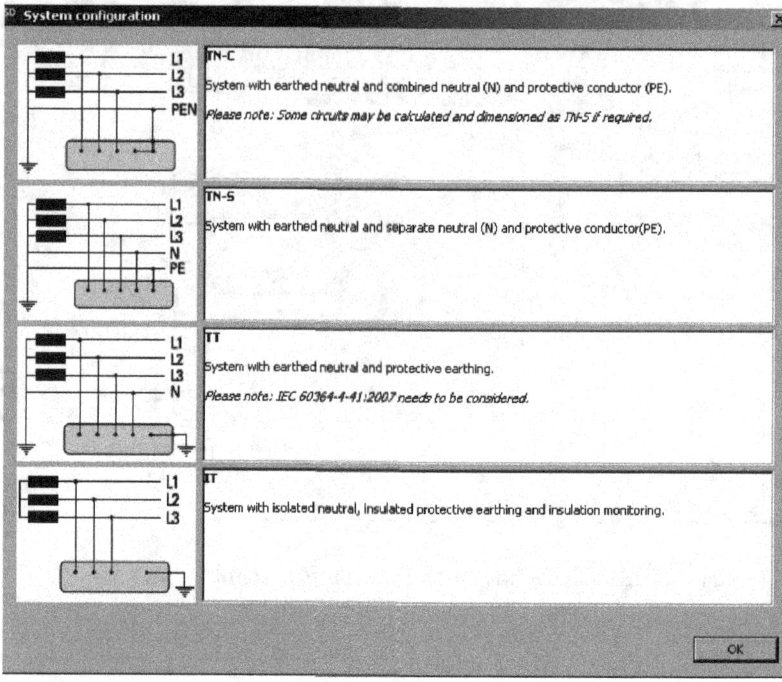


Fig. 3.23 – System configuration information

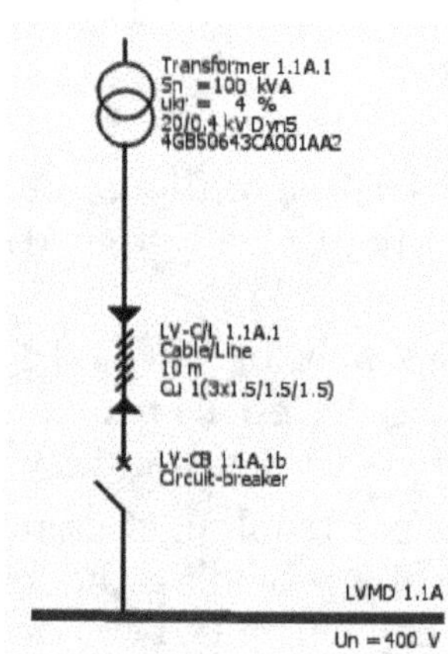


Fig. 3.24 - Example of a transformer bay formed using SIMARIS DESIGN

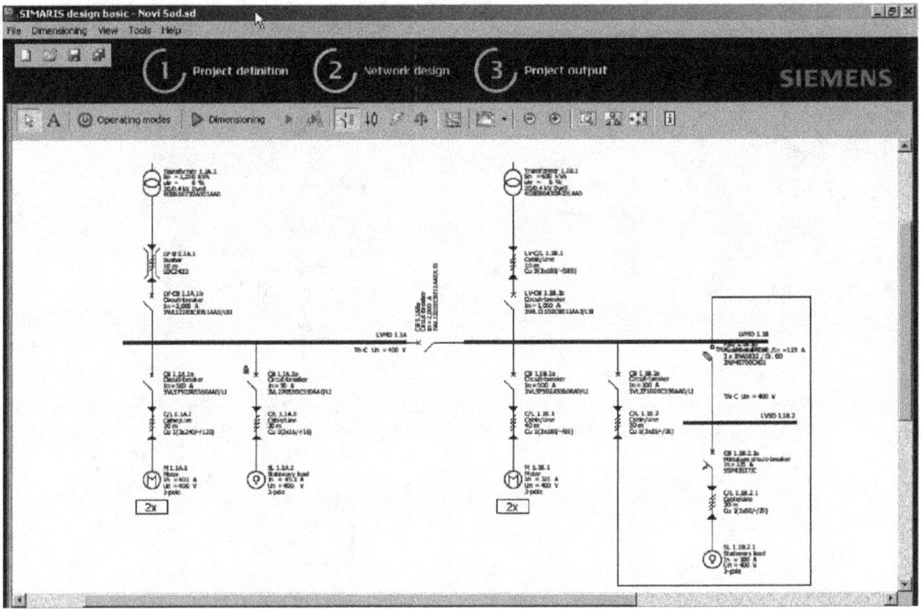


Fig. 3.25 – Complete single-pole diagram

For the selected feeder, the program SIMARIS DESIGN enables the display of various parameters such as: rated values (Fig. 3.26a), voltage drop (Fig. 3.26b), maximum and minimum values of short-circuit current (Fig. 3.26c) and the power report (Fig. 3.26d).

For the selected circuit breaker in the motor feeder, a window appears with data as displayed in Fig. 3.27. By activating the Catalog icon, the complete information about the selected circuit breaker is obtained (Fig. 3.28). Fig. 3.29 shows the selectivity of protection of the member of the considered circuit breaker. The complete selectivity of protection of the members of all circuit breakers in the considered single-pole diagram is shown in Fig. 3.30. Full selectivity is displayed in green while yellow indicates conditional selectivity. The theoretical bases of the aspect of designing medium- and low-voltage switchgear are covered in detail within the selected literature [15, 16].

Fig. 3.31 shows the dialog box for cable in the motor feeder. The designer is given the ability to provide all of the relevant factors for the proper selection of cable. By selecting the $i$ icon, a detailed description is provided for the type of installation and the method for installation (Fig. 3.32). Fig 3.33 shows the dialog box for selection of correction

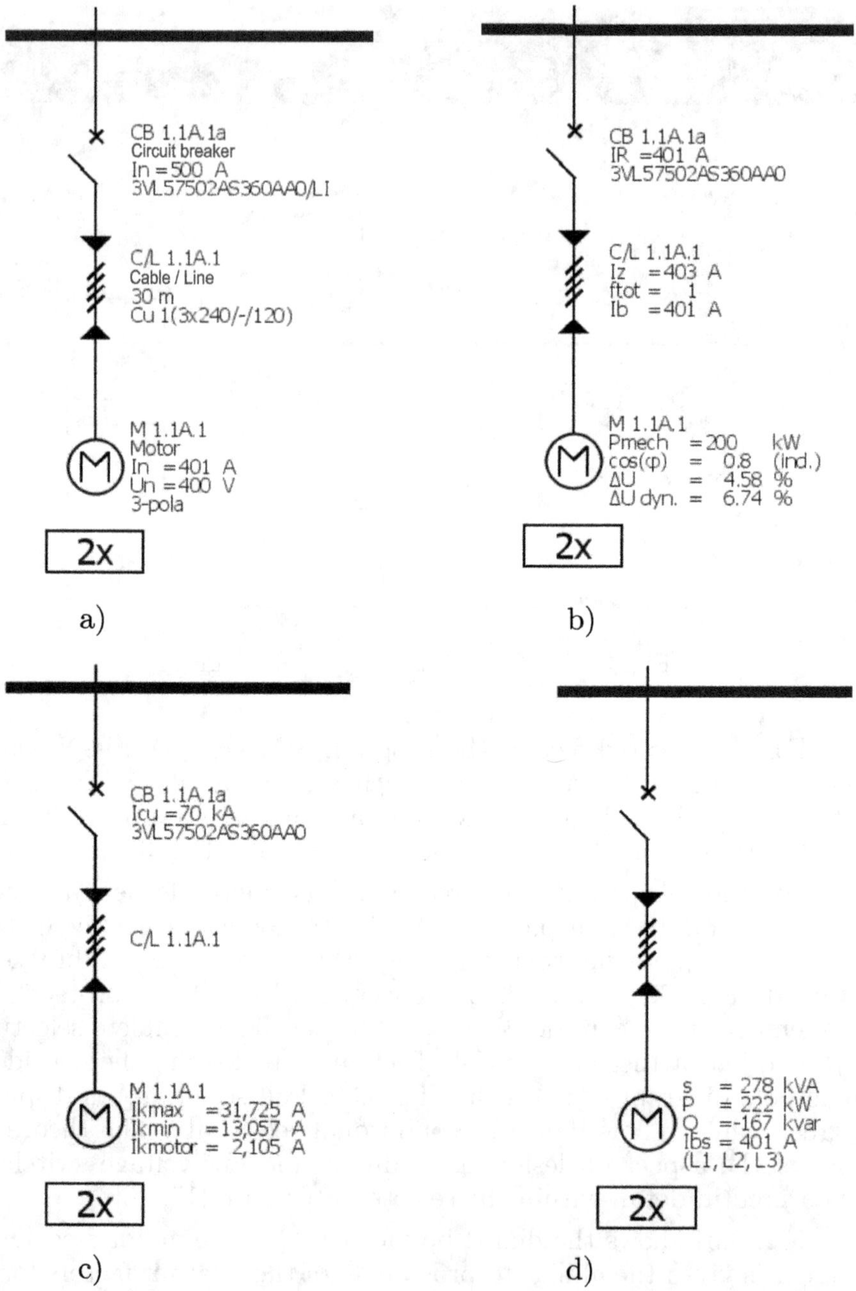


Fig. 3.26 – Display of various parameters for motor feeder;
a) Rated values; b) Voltage drop; c) Maximum and minimum values of short-circuit current; d) Power report

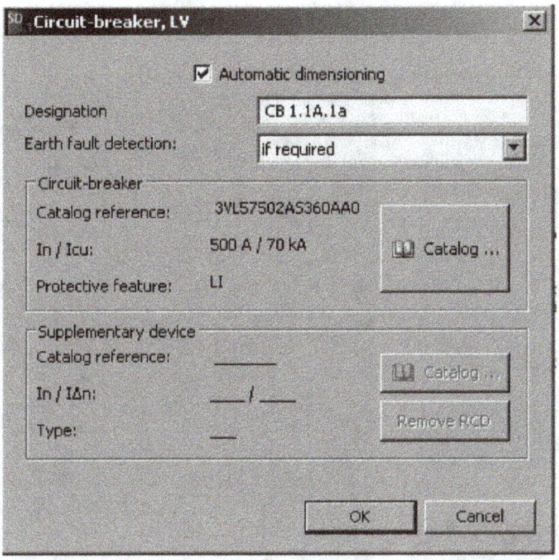


Fig. 3.27 - Window for data about the selected circuit breaker in the motor feeder

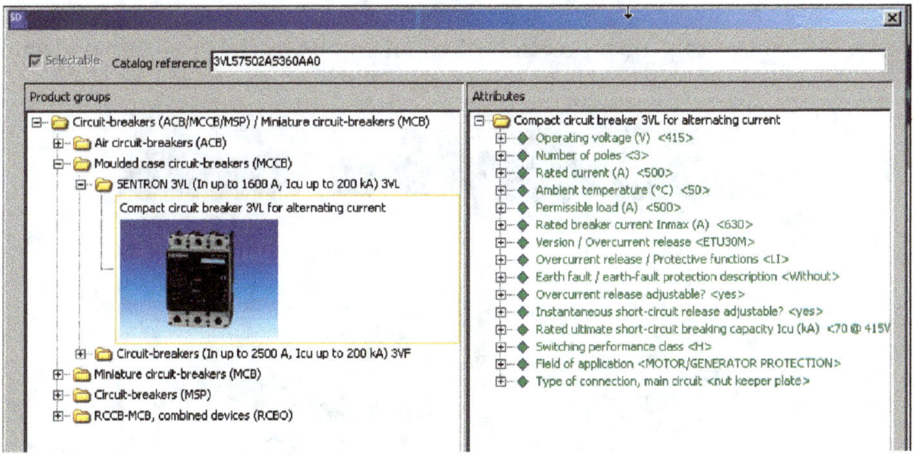


Fig. 3.28 - Complete information about the selected circuit breaker in the motor feeder

factors depending on the ambient temperature and number of cables.

The final option relates to the forming of documentation with the output data of the project. It is possible to create:

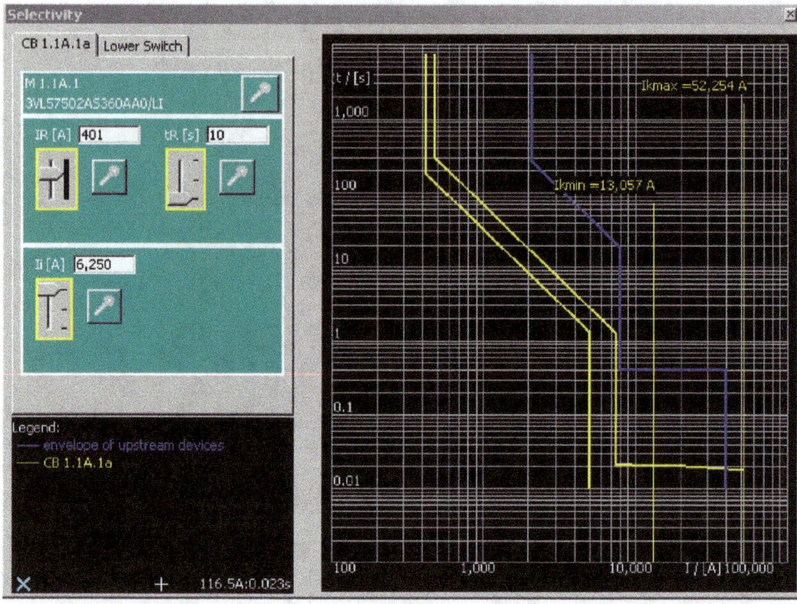


Fig. 3.29 - Selectivity of protection of the member of the considered circuit breaker

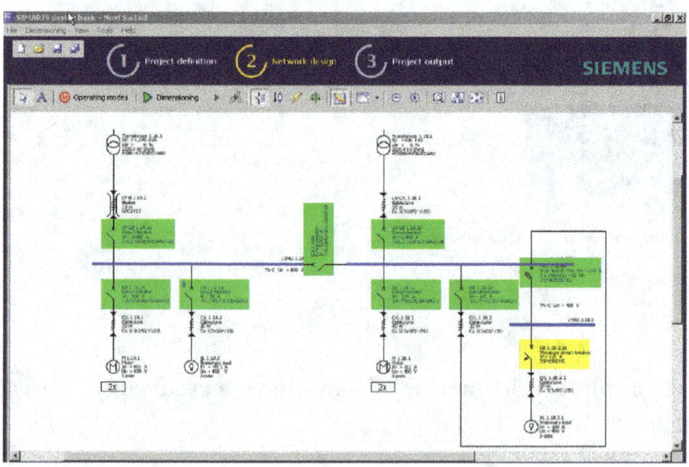


Fig. 3.30 - Complete selectivity of protection of the members of all circuit breakers in the considered single-pole diagram

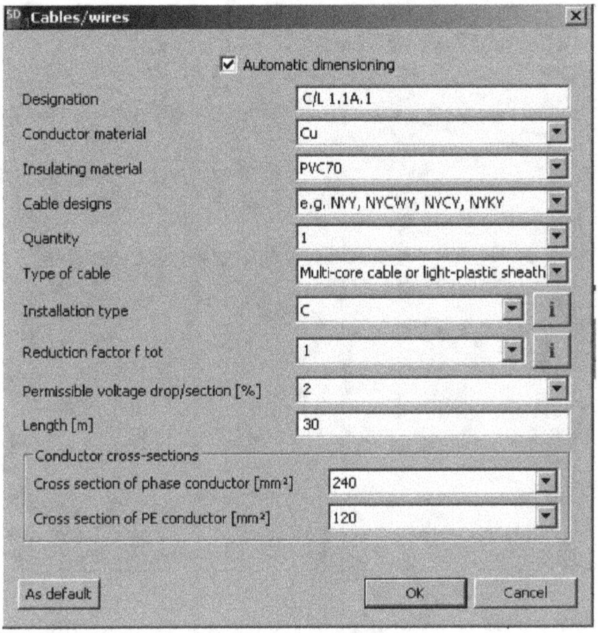


Fig. 3.31 – Dialog box for cable in the motor feeder

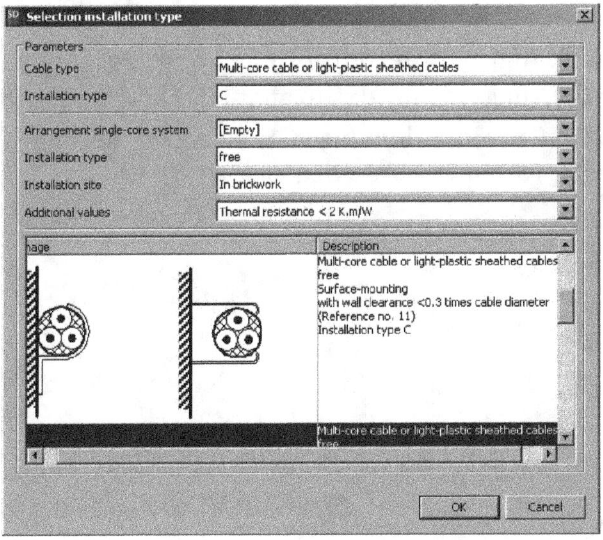


Fig. 3.32 - Detailed description of the type of installation and installation method

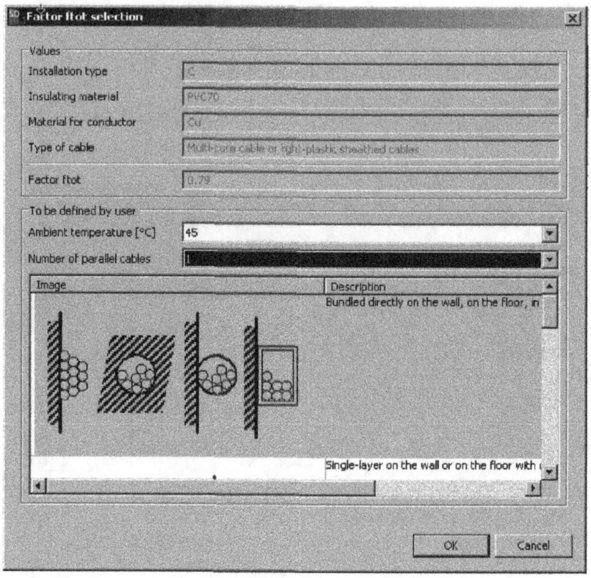


Fig. 3.33 - Window for selection of correction factors depending on ambient temperature and number of parallel cables

– project documentation,
– documentation on selectivity,
– network diagram in .pdf format,
– network diagram in .dxf format.

A report on project documentation which was formed using the aforementioned program is provided below.

**Project Documentation**
**created with**
**SIMARIS design basic**
**Version: 4.1 (14-12-2007)**
**Subrevision: 0591**

**http://www.siemens.com/sdbasic**

## Master data

| | |
|---|---|
| Project name: | Novi Sad |
| Short description of project: | Transformer station 20/0,4 kV |
| Created by: | Z. S. |
| Planning office: | |
| Created on: | Friday, April 03, 2009 |
| Changed on: | Friday, April 03, 2009 |

## Customer data

| | |
|---|---|
| City: | |
| Customer: | |

## Comments:

| |
|---|
| |

**Network parameters:**

| General | |
|---|---|
| Standard | IEC |
| Altitude of site | < 1000 m |
| Medium voltage | |
| Rated voltage | 20 kV |
| Ambient temperature | 40° C |
| Max./Min short-circuit power | 250 / 100 MVA |
| Neutral System | Low-resistance |
| Relation R1/X1 min | 0,2 |
| Conductor temperature of MV cable | |
| on beginning of short circuit | 20° C |
| for disconnection | 80° C |
| for voltage drop | 55° C |
| Low voltage | |
| Rated voltage | 400 V |
| System configuration | TN-C |
| Frequency | 50 Hz |
| Tolerable touch voltage | 50 V |
| Ambient temperature of devices | 45° C |
| Voltage Factor c max | 1,05 |
| Voltage Factor c min | 0,95 |
| Conductor temperature of LV cable | |
| on beginning of short circuit | 20° C |
| for disconnection | 80° C |
| for voltage drop | 55° C |

**Defined Network Modes for Calculation and Dimensioning:**

Operating mode: 1

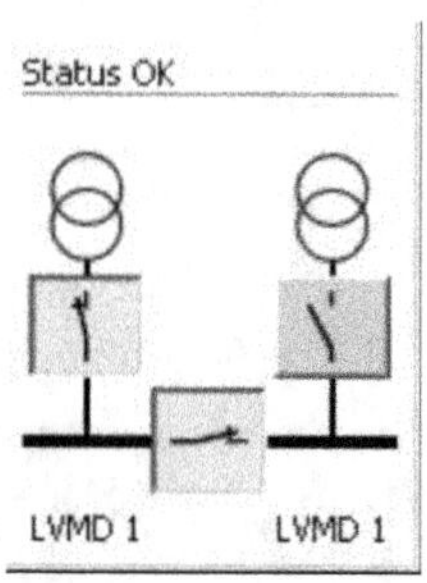


Operating mode: 2

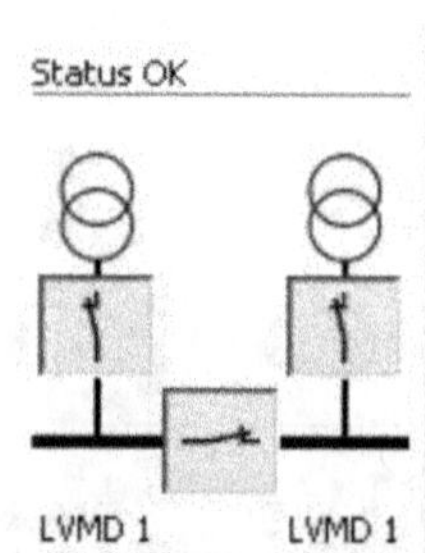

## List of Devices: Sources of Feeding: Transformers:

| Designation | Type | Sn [kVA] | ukr [%] | Uprim [kV] | Usec [V] | Pk [kW] | P0 [kW] | Vector Group | Catalog Reference |
|---|---|---|---|---|---|---|---|---|---|
| Tr. 1.1A.1 | GEAFOL | 1.250 | 6 | 20 | 400 | 10,6 | 2,7 | Dyn5 | 4GB61673D A001AA0 |
| Tr. 1.1B.1 | GEAFOL | 630 | 6 | 20 | 400 | 7,2 | 1,65 | Dyn5 | 4GB58643D A001AA0 |

## Switches/Fuses: Circuit Breaker/Miniature Circuit Breaker:

| Place | Desig-Desig | Catalog Reference | In [A] | Icu/Icn [kA] | Ik"max[A] Location | Release Type / Characteristics | Qty. |
|---|---|---|---|---|---|---|---|
| LVMD 1.1A.1 | LV-CB 1.1A.1b | 3WL12202 CB311AA2 | 2.000 | 66 | 30,782 | ETU25B | 1 |
| M 1.1A.1 | CB 1.1A.1a | 3VL57502 AS360AA0 | 500 | 70 | 52,254 | ETU30M | 2 |
| SL 1.1A.2 | CB 1.1A.2a | 3VL27052 DC330AA0 | 50 | 70 | 54,322 | TM | 1 |
| Coupling 1.1AB | CB 1.1ABa | 3WL12202 CB311AA2 | 2.000 | 66 | 30,782 | ETU25B | 1 |
| LVMD 1.1B.1 | LV-CB 1.1B.1b | 3WL11102 CB311AA2 | 1.000 | 55 | 16,117 | ETU25B | 1 |
| M 1.1B.1 | CB 1.1B.1a | 3VL57502 AS360AA0 | 500 | 70 | 52,667 | ETU30M | 2 |
| LVSD 1.1B.2 | CB 1.1B.2a | 3VL27102 DC330AA0 | 100 | 70 | 54,322 | TM | 1 |
| SL 1.1B.2.1 | CB 1.1B.2.1a | 5SP43927 | 125 | 10 | 9,505 | C | 1 |

## Switch Disconnector with Fuses:

| | |
|---|---|
| Place | LVSD 1.1B.2 |
| Designation | FSD1.1B.2b |
| Catalog Reference Socket / Fuse | 3NP40700CH01/3NA3832 |
| Quantity Socket / Fuse | 1/3 |

**Connections and Line Distribution:**
**Cable/Conductor Low Voltage:**

| Designation | C/L 1.1A.1 | C/L 1.1A.2 | LV-C/L 1.1B.1 |
|---|---|---|---|
| Type / Profile | 3x240/-/120 | 3x16/-/16 | 3x185/-/185 |
| Starting Point / Target Point | LVMD 1.1AM 1.1A.1 | LVMD 1.1ASL 1.1A.2 | LVMD1. 1B.1 |
| Ib [A] Iz [A] | 400,93 403 | 45,10 76 | 532,31 1.023 |
| Material | Cu | Cu | Cu |
| No. | 1 | 1 | 3 |
| Length [m] | 30 | 20 | 10 |
| Isolation | PVC70 | PVC70 | PVC70 |
| Installation Type | C | C | C |
| ftot | 1 | 1 | 1 |
| ΔU [%] / section /accumulated | 0,610 4,577 | 0,429 4,395 | 0,107 2,622 |
| Quantity | 2 | 1 | 1 |

| Designation | C/L 1.1B.1 | C/L 1.1B.2 | C/L 1.1B.2.1 |
|---|---|---|---|
| Type / Profile | 3x185/-/95 | 3x35/-/35 | 3x50/-/25 |
| Starting Point / Target Point | LVMD 1.1B M1.1B.1 | LVMD 1.1B LVSD 1.1B.2 | LVSD1.1B.2 SL 1.1B.2.1 |
| Ib [A] Iz [A] | 320,75 341 | 100 119 | 100 144 |
| Material | Cu | Cu | Cu |
| No. | 1 | 1 | 1 |
| Length [m] | 40 | 30 | 20 |
| Isolation | PVC70 | PVC70 | PVC70 |
| Installation Type | C | C | C |
| ftot | 1 | 1 | 1 |
| ΔU [%] section /accumulated | 0,770 4,737 | 0,684 4,651 | 0,332 4,983 |
| Quantity | 2 | 1 | 1 |

**Busbar Trunking Systems:**

| Designation | LV-B 1.1A.1 |
|---|---|
| Type / Material | LDC-2000 Cu |
| Starting Point / Target Point | LVMD1.1A.1 |
| In [A] | 2.000 |
| Icw 1s [kA] | 58 |
| Length [m] | 10 |
| Busbar Configuration | L1, L2, L3, PEN |
| Degree of Protection | IP34 |
| $\triangle$U [%] Section / Accumulated | 0,360 3,967 |
| Quantity | 1 |

**Load: Stationary Load:**

| Designation | SL1.1A.2 | SL1.1B.2.1 |
|---|---|---|
| Pn [kW] | 25 | 55,426 |
| In [A] | 45,105 | 100 |
| Un [V] | 400 | 400 |
| $\cos\varphi$ | 0,8 | 0,8 |
| ai | 1 | 1 |
| Phase Connection | L1-L2-L3 | L1-L2-L3 |
| Type of Load | inductive | inductive |
| Quantity | 1 | 1 |

**Motors:**

| Designation | M 1.1A.1 | M 1.1B.1 |
|---|---|---|
| Pn [kW] In [A] | 200 400,93 | 160 320,75 |
| Un [V] $\cos\varphi$ | 400 0,8 | 400 0,8 |
| ai | 1 | 1 |
| Phase Connection | L1-L2-L3 | L1-L2-L3 |
| Ia/In | 5 | 5 |
| Trip Class | Class 10 | Class 10 |
| Type of Load | inductive | inductive |
| Method of Starting | Direct-On-Line Starter | Direct-On-Line Starter |
| Ikre | 1 | 1 |
| $\eta$ | 0,9 | 0,9 |
| Quantity | 2 | 2 |

**Personal Protection Against Electric Shock**

All circuits of the project have a permissible break time ta-req > ta-cur and therefore fulfill the requirements for personal protection against electric shock.

**Symbols Legend:**

| Symbol [Unit] | Description |
|---|---|
| ai | Load factor |
| $\cos\varphi$ | Power factor |
| ftot | Derating factor |
| Ia/In | Inrush current relation |
| Ib [A] Iz [A] | Operating current / allowed load capacity |
| Ic base [kA] | Rated conditional short-circuit current – Socket |
| Icu/Icn [kA] | Rated ultimate short-circuit breaking capacity of ACB / MCCB<br>Rated ultimate short-circuit breaking capacity of MCB |
| Icw 1s [kA] | Rated short-time withstand current 1s |
| I$\triangle$n [mA] | Residual current RCD |

| Symbol [Unit] | Description |
|---|---|
| Ik"max [kA]Location | Maximum short-circuit current at installation location |
| Ik1max | Maximum single phase short-circuit current |
| Ik1min | Minimum single phase short-circuit current |
| Ik3max | Maximum three phase short-circuit current |
| Ik3min | Minimum three phase short-circuit current |
| Ik1D [kA] | Single-pole uninterrupted short-circuit current |
| Ik3D [kA] | Three pole uninterrupted short-circuit current |
| Ikmax/Ikmin | Relation of maximum and minimum short-circuit current |
| Ikre | Short-circuit current recovery factor |
| In [A] | Nominal current |
| P0 [kW] | No-load losses |
| Pk [kW] | Short-circuit losses |
| Pn [kW] | Active power |
| R0/R1 | Resistance relation zero / positive phase-sequence system |
| R1 [%] | Resistance in positive phase-sequence system |
| Sn [kV] | Nominal apparent power |
| ukr [%] | Short-circuit voltage |
| Un [V] | Nominal voltage |
| Uprim [kV] | Primary voltage |
| Usec [V] | Secondary voltage |
| X0/X1 | Reactance relation zero / positive phase-sequence system |
| xd" [%] | Subtransient reactance |
| Z1 max | Maximum impedance in positive phase-sequence system |
| Z1 min | Minimum impedance in positive phase-sequence system |
| ZS | Faulted circuit impedance |
| Zs max | Maximum faulted circuit impedance |
| Zs min | Minimum faulted circuit impedance |
| $\triangle$U [%] | per section / cumulative. Voltage drop between beginning and end of a section / cumulated voltage drop up to the specified point including voltage drop through transformer |
| $\eta$ | Efficiency |

**Standards for Computation:**

| Title | IEC | HD | EN | DIN VDE |
|---|---|---|---|---|
| Low-voltage electrical installations | 60364-1 ... 6 | 384 | | 0100 – 100...610 |
| Short-circuit currents in three-phase a.c. systems – calculation of currents | 60909 | | 60909 | 0102 |
| Short-circuit currents – calculation of effects, definitions and calculations methods | 60865 | | 60865 | 0103 |
| Low-voltage switchgear and controlgear - circuit-breakers | 60947-2 | | 60947-2 | 0660 – 101 |
| Low-voltage switchgear and controlgear assemblies | 60439 | | 60439-1 ... 5 | 0660 – 500 ... 505 |
| Method of temperature-rise assessment by extrapolation for partially type-tested assemblies (PTTA) of low-voltage switchgear and controlgear | 60890+C | 528 S2 | | 0660 – 507 |
| Application of cables and cords in power installations - recommended current-carrying capacity for sheathed and non-sheathed cables for fixed wirings in and around buildings and for flexible cables and cords | — | 384 | | 0298 -4 |
| Electrical accessories – circuit-breakers for overcurrent protection for household and similar installations – circuit-breakers for a.c. operation | 60898-1 | | 60898-1 | 0641 -11 |
| High-voltage switchgear and controlgear – alternating current switch-fuse combinations | 62271 | | 62271 | 0671 – 105 |

The documentation on selectivity contains the complete information related to the selectivity of all considered circuit breakers. The following are given: circuit name, switch name and value of selectivity (full or partial). The documentation also contains the corresponding current - time diagrams.

The final output data refers to the network diagram which represents the single-pole diagram in .pdf or .dxf format.

### 3.4.3. Application of the program SIMARIS SIVACON

This section presents the application of the new version (R.6.1) of the program SIMARIS SIVACON for the design and installation of 8PT SIEMENS SIVACON low-voltage switchgear [17-20].

After opening the program, the introduction window appears which offers the following options (Fig. 3.34):

- create a project for an 8PT busbar position on rear (loads model for 8PT with busbar on the rear),
- create a project for an 8PT busbar position on top (loads model for 8PT with busbar on the top),
- load an existing project (model will be loaded according to the project),
- load example project (using this function allows the loading of an example from the group of projects located in the defined directory).

After opening the program, the model with the busbar on top is selected (upper right corner on Fig. 3.34). Then the window for forming the switchgear is activated. The following data must be filled in (Fig. 3.35):

- name of switchgear,
- frequency (for all 8PT switchgear is 50 Hz),
- rated voltage,
- rated current,
- short circuit current,
- degree of protection (IP 40),
- ventilation type,
- ambient temperature,
- type of construction and cross section of the busbar.

By selecting the busbar construction, the height of the switchgear is automatically determined and it amounts to 2200 mm. This value cannot be changed.

Further selection of the switchgear characteristics is performed in the system of drop-down menus and consists of:

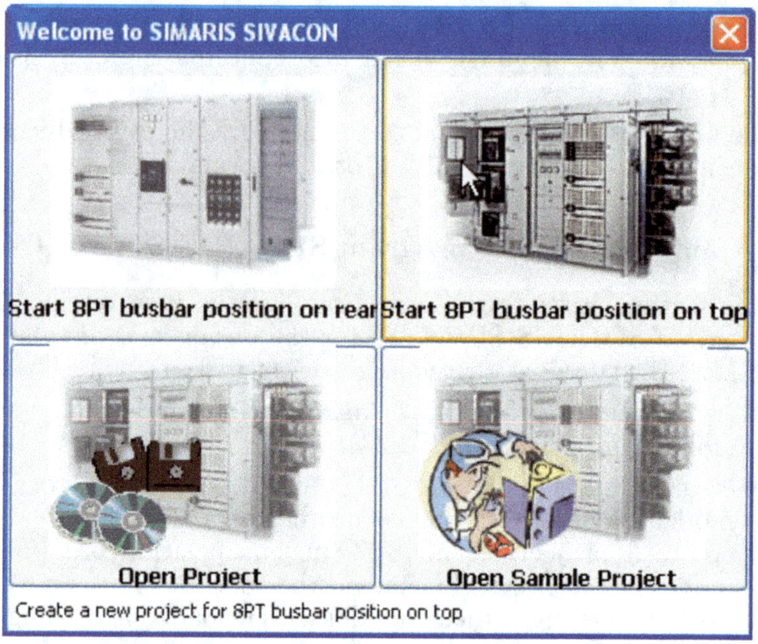


Fig. 3.34 – Introduction window of the program SIMARIS SIVACON

- selection of feeding system, as well as cross section of zero conductor,
- selection of connections to switchgear, by which depth of switchgear is automatically determined,
- selection of frame profile,
- selection of type of termination cubicle,
- selection of space for wiring,
- selection of option of switchgear,
- selection of existence of transport palettes.

Fig. 3.35 shows the complete selected characteristics of the switchgear. These characteristics will automatically be valid for all cubicles and feeders which are located within it and, when creating a cubicle, this data appears within the windows as unchangeable.

After defining the characteristics of the switchgear comes the forming of the first cubicle. It is necessary to define the following elements:

- name of cubicle,
- design and type of cubicle.

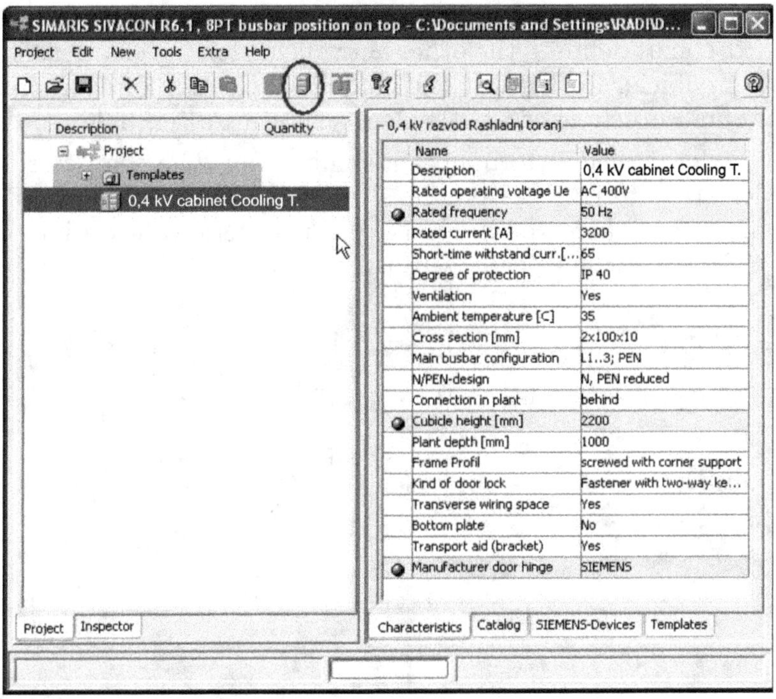


Fig. 3.35 – Complete selected characteristics of the switchgear

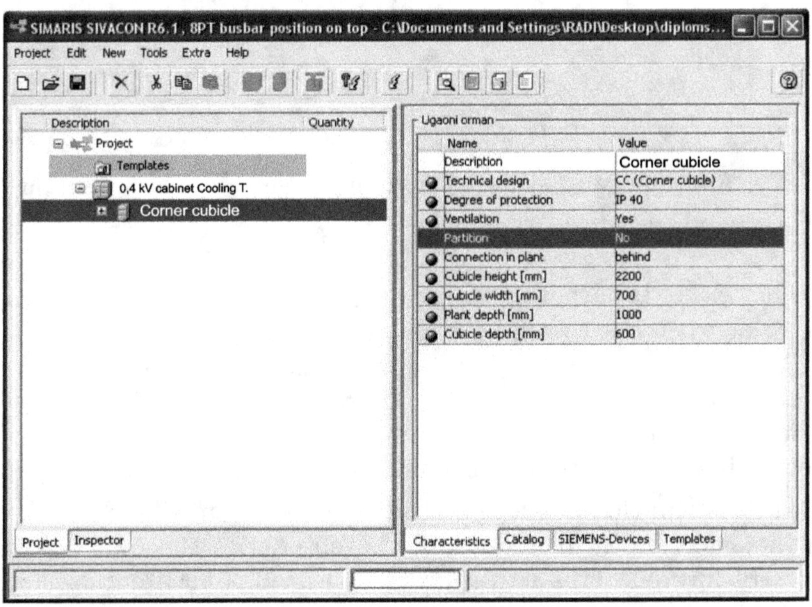


Fig. 3.36 – Complete selected characteristics of the corner cubicle

In the aforementioned case a corner cubicle was selected (symbol CC). Next comes defining the other entry fields for the characteristics of the corner cubicle. In this way the complete selection of characteristics of the corner cubicle is created which is shown in Fig. 3.36.

A feeder cubicle is formed in a similar way. The characteristics of the feeder are entered in as previously described. A 3WL1232 (3200A) circuit breaker was selected. By making this selection the options circled in Fig. 3.37 are automatically selected.

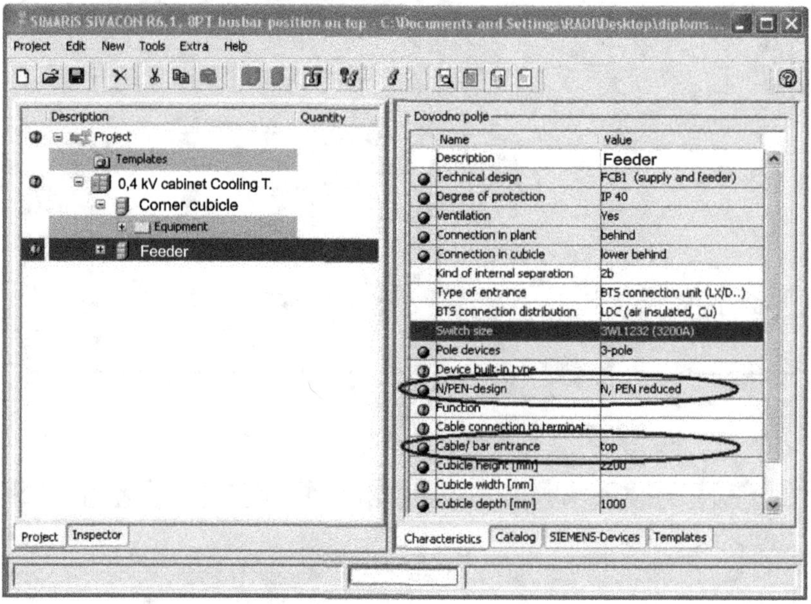


Fig. 3.37 - Automatically selected options based on the selection of a 3WL232 (3200A) circuit breaker

The selected characteristics of the feeder are displayed in Fig. 3.38. Once all characteristics of the bay (cubicle) have been selected, next comes the forming of the feeder. Fig 3.39 shows the feeder of the main circuit breaker. The name and the characteristics of the feeder which are not automatically selected are entered in on the right side of the window. This completes the formation of the feeder.

The distribution bay and motor feeder are formed in a similar way. After the selection of all requested characteristics the motor feeder is completely formed. This is displayed in Fig. 3.40. All cubicles within

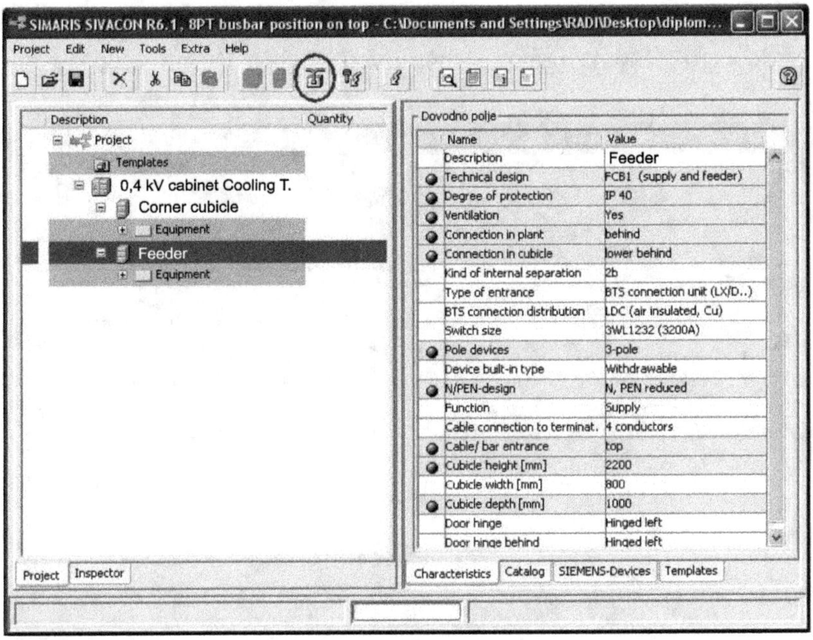


Fig. 3.38 – Complete selected characteristics of a feeder

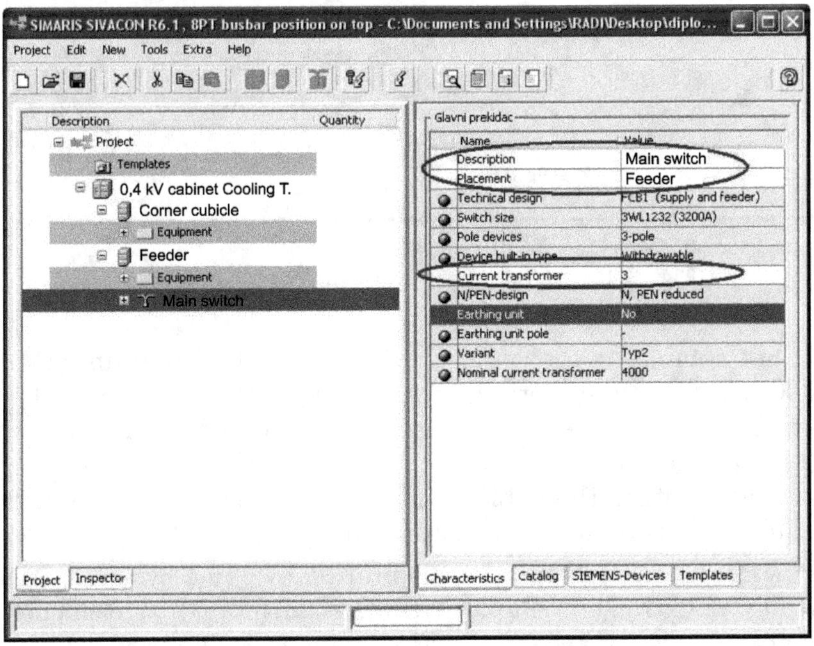


Fig. 3.39 – Entry of characteristics of the feeder (main switch)

the switchgear are entered in using the aforementioned method. A completely described switchgear is displayed within Fig. 3.41.

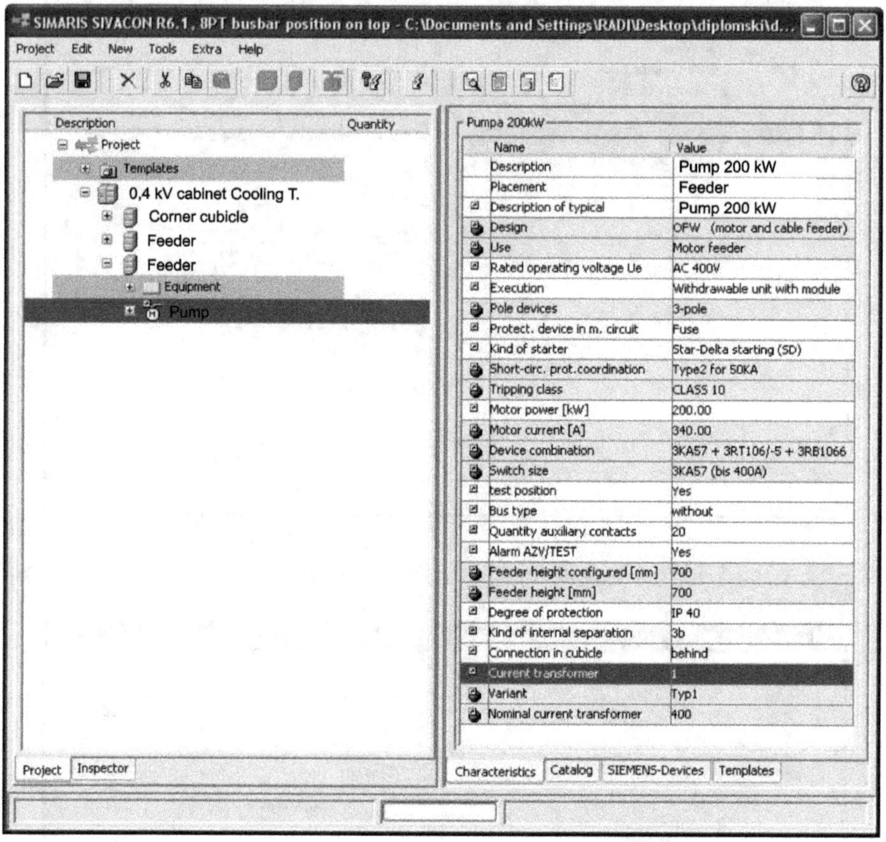

Fig. 3.40 – Completely described motor feeder

This completes the entry of data into the program SIMARIS SIVACON R.6.1. When forming the cubicle in the left part of the window, below the cubicle within the project tree itself, the Equipment folder is formed which holds all the mechanical components necessary for putting together the cubicle including the selected characteristics. The equipment is organized according to its code.

Using the catalog and documentation overview within SIMARIS SIVACON is extremely simple. Drawings of individual elements of a cubicle can be easily accessed in this manner. It is possible to review

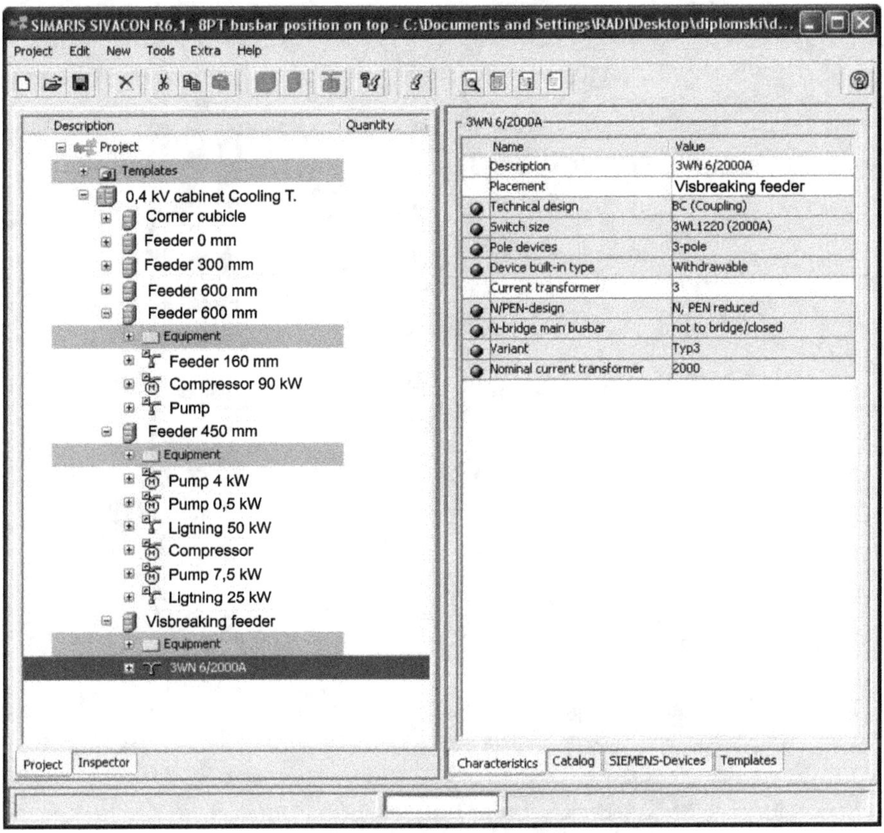


Fig. 3.41 – Complete described switchgear

the necessary parts for a complete switchgear, for each cubicle (bay) or for each individual feeder.

An exceptional convenience when creating parts from both tin and copper is the ability to use these drawings in computer controlled presses, which significantly decreases the engagement of human resources. Many of the parts are available in the form of 3D models which can be viewed from any angle, moved and rotated continuously in real time, as well as enlarged. A few positions of the 3D model of the 8PT11358 three-pole module FCB1 1600A are provided in Fig. 3.42.

Finally, it is necessary to note that within the program it is possible to select the options to display prices and generate reports. Reports come in the form of tables with a list of equipment and/or devices

Fig. 3.42 - Display of 3D model of an 8PT11358 three-pole module FCB1 1600A

compared to the desired parameters: according to switchgear, cubicles, feeders or just according to typical feeders. At the same time, characteristics which describe a device or part of the equipment may be selected.

# LITERATURE


[1] G. Brechmann, C.W. Dzieia, R.E. Hornemann, H.H. Hubscher, L.D. Jagla, N.J. Klaue: Westerman's Electrical Engineering Handbook, Građevinska Knjiga, Belgrade, 2000.

[2] G. Dotlić: Power Engineering Through Standards, Laws, Rulebooks and Technical Recommendations, revised and amended edition, SMI-ETS, 2006.

[3] IEC 617-7:1996: Graphical Symbols for Diagrams. Part 7: Switchgear, Controlgear and Protective Devices.

[4] IEC 617-8:1996: Graphical Symbols for Diagrams. Part 8: Measuring Instruments, Lamps and Signaling Devices.

[5] Main Project of the Adaptation of the 110 kV/ 35 kV "Vreoci" Transformer Station and Transmission Lines 110 kV of the Substation of the TPP "Kolubara", Faculty of Electrical Engineering, Belgrade, Orderer Electrical Utility of Serbia, Mining Kolubara – Lazarevac, 2005-2007, (Chief of project Z. Stojković).

[6] E. Finkelstein: AutoCAD 2002 Bible, Mikro Knjiga, Belgrade, 2002.

[7] H. J. Engelke: 3D – Konstruktion mit AutoCAD 2002, Volumen-Modellieren fur Einsteiger, Hanser Verlag Munchen Wien, 2002.

[8] Z. Stojković, J. Mikulović, Z. Stojanović: Workshop for Software Tools in Power Engineering, Faculty of Electrical Engineering, Belgrade, Academic Mind, Belgrade, June 2006.

[9] IEC 1082-2:1993: Preparation of Documents Used in Electrotechnology. Part 2: Function - Oriented Diagrams.

[10] As-Built Design of the Adaptation of 110 kV/ 35 kV "Vreoci" Transformer Station, 110 kV Substation, Volume E1: Wiring Diagram and Connections Diagram for Transformer Bay T1, Faculty of Electrical Engineering, Belgrade, Orderer Electrical Utility of Serbia, Mining Kolubara – Lazarevac, 2008, (Chief of project Z. Stojković).

[11] EPLAN Electric P8, Catalog, EPLAN Software & Service, 2008.

[12] Engineering Base 2004, The Electrical Engineering Office, Katalog, AUCOTEC, 2004.

[13] http://www.caddy-electrical.com

[14] SIMARIS DESIGN, Basic 4.1, Siemens, 2007.

[15] Totally Integrated Power – Application Manual - Basic Data and Preliminary Planning, Siemens.

[16] Totally Integrated Power – Integrated Solutions for Power Distribution in Commercial and Industrial Buildings – Application Manual – Part 2: Draft Planning, Siemens.

[17] Help Menu Program SIMARIS – SIVACON 8PT, Siemens.

[18] SIVACON 8PT Type-Tested Low-Voltage Switchboard, Startup Training, Siemens.

[19] SIVACON Catalog – Multi-purpose Switchboards, Typically Tested Switchboards for Power Distribution 8PT, Montprojekt, Belgrade.

[20] V. Sovrlić, Z. Stojković: SIMARIS SIVACON 6.1 Project for Cooling Tower, Faculty of Electrical Engineering, Belgrade, 2006.

# 4. APPLICATION OF SOFTWARE TOOLS IN POWER ENGINEERING PROJECT MANAGEMENT

## 4.1. About projects

### 4.1.1. Definition of a project

A project, in its essence, is every process through which a goal or group of goals is reached [1]. An investment project is a complex technical-technological, organizational, financial and legal undertaking with the goal of constructing and outfitting a structure or structures which have been shown and proven, through previously performed investment studies, to be necessary for the investor or as such provide the investor with a profit.

A project can also be defined as an undertaking which must be realized with the necessary level of quality, within provided time period and within the envisaged budget. In accordance with that, the following four parameters define a project: time period ($T$), scope of work ($S$), performance ($P$) and cost ($C$). The mutual relation of these parameters is often expressed through the formula $C = f(T, S, P)$.

### 4.1.2. Types of designs

According to the Law on planning and construction of structures, the following types of designs exist [2, 3]:

- general design,
- principle design,
- main design,
- contractor design,
- as-built design.

Definitions of the individual types of designs are provided below.

Z. Stojković, *Computer-Aided Design in Power Engineering*,
DOI: 10.1007/978-3-642-30206-0_4, © Springer-Verlag Berlin Heidelberg 2012

*a) General design*

The general design establishes the following:

- general concept,
- macrolocation of the structure,
- technical-technological and economic characteristics of the structure,
- method of securing infrastructure,
- possible variants of spatial-technical solutions from the perspective of:
  - fitting into space,
  - natural conditions,
  - evaluation of environmental effects,
  - protection of natural and immobile cultural heritage,
  - feasibility for construction of the structure.

*b) Principle design*

The principle design is the design which establishes:

- position (microlocation of the structure),
- capacity,
- technical-technological and functional characteristics of the structure,
- organizational elements of construction of the structure,
- elements of structure maintenance,
- value of the structure,
- measures for avoiding or decreasing negative effects on the environment,
- feasibility for construction of the structure, which includes a comparative analysis of the variant solutions from the perspective of characteristics of the location and soil, functionality, stability, evaluation of environmental effects, natural and immobile cultural heritage, rationality of construction and exploitation,
- supplying energy.

Based on the general design and principle design it is possible to establish the concept of the structure.

*c) Main design*

The main design establishes the following:

- construction-technical, technological and exploitation characteristics of the structure with equipment and installations,
- technical-technological and organizational solutions for construction of the design,
- value of the structure,
- conditions for structure maintenance.

The main design is drafted for the purposes of construction of the structure and acquiring construction permits, and is done so in accordance with the urbanistic permit.

*d) Contractor design and as-built design*

All details necessary for constructing the structure according to the main design are elaborated in the contractor design.

A display of the as-built state of the structure for the purposes of exploitation and structure maintenance are provided within the as-built design. Basically, the as-built design is the main design with changes which occurred during construction of the structure. In the case that during construction of the structure there were no deviations from the main design, the investor and contractor can state on the main design that the as-built state is identical to that which was designed.

### 4.1.3. Participants in realization of the project

The participants in the realization of the project are [1]:
- the investor,
- the general contractor or contractor,
- subcontractors and collaborators (including delivery of equipment and materials),
- consultant,
- designer,
- those who carry out expert and technical checks,
- expert supervisor and, within that, engineering supervisor,
- administrative supervisor.

*a) Investor and contractor*

The investor is an organization, individual or company which requires the construction of the structure and possesses the financial resources necessary for realization of the design. For the purpose of realization of works on the construction site, the investor hires one or more contractors. Their role is to carry out individual tasks or all tasks which relate to construction of the structure, where construction of the structure entails:

- drafting of technical documentation,
- carrying out preliminary works related to construction,
- constructing or refurbishing the structure.

Generally speaking, the designer also has the status of a contractor in relation to the investor.

*b) Consultant*

The consultant primarily has an advisory function but may also be authorized to represent the employer in certain tasks. The investor and contractor may hire a consultant if they feel that in a certain field they require expert knowledge or highly-qualified services. The area of expertise, rights and responsibilities of the consultant are the subject of consideration of the international Federation of Consulting Engineers (FIDIC) and the United Nations Economic Commission for Europe (UN ECE) [4, 5]. Additionally, the Law on public procurement defines the process of selection of consultants and awarding of consultant services [6].

The expert services which are offered by the consultant, according to the FIDIC "the guide for offering engineering services of independent consultants", can generally be divided into five categories [1, 4]:

- Counseling services – CS,
- Pre-investment studies – PIS,
- Design and supervision services for construction of works – DSSCW,
- Specialized design and development services – SDDS,
- Project management – PM.

According to the FIDIC, early inclusion of a consultant in the task is of key significance, i.e. in the drafting of the pre-investment study, because it is considered that 85 % to 90 % of possible savings in a project are achieved in the phases of work which lead up to the drafting of the tender (bidding conditional documentation) for selection of the contractor, and that remaining savings can be achieved during construction and outfitting of the structure, that being through strict control over expenses, quality of works and deadlines.

In reference [5] the term "engineering consulting" is used which exclusively refers to intellectual services. Four main methods of approach to the offering of consulting services have been adopted:

- conventional approach,
- in-house approach – IH,
- project management,
- turn-key – TK.

The aforementioned methods are briefly described below. The conventional approach implies that the investor takes complete responsibility for the realization of the project, leaving the consultant to perform professional services such as:

- drafting plans and specifications,
- preparation of bidding conditional documentation (tender)
- cost analysis,
- representing the interests of the investor during submission of bids or during negotiations and selection of a contractor,
- supervising construction of the structure.

The conventional approach is applied in a majority of projects. Using the in-house method is a method of limited consulting because, when applied in full measure, it leads to a minimum use of the services of independent consultants and a maximum use of the investor's experts. This specifically implies that the investor's experts oversee management of the project, prescribe the methodology of drafting and content of bids, and even the procedure of construction, while the role of consultants is limited to help in specific aspects of the project for which the investor's experts are not trained. This method has a series of advantages which are described within the literature [1, 5].

The third method, the project management method, entails the hiring of a consultancy company or several companies for carrying out the following activities within the framework of the realization of the project:

- planning the realization of the project,
- designing and/or managing the drafting of project documentation,
- managing construction.

Companies which deal with project management prepare and consider contracts and all important elements which are related to the process of construction of the structure. Along with that, these companies act as representatives of the investor as opposed to the "turn-key" method where they act as representatives of the contractor. The project management method has a series of advantages which are described in detail within the literature [1].

The "turn-key" method entails a contract with one company for planning and construction of the complete structure where the consultant may have a dual role, i.e. he may be hired by the investor to protect the investor's interests on the project or by the contractor. The advantage of this method is that a minimum number of the investor's personnel are engaged in the project and construction can begin before the entire project documentation is completed.

Within the work of the consultant, two fundamental principles are applied:

- principle of professionalism (abiding by the rules of the trade),
- principle of impartiality (protection of the interests of the project, and not of the investor or contractor).

The consultant is selected based on qualifications. Payment for the services of a consultant is most often based on the hours of engagement with additional expenses which are acknowledged, or through a percentage of the total expenses on the project. In the case where the volume of the job and plan of work are clearly established, payment is regulated in the form of a fixed rate [1].

*c) Designer*

The role of the designer is to, according to the requirements of the investor, conceptualize and combine the appearance, content and functionality of the structure with the dimensions of its individual components, materials and technology of construction. The conditions for performing the activities of design and construction are defined by the Law on planning and construction of structures [2]. According to this law, designing and/or carrying out works for construction of power system structures and installations may be conducted by companies or other legal entities which are registered in the appropriate registrar for designing and/or carrying out works for construction of power system structures.

Designing and carrying out works for construction of power system structures, for which approval is issued by the ministry in charge of construction matters or by the autonomous province, may be conducted by a company or other legal entity which is registered in the appropriate registrar for designing and/or carrying out works for construction of those types of structures and which has an employee who is licensed as an authorized designer or authorized contractor.

Establishing the fulfillment of the conditions for issuance of a license for an authorized designer or authorized contractor is conducted by the Engineering Chamber of Serbia in accordance with the rulebook on the conditions and procedure for issuing and revoking licenses for authorized city planners, designers and contractors [7, 8].

According to the aforementioned rulebook, the Engineering Chamber issues those who have completed post-secondary education in electrical engineering with the following licenses:

- for authorized designers:
  - License 350 for authorized designers of low- and medium-voltage power system installations,
  - License 351 for authorized designers of high- and medium-voltage power system installations – switchgear and transmission of electrical energy,
  - License 352 for authorized designers of management of electromotor drives – automatic, measurement and regulation,
  - License 353 for authorized designers of telecommunications networks and systems.
- for authorized contractors:
  - License 450 for authorized contractors of low- and medium-voltage power system installations,
  - License 451 for authorized contractors of high- and medium-voltage power system installations – switchgear and transmission of electrical energy,
  - License 453 for authorized contractors of telecommunications networks and systems.

*d) Expert and technical inspector*

The designer is responsible for the drafting and inspection of technical documentation. For the principle and main designs, an expert inspection and technical inspection are conducted [2]. Those who carry out these inspections are licensed designers who are responsible for verifying the adherence to laws and other regulations, standards, technical norms, as well as quality standards which are mandatory in the planning and construction of power system structures.

*e) Expert supervision*

Expert supervision includes the following:

- verifying the quality of works,

- ensuring adherence to regulations, standards, technical norms and standards of quality,
- inspecting materials which are installed,
- ensuring the congruence of carried out works and the technical documentation on which approval of construction is based,
- verifying the dynamic of construction in stages,
- verifying the realization of financial constructions envisaged by the investment program.

Expert supervision is provided by the investor. In the case of more complex projects, where planning and construction are carried out simultaneously, it is possible, and even desirable, that the expert supervision be divided into supervision of engineering and supervision of contracting.

*f) Administrative supervision*

Each of the aforementioned functions (investor, designer, supervisor, expert and technical auditor, contractor) is precisely defined, so to ensure that the conditions of the defined functions are respected, the administrative supervisor is defined which, in the name of the administrative body (i.e. the state), conducts inspections and verifies the application of all prescribed conditions for designing and building the structures.

All of the aforementioned participants in the realization of the project have their own obligations and framework within which those obligations can be performed.

They are included within the realization of the project in various time intervals and with various intensities.

Fig. 4.1 shows the dependency of the level of engagement of key participants (investor, contractor and designer) during realization of the project [1].

Based on the diagram in Fig. 4.1, it can be clearly noticed that the level of engagement of individual participants in the realization of the project changes over time, depending on their expertise, interests and the needs of the project. It is obvious that the investor represents the only engaged party at the very beginning. Following a certain amount of time the designer is included whose fundamental task is the drafting of the project documentation. During the phase of construction, the contractor is responsible for the realization of the project, while the remaining participants conduct control and verification of changes.

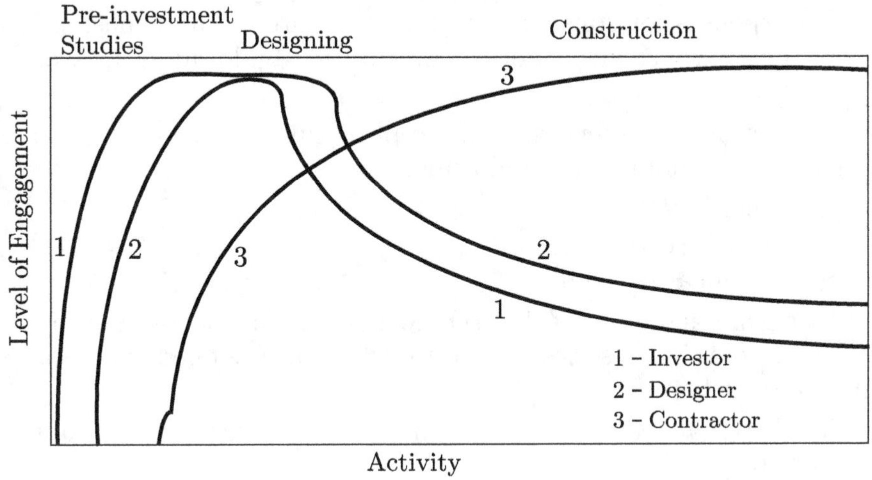


Fig. 4.1 - Dependency of the level of engagement of key participants (investor, contractor and designer) during realization of the project

### 4.1.4. Content of main design

The main design is of special significance considering that it is drafted for the purposes of construction of the structure and acquiring construction permits, and done so in accordance with the urbanistic permit. It is necessary to note that the content of the main design of electrotechnical documentation has not been established by a single valid document, but rather within legal regulations [2, 9, 10]. The main design must contain the following components [2, 3]:

- additional investigative works,
- elaboration of technical-technological and exploitation characteristics of the structure with equipment and installations,
- calculation of stability and security of the structure and calculation from the area of construction physics,
- foundation schematic of the structure,
- technical-technological and organizational solutions for construction of the project,
- elaboration of connections of the structure to the appropriate trafficways and other infrastructure, as well as the layout of open space,
- conditions for protection of neighboring structures,
- elaboration of measures for avoiding or decreasing the negative effects on the environment,

- expenses for construction and structure maintenance,
- geodetic-technical documentation for building and observing the structure.

According to the occasionally applied rulebook [10] the technical documentation contains the following:

- general section,
- textual section,
- graphical section.

The general section of the technical documentation goes into great detail and contains a series of details which are defined in the aforementioned rulebook.

The textual section of the technical documentation contains the following components:

- project task,
- technical description of works,
- preliminary calculation and specification of works,
- necessary calculations.

The graphical section of the technical documentation contains the following components:

- situational plan,
- characteristic horizontal and vertical cross sections and appearances which are determined by the function, scope, appearance and construction of the structure in the appropriate proportion,
- regulation protocol or act which replaces it,
- excerpt from the cadastre of installations,
- excerpt from the land cadastre with a geodetic display of the current state.

Based on legal regulations, the following components are most often submitted with the main design [3].

- general data,
- project task,
- technical description,
- general and technical conditions,
- specification of materials,
- bill of quantities,
- attachment or separate elaboration of the applied prescribed measure of protection at work,

- calculations,
- graphical documentation which includes:
  - situation,
  - installation plans,
  - dispositions,
  - single-pole diagrams,
  - wiring diagrams,
  - connection diagrams,
  - details for carrying out electrical installation works.

Supplement 4.1 shows the typical content of a main design book for adaptation of a power system structure [11].

Supplement 4.1 Typical content of a main design book for adaptation of a power system structure [11]

## CONTENT
## VOLUME E0: TEXT AND GRAPHICAL DOCUMENTATION

**I LIST OF BOOKS**

**II GENERAL DOCUMENTATION**

1. DECISION ON REGISTRATION OF THE INVESTOR

2. DECISION ON REGISTRATION OF THE ENGINEERING ORGANIZATION

3. DECISION ON APPOINTING THE AUTHORIZED DESIGNER

4. LICENSE OF THE AUTHORIZED DESIGNER

5. CONFIRMATION FROM THE DESIGNER REGARDING CONFORMITY OF DOCUMENTATION

6. DECLARATION OF THE DESIGNER REGARDING USE OF REGULATIONS

7. PROJECT TASK

**III TECHNICAL DOCUMENTATION**

1. TECHNICAL DESCRIPTION

2. TECHNICAL CONDITIONS

3. TECHNICAL CALCULATIONS

3.1 Introduction

3.2 Data on installation

3.2.1 Normal operation

3.2.2 Short circuit

3.3 Selection and verification of high-voltage equipment

3.3.1 110 kV circuit breaker

3.3.2 110 kV disconnector

3.3.3 110 kV disconnectors with earthing switches

3.3.4 110 kV current measuring transformer

3.3.5 110 kV voltage measuring transformer

3.3.6 110 kV surge arrester

3.4 Equipment specifications

3.4.1 110 kV circuit breakers

3.4.2 110 kV disconnectors

3.4.3 110 kV current measuring transformers

3.4.4 110 kV voltage measuring transformers

3.4.5 110 kV surge arresters

3.5 Installation grounding

3.5.1 Distribution of current within the grounding system

3.5.1.1 Current which is released into the soil through the grounding system

3.5.1.2 Current which is released into the soil through the "Vreoci" transformer station (TS) grounding system

3.5.1.3 Selection of cross section of the grounding grid conductor

3.5.2 Installation grounding grid

3.5.3 Realization of grounding system

3.6 Description of protective-control system within the 110 kV/35 kV "Vreoci" TS

4. PROTECTION AT WORK

4.1 General information on measures

4.2 Special technical instructions and protective measures

4.3 Protective measures from the effects of electrical energy

5. BILL OF QUANTITIES

5.1 High-voltage equipment

5.2 Protective equipment and control equipment

5.3 Material for grounding grid

5.4 Recapitulation

**IV GRAPHICAL DOCUMENTATION**

1. Single-pole diagram 110 kV substation E0-01
2. Disposition 110 kV substation E0-02
3. Intersection and basis 110 kV substation E0-03
4. Disposition of grounding grid E0-04

On the basis of supplement 4.1 it can be concluded that the project task is an integral component of the general documentation (item II.7). This important document is created and signed by the investor, but in practice is most often created through the joint work of the investor and designer. The project task determines the content of item III.3 Technical calculations.

Supplement 4.2 shows the project task of the main design of adaptation of the power system structure, the contents of which are defined in Supplement 4.1 [11].

Supplement 4.2 Project task of the main design of adaptation of the power system structure, the contents of which are defined in supplement 4.1 [11]

## PROJECT TASK FOR DRAFTING OF THE MAIN DESIGN OF ADAPTATION OF THE 110 kV/ 35 kV "VREOCI" TS

### 1.0 GENERAL INFORMATION

1.1 Project title: Main design of adaptation of the 110 kV/35 kV "Vreoci" TS

1.2 Investor: Electrical Utility of Serbia, mining Kolubara – Lazarevac

1.3 Designer: Faculty of Electrical Engineering, Belgrade, Serbia

### 2.0 TECHNICAL DATA

2.1 110 kV/ 35 kV "VREOCI" TS

2.1.1 110 kV substation

1. Provide a solution for a 110 kV busbar system through which reliably feeds the TS within the possible operational variants. This solution should envisage the installation of a sectional circuit breaker in the 110 kV busbars.

2. Replace the existing circuit breakers within transformer bays with three pole circuit breakers featuring motor-spring operation with the appropriate circuit breaking characteristics.

3. Replace the existing disconnectors with disconnectors of the appropriate rated current with motor operation.

4. Envisage new voltage measuring transformers in sections of the busbar which conform to the valid internal recommendations and standards.

5. Envisage new current transformers in the transformer bays which conform to the valid internal recommendations and standards.

6. Envisage new, metal-oxide surge arresters in transformer bays on the 110 kV side.

7. Replace disconnectors within the neutral point of the transformers on the 110 kV side with the appropriate motor-driven disconnectors.

8. Conduct the selection of equipment based on the analysis of perspective current loads.

2.1.2 35 kV switchgear

1. Conduct an analysis of perspective current loads.

2. Completely replace circuit breakers, disconnectors and corresponding equipment in transformer and feeder bays with modular blocks, of a modern construction, with vacuum contacts in $SF_6$ environment.

3. Envisage reconstruction of the 35 kV busbar with the goal of enabling the parallel operation of the transformers on the 35 kV side.

4. Verify the equipment at perspective current loads and, when necessary, replace the remaining power system equipment on the 35 kV side.

5. Verify the condition of the TS grounding grid and, if necessary, draft a restoration plan for the grounding grid.

6. Ground the neutral points of the transformer on the 35 kV side through a separate grounding resistor for a current of 300 A.

7. Remove the existing link between the neutral points of the transformer.

2.1.3 Protection and control system

1. Replace all existing electromechanical protection with new microprocessor protection with the following characteristics:

- complete microprocessor platform,
- devices for medium-voltage protection placed as close as possible to the substation (within the distribution board),
- devices for protection and control of transformers and transmission lines in the substation of the "Kolubara" thermal power plant (TPP) placed in a separate cabinet,
- devices for high-voltage protection must have separated devices for protection and control and with medium-voltage protection those two functions must be combined in one device,
- the connection with the local SCADA in the TS must be achieved according to protocol IEC 61850 through an optic line,
- the devices must contain an events list and digital register of events,
- adjustment of the device must be through a PC or directly on the device itself,

- remote access by the engineer to the devices must be through an Ethernet line according to protocol IEC 61850,
- the devices must contain the option of measurement of electrical energy,
- protective devices must contain all protective functions envisaged by the technical recommendations of Electrical Utility of Serbia and the conditions in the TS,
- the devices must have LEDs for signalization of protective operation,
- the control devices must enable the setting of all blockages through software on both the level of the bay as well as on the level of the TS through protocol IEC 61850,
- a portable laptop computer must be envisaged with a program for adjusting and viewing the protection and control devices.

2. Install a new control system with the following characteristics:
- the new system must be based on microprocessor technology,
- it must have the ability to display the actual condition of equipment in the TS,
- it must enable the command of all switching devices with motor drive,
- it must enable command of auxiliary devices (ventilators, regulating power switches, 0.4 kV circuit breakers, lights, etc.),
- it must be able to collect and display data on the reaction of protection through alarm lists and events lists,
- communication with devices according to bay must be through a fiber optic (FO) line according to protocol IEC 61850,
- communication with the dispatcher center must be according to protocol IEC 101,
- registration and archiving of measurement data,
- display of all available measurements in the TS,
- the system must be equipped with a device for synchronization of the exact time.

3. Control is envisaged from one location within the "Kolubara" system. The entire command should be realized using direct voltage.

2.1.5 Self consumption of the TS

1. For the self consumption within the TS, envisage the use of an auxiliary alternating voltage $3 \times 400/230$V, 50 Hz and direct voltage of 110 V. It is necessary to conduct revitalization of the self consumption by placing the entire equipment within cabinets.

2. Supply the alternating-voltage self-consumption feeder is envisaged from the 35 kV/0.4 kV self-consumption transformer. The self-consumption transformer feeds the main alternating voltage feeder through the corresponding 1 kV cable. The alternating-voltage self-consumption feeder is located in a separate cabinet. The cabinet has been envisaged in the self-consumption facilities. From the self-consumption feeder – the alternating sections feeds: the TS management system, rectifiers which feed accumulator batteries, self-consumption feeder - DC part, cooling cabinet of the power transformer, regulation cabinet, etc.

3. Supply the direct-voltage self-consumption feeder is envisaged through the rectifier and, in the case of a loss of alternating voltage, from the stationary hermetically sealed battery of a proper capacity. The self-consumption direct-voltage distribution cabinet is located beside the alternating-voltage distribution cabinet. Signalization of the loss of voltage and drop of individual automatic devices has been envisaged on separate signalization devices which are located in the cabinet together with the station computer. The connection between the battery and rectifier is by cable, through the cable duct in the floor of the facility.

**3.0 SPECIAL REQUIREMENTS**

1. Abide by existing regulations and internal standards in the selection of equipment and other solutions. A detailed justification must be provided for any possible deviations from these norms.

2. Carry out the works in stages within the substation so that during works the entire operation of the TS does not have to be shut down.

**4.0 CONTENT OF DESIGN**

The design must contain the following components:

1. Technical description.
2. Technical conditions.
3. Calculations.
4. Protective measures.
5. Bill of quantities.
6. Graphical documentation.
7. List of applied regulations and measures.

INVESTOR

### 4.1.5. Phases within the realization of the project from the perspective of the investor

Based on the definition of the participants in the realization of the project, it is clear that each of the participants has expressed goals. The fundamental goals, common for all participants, relate to:

- striving toward the successful realization of the project,
- achieving a profit,
- acquiring references significant for future work.

The investor bears the greatest consequences in the case of an unsuccessful project, considering that the idea and financial support for realization of the project come from the investor. For this reason, the investor has the greatest interest in ensuring that the project is managed with the highest level of quality, especially during the initial works on realization of the project. These works relate to the conducting of the pre-investment studies, drafting of the principle design and elaborating the specification of equipment. Based on all of this the decision is made of whether or not the structure should be built.

For the aforementioned reasons, the investor alone, or with the assistance of a consultant and/or designer, must consider and formulate the requirements which relate to the following issues [1]:

- functionality of the structure,
- quality of works,
- time,
- expenses and payment,
- special requirements.

Each of the participants in the realization of the project has their own interests and periods of engagement and, on the basis of that, develops their own management methodology. Effective management of the project entails the defining of the basic phases within the realization of the project. Within the literature there are various approaches to defining the individual phases of a project from the perspective of the investor [1]:

- general approach,
- W. Hughes approach,
- R. Pilcher approach.

*a) General approach*

The general approach entails two phases:

- construction phase, which is divided into:
    - phase of forming the concept, which ends with the decision on investment, and
    - phase of realization, which definitively concludes upon the expiration of the warranty period,
- exploitation phase of the investment structure.

The phase of forming the concept is the most delicate phase from the perspective of the investor. In this phase the investor unavoidably hires a consultant for individual areas, as well as an independent designer for drafting the conceptual solution and conceptual design.

The strategy of the project is defined entirely on the basis of:

- drafted project documentation,
- opportunity study – OS,
- prefeasibility study – PFS,
- feasibility study – FS.

The opportunity study identifies the general opportunities of constructing a future investment and estimates the number and content of studies which must still be conducted. The prefeasibility study provides a provisional decision on the feasibility of carrying out the project and the decision on whether it is necessary to conduct the feasibility study. The feasibility study contains the final selection of the future characteristics of the project which are defined on the basis of previous studies and determine the justification of the project in accordance with the set criteria.

Along with the aforementioned studies, the following is also normally done:

- support study – SS, through which research is done on individual aspects of the future project such as:
    - marketing,
    - raw materials,
    - financing.
- evaluation study – ES, which essentially contains the final decision on investing in the project.

*b) W. Hughes approach*

According to the W. Hughes approach, the following phases are defined within the realization of the project from the perspective of the investor [1]:

- inception stage – IS,
- feasibility stage – FS,
- scheme design – SD,
- detail design – DD,
- contracting,
- construction,
- commissioning.

The contents of the individual phases according to this approach are described in detail within the selected literature [1].

*c) R. Pilcher approach*

According to the R. Pilcher approach, the fundamental phases which are important for the investor are [1]:

- feasibility and design – FD,
- purchasing/ordering,
- construction,
- commissioning and operating – CO.

### 4.1.6. Types of contracts

The contract represents the legal document which regulates the rights and obligations of the participants in the realization of a project. A contract, from a general perspective, is the free will agreement of two or more entities through which a legal action is achieved. The legal action may consist of or be based on a legal relation or on the exchange of an existing legal relation or the termination of such a relation. According to their theoretical structure, plural form and application in practice, the most significant are obligational contracts and among those, the most important are those which create obligations - binding responsibilities, duties [1].

Contracts result from one side – the bidder, submitting a bid to the other side - the investor, with the acceptance of the bid by the second side. The bid and acceptance of the bid are therefore the two acts from which the concluding of the contract is composed. The process of negotiation is in fact the period of mutual harmonization.

Contracts are divided into [1]:

- classic contracts,
- complex contracts.

The three basic groups for the new division of contracts are [1]:

- division of contracts according to the subject of the obligations of the contractor,
- division of contracts according to the method for establishing the price, and
- division of contracts according to method of financing.

The basic elements of the aforementioned divisions are briefly covered below.

*a) Division of contracts according to the subject of the obligations of the contractor*

According to the subject of the obligations of the contractor, contracts can be divided into contracts for:

- carrying out construction works,
- installation of electrical-machine equipment within the framework of which the following can be specified:
    - only delivery of equipment,
    - only installation of equipment,
    - delivery and installation of equipment, or
    - all three previous tasks with supervision of works.
- turn-key construction and outfitting of the structure,
- construction and outfitting of industrial structures with a performance guarantee, i.e. for achieving the planned capacities, and
- construction management – CM.

*b) Division of contracts according to the method for establishing the price*

Along with the subject of the contract, another important element is the price for realization of the contracted work. These prices can be established within contracts as [1]:

- Lump sum – LS,
- Unit prices – UP,
- Cost plus – CP,
- Sealing price – SP,
- Fixed price – FP.

The methods for determining the price for services according to the Law on planning and construction of structures [2] have been published and generally accepted within the so-called blue book [12]. Forming of prices for services during the drafting of technical documentation from the field of power engineering is most often done using the percentage of investment value method [3]. The application of the method is possible after realizing the following steps in the calculation of the price for services:

- establishing the investment value,
- classifying the project,
- establishing the percentage for developing the electrical project.

The established investment value implies introducing a coefficient of correction to the actual investment value. This coefficient is determined on the basis of the average annual index of price increases for construction works, trade works and equipment in relation to the considered year, established according to data from the Republic Institute for Statistics [3].

The classification of the project is from the perspective of the services for drafting the technical project documentation conducted in the following manner:

- projects for electrotechnical installations in structures,
- industrial plant electrical projects,
- electrical power and communications projects,
- electrical projects for electrical facilities in railway traffic.

Establishing the percentage for drafting the electrical project is done for each type of project depending on the uncorrected investment value and class of project. These dependencies are displayed in the form of tables and diagrams [3].

Forming the price for services of constructing power system structures is done by applying a method which is based on the recognition of the appropriate norms of work [13]. Most often the method of combined norms for the volume of work is applied which, along with the percentage of the investment value, also contains expenses for operations, revenue and the norm for necessary consumption of working hours. The norm for work is a calculation of the necessary consumption of working hours for completing a unit of measure of a certain item during which the following criteria of complexity are recognized:

- per number of technological balance operations,
- per share of non-standard and non-cataloged equipment,
- per characteristics of work and level of outfitting,
- per share of its own technology.

The procedure for calculation of the norm of work is displayed in references [12, 13].

When forming the price for realization of the contracted work, it is necessary to recognize the effect of the specific conditions in which the project is realized. This, above all, relates to recognizing the effect of the following two conditions:

- inflation,
- unplanned cost increases.

The effect of inflation can be recognized in one of the following two ways:

- factoring inflationary movement into the estimation of expenses based on official periodic publications of the government,
- specifying a clause within the contract which serves to harmonize the actual expenses at the time they are incurred with the actual price increases since the time when those expenses were estimated (sliding scale).

The second, very common case of changed estimated expenses is unplanned price increases which can occur due to the following reasons [1]:

- poor-quality tender documentation and initial project solution,
- inadequately trained workforce carrying out works,
- failures by the contractor and supplier to complete their works within the planned frameworks and deadlines,
- surprise moves by the government within the country where works are carried out,
- failures to properly factor in customs regulations during the estimation of expenses,
- additional expenses of transport due to poorly analyzed sources for materials,
- sudden changes in the offer and prices of the local workforce,
- natural catastrophes,
- additional expenses due to the lack of geological foundation or inadequately conducted investigative works,
- additional expenses for preparation of the construction site due to unplanned shifting of underground installations, etc.

*c) Division of contracts according to method of financing*

The division of contracts according to the method of financing is done in the following manner:

- payment in hard currency,
- payment partially in cash, partially in credit,
- establishing a combined company which is responsible for payment matters,
- payment in the counter value of delivered goods and services with that which is produced by the newly constructed or reconstructed structure,
- payment in goods and services through counter deliveries of the investor according to the specific market circumstances and depending on the relations of the partner and the adopted concept of work.

Details about the methods of financing are displayed in the selected literature [1].

### 4.1.7. Example contract

Supplement 4.3, with the consent of the orderer and executor, shows an example contract for the drafting of the main design of adaptation of the considered power system structure [11]. For understandable reasons, the price has been left out.

On the basis of the displayed example, it can be concluded that the basic elements of a contract are:

- subject of the contract,
- completion deadline,
- obligations of the orderer,
- obligations of the executor,
- price and method of payment,
- engagement of external collaborators,
- contracted fines,
- transitional and closing regulations.

It must be noted that the estimation of the completion deadline is possible through the application of the following expression [14]:

$$T_e = \frac{T_0 + T_p + 4 \cdot T_m}{6} \tag{4.1}$$

where: $T_e$ – estimated time – ET, $T_0$ – optimistic time – OT (shortest), in which just 1 % of similar projects are finished , $T_p$ – pessimistic time – PT (longest), in which 99 % of similar projects are finished, $T_m$ – most probable time – MPT.

Supplement 4.3 Example contract for drafting of the main design of adaptation of the considered power system structure [11]

Electrical Utility of Serbia
Mining Kolubara - Lazarevac

CONTRACT

CONCLUDED BETWEEN:

ORDERER: Electrical Utility of Serbia, Mining Kolubara - Lazarevac

EXECUTOR: Faculty of Electrical Engineering, Belgrade, Serbia

SUBJECT OF THE CONTRACT

Article 1.

The subject of this Contract is the drafting of the main design for adaptation of the 110 kV/ 35 kV "Vreoci" TS and 110 kV transmission line bays numbers 23 and 24 of the substation of the "Kolubara" TPP according to Bid No. 2292 from 27.09.2005.

COMPLETION DEADLINE

Article 2.

The completion deadline for the project is 9 (nine) months from the day on which the Contract comes into force.

OBLIGATIONS OF THE ORDERER

Article 3.

The Orderer is obligated to place all information and documents necessary for the drafting of the subject design at the disposal of the Executor. Information which is not at the disposal of the Orderer will be acquired or determined jointly by the Executor and Orderer.

The Orderer is obligated to provide the execution of work related to possible preceding checks of the conditions of the installation (test excavation of the grounding grid) on the basis of the expert instructions and supervision of the Executor.

The Orderer is obligated to form a commission for the final calculation and inform the Executor of such.

OBLIGATIONS OF THE EXECUTOR

Article 4.

The Executor is obligated to, in a timely manner and with a high level of quality, draft the subject design in accordance with the adopted project task while abiding by all valid regulations and standards.

The Executor is obligated to submit the specifications of the most important necessary power system equipment to the Orderer within a deadline of 30 days from the day on which the Contract comes into force.

Article 5.

The Executor is obligated to submit 6 (six) copies of the main design to the Orderer.

PRICE AND METHOD OF PAYMENT

Article 6.

The price for drafting the main design is the net amount of:

Change of the contracted price will be conducted on the basis of movement of the middle exchange rate of the Euro according to the National Bank of Serbia on the day of invoicing. The base date for calculation is 27.09.2005. VAT is not calculated into the price.

Article 7.

The Executor will issue invoices for completed works on a monthly basis, temporary situations and in accordance with the mutually signed records on the level of completion of the contracted work. The actual value of the completed work will be established upon the completion of the final calculation.

Payment per temporary situations and the final situations will be completed within 15 days from receipt of the same by the Orderer.

EXTERNAL COLLABORATORS

Article 8.

For the purpose of executing the portion of works from Article 4 and Article 5, the Executor may hire external collaborators. Their mutual rights and obligations will be regulated by separate copyright contracts concluded between the Executor and the external collaborators.

CONTRACTED FINES

Article 9.

For failure to complete the contracted work, the Executor will pay the Orderer the contracted fine in the amount of 1 ‰ (one per mil) daily, but no more than 5 % (five percent) of the contracted value.

The Executor is responsible for fixing/removing any short comings within the main design within a deadline of three weeks from the day when complaints are submitted.

If the Orderer fails to submit complaints to the Executor within a deadline of three weeks from receiving the main design, it will be considered that the main design has been accepted.

TRANSITIONAL AND CLOSING OBLIGATIONS

Article 10.

For all matters which are not defined by this Contract, the Law on obligational relations will be applied.

Article 11.

All disagreements between the contracted parties will be resolved by mutual consent. If it is not possible to reach an agreement, the Commercial Court of Belgrade has jurisdiction over all possible disputes.

Article 12.

This Contract comes into force on the day when it is signed by both parties.

Article 13.

This Contract is composed of 6 (six) identical copies. Three signed and certified copies belong to each of the contracted parties.

| FOR THE EXECUTOR | FOR THE ORDERER |
|---|---|
| DEAN OF THE | DIRECTOR |
| FACULTY OF ELECTRICAL | |

## 4.2. Basics of project management

Project management, according to the FIDIC guide [1, 4] is defined as the mobilization of various sources and coordination of their activities with the goal of the task which is executed by each individual accumulating into a multidisciplinary team effort for achieving the investor's goal within the framework of the agreed plan, budget and level of quality.

Project management can also be defined as the efficient use of resources in order to realize a project as planned, within the deadline, with the envisaged method of execution and within the framework of the planned budget. Project management is the application of knowledge, skills, tools and techniques in order to achieve the project requirements, too.

### 4.2.1. Project cycle

The project cycle broadly describes the phases through which a project goes from start to completion. Every expert in project management has their own description of the project cycle. Below are the aforementioned phases of one six-step project cycle which is most commonly applied in practice:

- identifying the problem,
- defining the problem,
- planning the project,
- realizing the project,
- applying the project,
- evaluation.

Regardless of the number of phases in the project cycle, the order remains the same. The aforementioned phases are discussed in Chapter 1.

It is necessary to note that planning the project represents a crucial phase of the project. For the planning of the three fundamental parameters of the project (quality, price, duration), it is necessary to take the following steps [14]:

- establish the goal of the project,
- select the basic strategy for achieving the goal,
- separate the project into segments,
- determine the standards for carrying out each segment,
- determine the necessary time for completing each segment,
- determine the proper order for executing the segments,
- determine the price for each segment and include the expenses into the project budget,
- determine whether and what kind of training is necessary for the members of the project team.

### 4.2.2. Phases of project management

Project management consists of several phases, of which the most important are:

- project integration management,
- project goal management,
- project time management,
- project cost management,
- project quality management,
- project human resource management,
- project communications management,
- project risk management,
- project procurement management,

Each of the aforementioned phases is composed of the following five processes:

- initialization,
- planning,
- execution,
- control,
- completion.

For the realization of the aforementioned phases of project management, it is necessary to recognize the input data and apply the proper tools and techniques for the purpose of obtaining the proper output data. The most commonly used tools and techniques in project management are:

- Monte Carlo Analysis – MCA,
- Project Management Information System – PMIS,
- Sensitivity Analysis – SA,
- Benefit / Cost Analysis – BCA,
- Decomposition,
- methods, techniques and analyses based on a network diagram such as:
  - GERT – Graphical Evaluation and Review Technique,
  - PERT – Program Evaluation and Review Technique,
  - CPM – Critical Path Method,
  - What-If Analysis.
- flow charts (cause-and-effect diagrams (Ishikawa diagrams)),
- statistical methods for identifying influential factors,
- control diagrams,

– Pareto charts,
– Delphy technique.

Gantt charts and PERT diagrams, which are created during the project planning phase, are very useful to the project manager for control of the project in the sense of quality, time and budget. A Gantt chart is a horizontal display of activities within the function of their duration. A PERT diagram enables the functional display of individual activities and their mutual connection.

For control of the project, as the center activity during the realization of the project, the following four diagrams are useful:

– CPIC – Control Point Identification Charts,
– PCC – Project Control Charts,
– MC – Milestone Charts,
– BCC – Budget Control Charts,

Details of the tools and techniques for project management are provided within the selected literature [14-17].

## 4.3. Examples of project management

### 4.3.1. Application of the program MS Excel

Through the application of the program MS Excel, one can form [18]:

– static Gantt charts for visual presentations,
– dynamic Gantt charts which calculate duration and enable automatic graphical representation,
– a combination of Gantt charts with other diagrams and formulas.

Due to the fact that all projects are carried out over a certain period of time, the use of MS Excel can calculate the duration of processes in production, sales, finance or any other business segment. If the timeline is properly set, to the levels of monitoring and envisaging which are necessary for the concrete circumstances, it is possible to connect them or address the formula.

An illustrated example of the forming of a Gantt chart for a technological development project is provided below [19]. A similar procedure is required for management of later technological development projects [20, 21].

#### 4.3.1.1 Forming dynamic Gantt charts

Through the technological development project [19], the realization of a prototype fiber optic device for measuring current intensity (Fiber Optic Current Sensor - FOCS) has been envisaged. Fig 4.2 shows the appearance of this device which is able to be applied for measurement of current intensity and harmonic analysis, as well as for measurement of the impulse current provided in the selected literature [22-24].

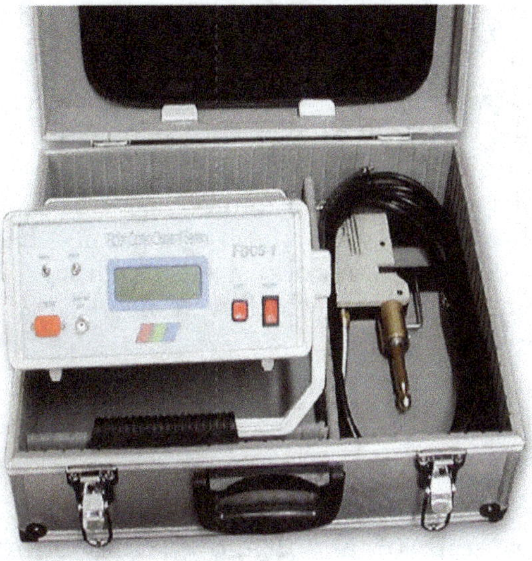

Fig. 4.2 – Fiber Optic Current Sensor – FOCS [19]

Dynamic Gantt charts are activated by timelines and start and finish dates. The duration of the project is the period between its start date and completion date, and the week value shows the weekly output. The week value can be a percentage, a unit of production, an individual, etc. The essence is in the fact that a Gantt dynamic timeline is based on a simple formula which references the week and week value. The week value can be a name, letter or number. It is necessary to pay special attention to how the formula addresses four key cells:

- first cell of the timeline,
- second cell which contains the start date,
- third cell which contains the completion date,
- fourth cell which displays the week value.

The duration from the start date to the completion data is obtained through extraction which is the key difference between MS Excel and other project management programs.

The following example shows the application of MS Excel's basic characteristics and functions in project management. The goals of this project are:

- development of voltage divider (VD),
- development of electronic circuit,
- development of optical system,
- forming of an impulse current generator and testing the measurement characteristics of the device.

The project was realized within the microelectronic laboratory of the Faculty of Electrical Engineering, University of Belgrade and within the laboratory of the Institute for High Voltage and Power Systems of the Faculty of Electrical Engineering, University of Karlsruhe, FR Germany. The project was composed of nine phases, each lasting one month. Within the framework of each phase of the project, certain activities were carried out which were necessary for the successful realization of the project.

#### 4.3.1.2 Description of activities

The activities on which the forming of the Gantt chart was based are described below.

**1. Development of a voltage divider model (VD).** This activity envisages the forming of a mathematical model for analyzing the voltage divider. Based on the mathematical model, the characteristics of individual components of the voltage divider will be determined from the aspect of measuring alternating and impulse current. Persons A, B, C and D took part in the realization of this activity with 100 % engagement (Fig. 4.3). The activity was completely achieved.

**2. Procurement and/or manufacturing of individual VD components.** This activity envisages the recognition of the need for procurement of individual components of the voltage divider, as well as the possibility of manufacturing individual components. Persons E, F, G and H took part in the realization of this activity with 100 % engagement. The activity was fulfilled completely according to the envisaged plan.

**3. Installation of the VD measurement system.** Based on the procured and formed components, the voltage divider was composed in the corresponding laboratory for high voltage. Persons A, B,

C, D, E, F, G and H took part in the realization of this activity with 100 % engagement. The activity was fulfilled completely according to the envisaged plan.

**4. Testing of the VD measurement system.** Within the framework of this activity, testing was conducted on the measurement characteristics of the formed voltage divider. The relation of division and transmission characteristics of the voltage divider was determined.

Persons A, B, C and D took part in the realization of this activity with 100 % engagement.

The activity was fulfilled completely according to the envisaged plan.

**5. Development of electronic circuit.** Within the framework of this activity, the manufacturing of individual components of the electronic circuit for the purposes of measuring current intensity was envisaged. The activity was conducted within the microelectronics laboratory of the Faculty of Electrical Engineering, University of Belgrade. Persons E, F, G and H took part in the realization of this activity with 100 % engagement. The activity was fulfilled completely according to the envisaged plan.

**6. Testing of electronic circuit.** Within the framework of this activity, the testing of the electronic circuit was envisaged along with possible modifications in accordance with the obtained results. The activity was conducted within the microelectronics laboratory of the Faculty of Electrical Engineering, University of Belgrade. Persons A, B, C, D, E, F, G and H took part in the realization of this activity with 100 % engagement. The activity was fulfilled completely according to the envisaged plan.

**7. Development of optical system.** The development of an optical system is the subject of the seventh activity. The characteristics of individual components of the optical system are determined on the basis of the needs for measuring impulse currents with a nominal rise time front amounting to several $\mu$s. This means that the characteristics of the measurement component are adapted to a frequency range of several MHz. Persons A, B, C, and D, with 100 % engagement, participated in the realization of this activity. The activity was fulfilled completely according to the envisaged plan.

**8. Testing of optical system.** Based on the results of measurement which were obtained following the seventh activity, modifications were made along with the repeated testing of the optical system. These activities were conducted by persons E, F, G and H with 100 % engagement. The activity was fulfilled completely according to the envisaged plan.

**9. Forming of an impulse current generator and testing the measurement characteristics of the device.** In the aforementioned laboratory of the Institute for High Voltage and Power Systems, within the framework of the ninth activity, the forming of an impulse current generator was envisaged along with the testing of the measurement characteristics of the device. The components of this generator should enable the generating of impulse currents defined by regulations and verification of the measurement characteristics of the (Fiber Optic Impulse Current Sensor – FOICS) device [24].

In accordance with this activity, an optical-electronic system was composed with its components being determined in such a way that they completely enable the precise and reliable measurement of impulse currents of various forms and amplitudes.

Persons A, B, C, D, E, F, G and H took part in the realization of this activity with 100 % engagement. The activity was fulfilled completely according to the envisaged plan.

The creation of the Gantt chart was conducted on the basis of the aforementioned activities displayed in Fig. 4.3. The Gantt chart is illustrated in Fig. 4.4.

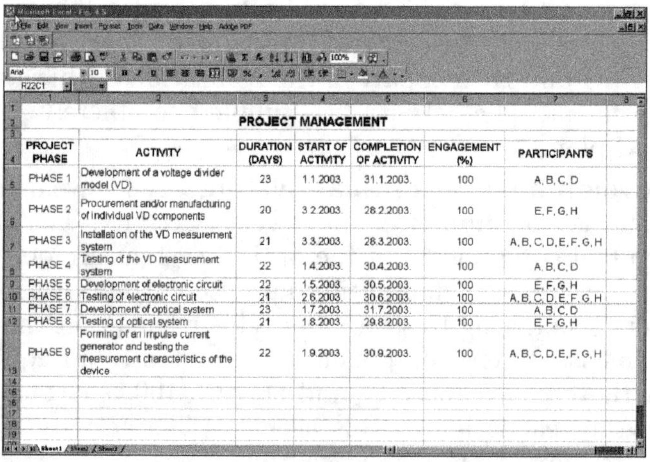

PROJECT MANAGEMENT

| PROJECT PHASE | ACTIVITY | DURATION (DAYS) | START OF ACTIVITY | COMPLETION OF ACTIVITY | ENGAGEMENT (%) | PARTICIPANTS |
|---|---|---|---|---|---|---|
| PHASE 1 | Development of a voltage divider model (VD) | 23 | 1.1.2003. | 31.1.2003. | 100 | A, B, C, D |
| PHASE 2 | Procurement and/or manufacturing of individual VD components | 20 | 3.2.2003. | 28.2.2003. | 100 | E, F, G, H |
| PHASE 3 | Installation of the VD measurement system | 21 | 3.3.2003. | 28.3.2003. | 100 | A, B, C, D, E, F, G, H |
| PHASE 4 | Testing of the VD measurement system | 22 | 1.4.2003. | 30.4.2003. | 100 | A, B, C, D |
| PHASE 5 | Development of electronic circuit | 22 | 1.5.2003. | 30.5.2003. | 100 | E, F, G, H |
| PHASE 6 | Testing of electronic circuit | 21 | 2.6.2003. | 30.6.2003. | 100 | A, B, C, D, E, F, G, H |
| PHASE 7 | Development of optical system | 23 | 1.7.2003. | 31.7.2003. | 100 | A, B, C, D |
| PHASE 8 | Testing of optical system | 21 | 1.8.2003. | 29.8.2003. | 100 | E, F, G, H |
| PHASE 9 | Forming of an impulse current generator and testing the measurement characteristics of the device | 22 | 1.9.2003. | 30.9.2003. | 100 | A, B, C, D, E, F, G, H |

Fig. 4.3 - Working MS Excel table with basic data for creating the Gantt chart for the technological development project [19]

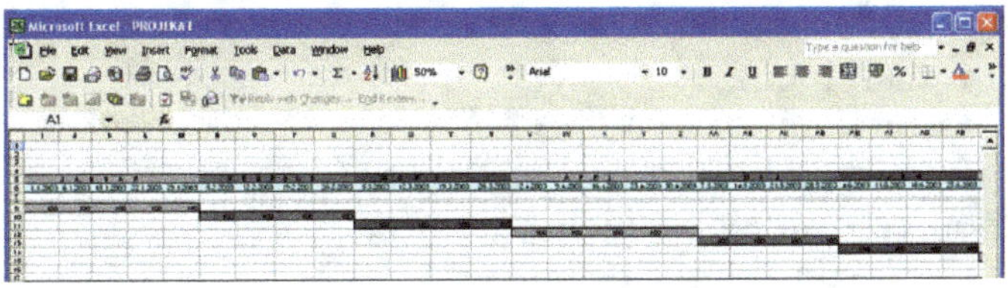

Fig. 4.4 – Gantt chart for the technological development project [19]

## 4.3.2. Application of the program MS Project

### 4.3.2.1 MS Project basics

In this section descriptions are provided for the most important elements of the program MS Project [25-27]. The primary task of this program is to enable the project manager to plan the timeline of the project and analyze financial indicators.

Upon opening the program MS Project, a working window is opened with the option for creating a Gantt chart already activated (Fig. 4.5). The calendar to the right side of the screen is displayed in days.

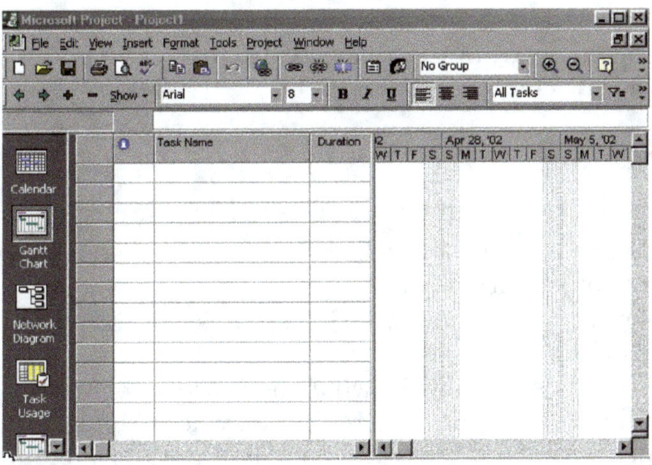


Fig. 4.5 - MS Project working window with option for creating a Gantt chart activated

Within the program MS Project there are two basic types of displays:

- activity displays, such as:
  - task name,
  - duration of task,
  - start data and completion data.
- resource displays, which refers to:
  - distribution of resources,
  - resource tracking.

*a) Activity displays*

MS Project contains five activity displays:

- calendar view which refers to a monthly calendar in which each activity is graphically represented by a line which crosses through the days planned for the realization of that activity,
- Gantt chart view which contains textual and graphical data about activities, as well as their mutual connections,
- task usage view based on which the actually consumed effort and expenses can be compared with that which was planned,
- tracking Gantt view based on which it is possible to compare the start date and completion date of the provided original plan with the start and completion dates determined by the current status of the project or compare the percentages of completed activities,
- network diagram view which contains the activities in the form of a flow chart.

*b) Resource displays*

The three most common resource displays are:

- resource graph view which graphically displays the arrangement of resources, work and expenses in time,
- resource sheet view which contains a table display of data about the resources,
- resource usage view in which activities are grouped according to their corresponding resources where the following is provided: quantity of work, schedule of work, availability for work for each resource.

The remaining views can be obtained by clicking on *more views* within the views palette.

The types of files which are supported by MS Project are displayed in Table 4.1.

Table 4.1 – Types of files supported by MS Project

| Type of file | Extension |
|---|---|
| MS Project documents | mpp |
| MS Project templates | mpt |
| MS Project workspace | mpw |
| MS Project database | mpd |
| MS Project exchange file | mpx |

#### 4.3.2.2 Application of MS Project in the drafting a complete project

*a) Basic data about the project*

When creating a new project, it is necessary to first enter in the general information about the project which includes the start date and completion date, as well as the project calendar. There are three types of calendars:

- standard calendar which is a calendar of working days with working hours from 08:00 to 17:00, beginning on Monday and ending on Friday,
- third shift calendar with working hours from 23:00 to 08:00, beginning on Monday and ending on Saturday,
- 24-hour calendar in which the working hours are 24 hours a day, all days of the week.

The calendar which most closely corresponds to the planned flow of works should be selected. This can be changed during the flow of the project. There is also the possibility of forming separate calendars.

*b) Entering in project activities*

When the material for the project is prepared, the goals of the project are defined and the method for measuring the success of the project is established, the next step is to enter in the project activities. There are two approaches to making the activities list. The first approach is according to the principle "from top to bottom" where a list of main activities or phases of the project is created. Each of these activities requires a certain number of lower level activities. A task which encompasses several subtasks (a task of a lower level) is called a collective task. Working according to the principle "from top to bottom", a list is created of collective tasks which are further expanded into subtasks. The list of subtasks can also be further expanded. However, it is necessary to ensure that tasks are not expanded too far by which they become insignificant from the aspect of tracking and managing. The goal is to reach a level of detail which allows one to see the progress of the project while not losing the global picture of the project.

The second approach to creating a list of tasks is according to the principle "from bottom to top". Here a list of all of the lower level tasks is created, normally in chronological order. Then the collective tasks are inserted and the tasks which are related to them are grouped together. This is a suitable approach for small projects.

Regardless of which method is used, the following must be respected:

- each task must have clear criteria for completion,
- tasks which last a relatively long time in comparison to the entire project should be expanded.

Every project normally contains the following four basic types of tasks:

- collective tasks,
- subtasks,
- repetitive tasks,
- significant tasks.

Collective tasks are those which contain subtasks. Tasks of a relatively smaller scope which fit into a collective task are subtasks. Significant tasks mark the completion of a certain phase of the project and normally do not have a duration.

*c) Using tools for classifying tasks into levels*

For the purpose of forming subtasks, it is necessary to enter the task into an empty cell in the task name column and then click on the indent button.

This causes the task directly above to turn into a collective task. Collective task titles appear in bold font with a contracting symbol (−) in front. The line in the Gantt chart which marks a collective task changes in appearance and becomes black with markings for the beginning and end (Fig 4.6).

If a certain task is of special significant then its duration is entered in as zero. In the Gantt chart that task is shown by a black rhombus

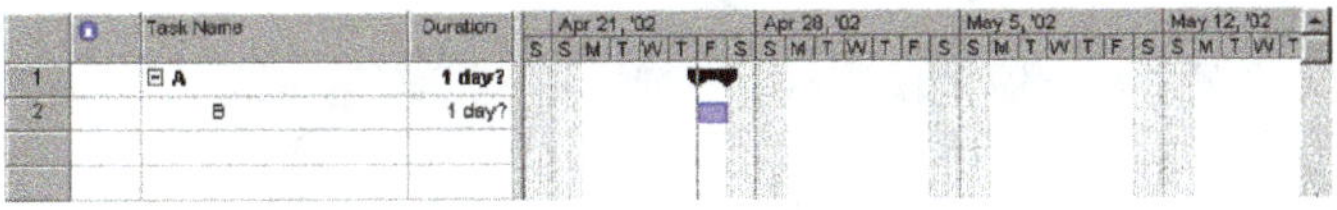

Fig. 4.6 – Forming a collective task

with the date above (Fig. 4.7). This point is called a Milestone Point (MP).

If a task occurs multiple times within the project, but not in subsequent time intervals, it is entered in multiple times. However, if the task does occur in subsequent time intervals, it can be entered in once and marked as a repetitive task.

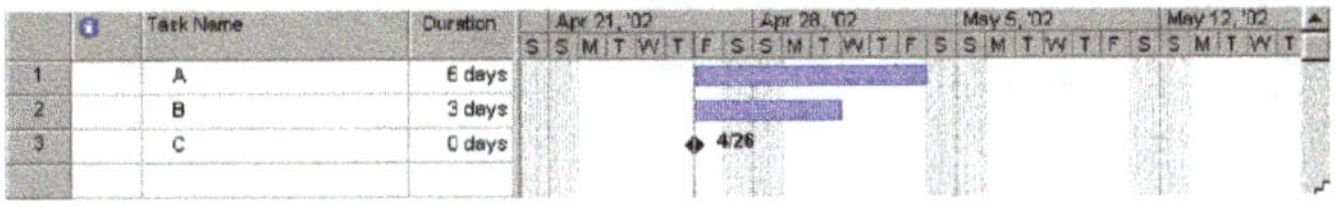

Fig. 4.7 – Example of a significant task (Milestone Point – MP)

*d) Drafting a dynamic plan and connecting activities within a project*

With interconnected tasks, the concepts of preceding task and subsequent task must be defined. A previous task is a task which must be started first or finished first. A subsequent task is a task which depends on the preceding task. Each task within the project is not required to have a preceding task or subsequent task. Defining those tasks which do have the aforementioned tasks will help the project manager form a realistic time schedule and efficiently control the realization of the project.

When creating a dynamic schedule and connection of tasks within a project, various types of connections are used.

*d.1 Use of various types of connections*

The program MS Project offers a selection of the following types of connections:

- Finish to Start – FS; This is the most commonly used type of connection which indicates that one task may not begin until the other is finished (Fig. 4.8).

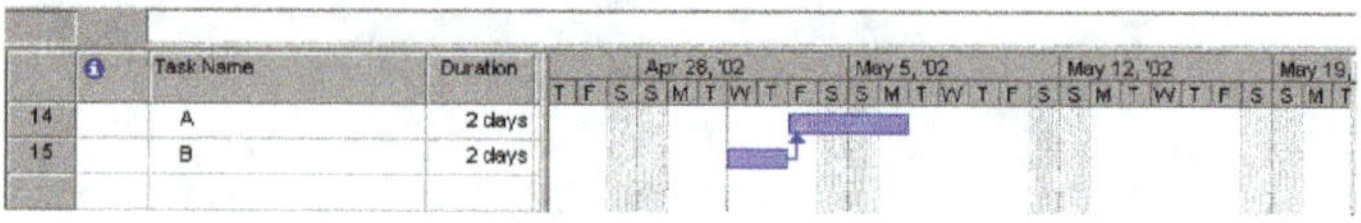

Fig. 4.8 – Connection type "Finish to Start"

– Start to Start – SS; With this type of connection, one task may not begin until another task begins. Instead of waiting for the completion of a previous task, the subsequent task may be started immediately upon or very soon after the start of the preceding task (Fig. 4.9).

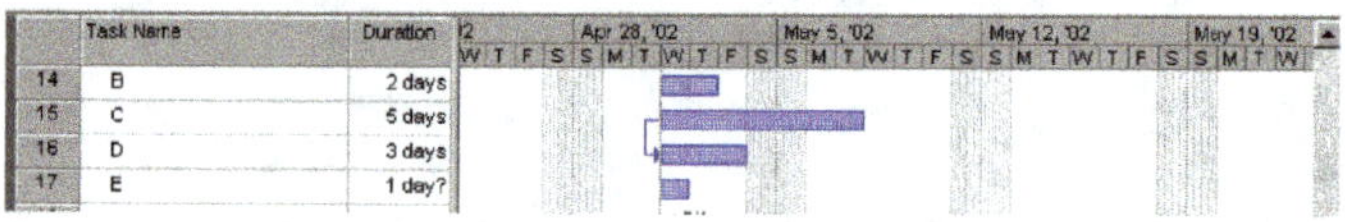

Fig. 4.9 – Connection type "Start to Start"

– Finish to Finish – FF; In this type of connection one task may not be completed until another task is completed. With this type dates are of no importance. It is important that the preceding task is finished before or at the same time as the subsequent task (Fig. 4.10).

Fig. 4.10 – Connection type "Finish to Finish"

– Start to Finish – SF; Here the completion date of one task is dependent on the start date of another task (Fig. 4.11).

*d.2 Establishing a connection*

The connection between tasks may be entered in at the same time as the tasks themselves, or connections can be created after entering in all tasks. It is only important that connections are established before arranging resources and expenses. Within the program MS Project, tasks can be connected in the following ways:

– using commands from the menu and tool palette,

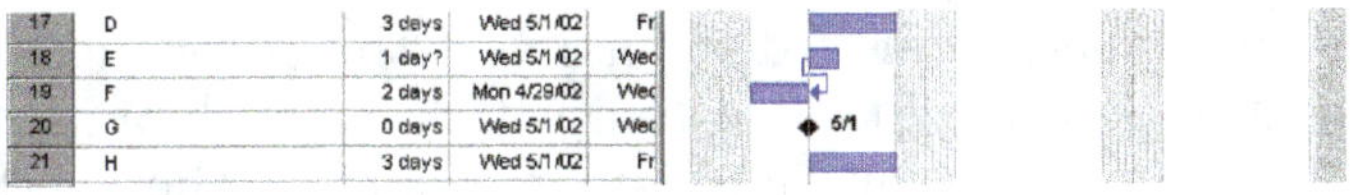


Fig. 4.11 – Connection type "Start to Finish"

– using the task information dialog box or tasks table,
– using the drag and release mouse function.

### *d.3 Establishing connections in tasks list which are classified by level*

When forming a list of activities which are classified by level, subtasks are automatically connected to the collective tasks which they belong to. However, there are several variants for the method of interconnecting subtasks within a collective task, as well as within groups of collective tasks.

The following options are available:

– connecting subtasks with a group of collective tasks and connecting collective tasks to other collective tasks,
– connecting subtasks which are located within various collective tasks,
– connecting collective tasks with subtasks from other collective tasks.

Connecting collective tasks with other collective tasks is the way in which MS Project itself connects tasks if they are all selected. By clicking on the link tasks button the aforementioned connection is created.

Connecting subtasks which belong to various collective tasks requires careful attention when shifting, moving or deleting subtasks. It is necessary to provide sustainable connections between subtasks of various groups.

Connecting collective tasks with subtasks is done for the case when one wants the completion of a certain collective task to be connected to a certain subtask in another collective task.

### *d.4 Delaying the start of a task and overlapping of tasks*

When two activities are connected, their planned start or completion may depend on various other factors. With the FS connection, the completion of the preceding task does not necessarily mean that the subsequent task may begin. Additionally, if the subsequent task may begin before the preceding task is completed an overlapping will occur (Fig. 4.12). The overlapping time is a measure of the overlapping of two interdependent tasks.

Delay time and overlapping time may be entered into the program in various ways. Delay time is entered in as a positive number and overlapping time as a negative number. This time can also be entered in as a percentage of the duration of the preceding task.

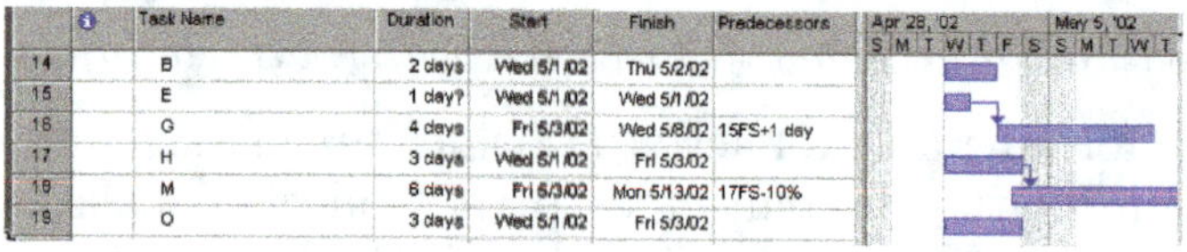


Fig. 4.12 – Example of delay time and overlapping time

*d.5 Division of a task*

In the case where an interruption occurs during a task, planned or unplanned, the task may be divided by displaying an interruption (Fig. 4.13). A task can be divided into an indefinite number of parts.

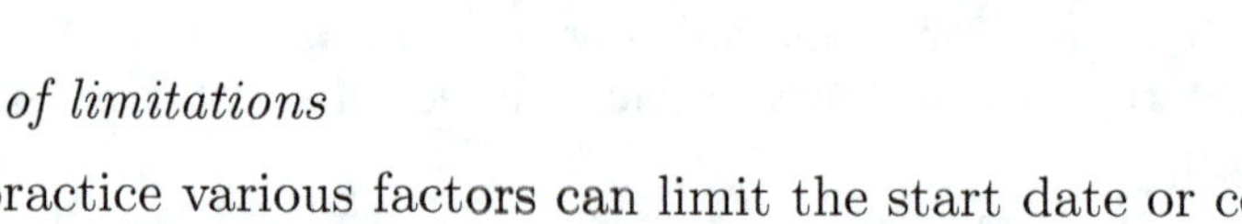


Fig. 4.13 – Example of a divided task

*d.6 Types of limitations*

In practice various factors can limit the start date or completion of individual tasks. Within the program MS Project there are eight options for setting limitations.

- *as late as possible* – used when the project is planned to begin from the completion date,
- *as soon as possible* – implied limitation when the project is planned to begin from the start date,
- *finish no earlier than* – task is not allowed to start before the given limitation date,

- *finish no later than* – task must be completed by the given limitation date,
- *must finish on* – task must be completed exactly on the given limitation date,
- *must start on* – task must begin exactly on the given limitation date,
- *start no earlier than* – task may not begin before the given limitation date,
- *start no later than* – task must begin on or before the given limitation date.

There are fixed and flexible limitations. Fixed limitations are those which limit scheduling by requiring a task to start or finish exactly within the given parameters. With flexible limitations, MS Project calculates the schedule and makes adjustments on the basis of the limitations. In practice it is possible for conflicts to arise due to limitations being placed on a task which do not conform to other parts of the time schedule of the project. These conflicts can be resolved by providing answers to the following questions:

- Is the limitation obligatory? Has it bet set intentionally or by accident? What must be changed within the time schedule in order for the limitation to be eliminated?
- Can the completion date of the preceding task be moved so that it comes before the limitation date?
- Can the duration of the preceding task be shortened so that it ends before the limitation date?
- Can the connection between the conflicting tasks be removed?
- Is it possible to replace the fixed limitation with a flexible limitation?

The other method for resolving a conflict is by not recognizing the limitation which is achieved [25-27].

### *d.7 Changing task parameters*

The typical tasks related to project management are:

- shifting resources from less critical tasks to more critical tasks,
- adding resources in order to complete a task faster,
- modifying the quantity of work necessary in order to successfully complete a task on the basis of information from members of the project team,

– setting the start date or completion date of a task in order to respect the project deadline.

There are three types of tasks:

– with fixed units,
– with fixed work,
– with a fixed duration.

The program MS Project enables the following five combinations of task types and options of fixed quantities of work:

– fixed units and fixed quantity of work (default option),
– fixed units,
– fixed work and fixed quantity of work,
– fixed duration,
– fixed duration and fixed quantity of work.

### *d.8 Scheduling resources*

There are various methods for scheduling the resources of work and materials within MS Project. Scheduling connects a task with the resource which is responsible for completing that task. Scheduling resources is conducted through:

– dialog box,
– Task table (Fig. 4.14).

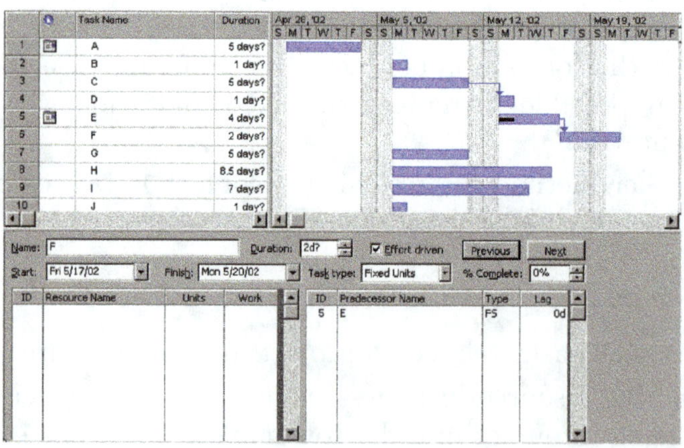

Fig. 4.14 – Task table and Gantt chart

*d.9 Locating overallocated resources*

A resource is considered to be overallocated when the sum of its works, over a certain period of time, surpasses the maximum number of units of that resource. Delays in tasks are a regular cause of overallocation because the work planned for the previous week is shifted into the next week. If a resource is planned to work on some other task it may become overallocated. Overallocation of resources cannot be seen within the Gantt chart. For this reason verification of resources is conducted within other displays before the project enters the phase of realization. In all resource displays, overallocated resources are displayed in red so that they are easily noticed. The report on overallocated resources contains a list of all overallocated resources and the engagements which contribute to the overallocation.

MS Project may suggest a method for correcting the overallocation. The following solutions are available:

- delay the task until the assigned resource is available,
- divide the task so that only a part of it is delayed,
- schedule additional tasks.

All three options are available while the project is still incomplete and unpublished.

One of the methods of resolving an overallocation is the leveling of resources. That is the procedure through which a task is delayed or divided. Leveling can be applied for a certain period of time or for the entire project. The factors which dictate which tasks are delayed and which are divided are:

- available free time,
- task priority, its dependencies and limitations,
- line number of the task,
- dates within the time schedule,

Fig. 4.15 shows the process of resource allocation.

*d.10 Tracking the realization of the project*

Brief descriptions are provided for all the columns which are used for tracking project data and the procedures for updating the data. For tasks which have not been allocated resources, the following data may be entered:

- actual start and completion date,
- task status, including:

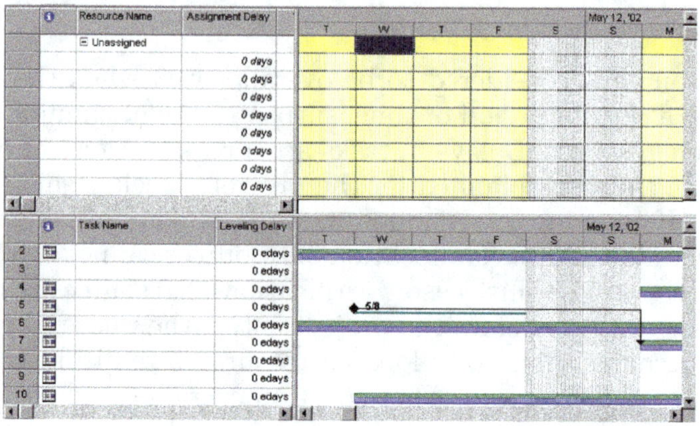

Fig. 4.15 – Resource allocation display

- percentage of completion,
- actual duration,
- remaining duration;

– costs.

For tasks which have been allocated resources, the following data can be entered manually:

– actual quantity of work,
– scheduled work.

The actual start date is the date when execution of the task actually begins. MS Project calculates that date on the basis of that which is entered into the percent complete column, actual work column or percent work completed column. The actual completion date is calculated when 0 % of work remains or when the value in the columns percent complete or percent work complete amounts to 100 %.

The difference between the actual duration and planned duration is located in the percent complete column. MS Project performs the calculation using the following formula:

$$\% \text{ complete} = \text{actual duration} \,/\, \text{duration} \cdot 100 \tag{4.2}$$

When the value for % complete is entered in, MS Project automatically calculates the actual and remaining duration. The actual duration is the total range of working time, without consideration of the actual number of working hours (quantity of work) consumed on a specific task.

The remaining duration is the difference between the planned duration and the actual duration. MS Project calculates the remaining duration on the basis of the value entered in for the percent completed.

The actual costs encompass the resource-related costs incurred up to the specified date, as well as fixed costs which are related to the task. For calculation of these costs MS Project uses the following formula:

$$\begin{aligned} \text{actual costs} = {} & (\text{actual work} \cdot \text{standard hourly rate}) + \\ & (\text{actual overtime work} \cdot \text{overtime hourly rate}) + \\ & \text{costs for using resources} + \text{fixed costs of the task} \end{aligned} \tag{4.3}$$

The actual costs can be entered in by the user or be automatically calculated by MS Project.

MS Project calculates the actual work on the basis of the quantity of work which the resource, allocated to the project, has completed. If the value for actual work is entered manually, MS Project calculates the percentage of completed work and remaining work according to the formulas:

$$\text{percentage of completed work} = \text{actual work}/\text{planned work} \tag{4.4}$$

$$\text{remaining work} = \text{planned work} - \text{actual work} \tag{4.5}$$

*d.11 Review of deviations from the initial plan and review of collective reports*

MS Project calculates deviations from the initial plan of the project for start and completion dates, duration, work and costs. These deviations can be viewed by opening the tracking Gantt display and the variance table. Within the tracking Gantt display, gray lines represent a task with start and completion dates according to the initial plan, and blue and red lines represent tasks according to the valid plan.

Within the variance table there are columns which display data from the initial plan as well as deviations from that data, and a column can be added in which the actual and planned dates, as well as dates from between-phase plans are compared.

The *task in progress, should have started tasks* and *slipping tasks* reports can be accessed through the view-reports menu, as well as by clicking on the current activities button.

The *cash flow, overbudgeted tasks* and *overbudgeted resources* reports can be accessed by clicking on the costs button within the reports option.

The *overallocated resources* report is accessed through the assignments button within the reports option, and the *task usage* and *resource usage* reports are accessed through the *workload* button.

### *d.12 Adding a progress line in the Gantt chart*

A progress line may be added in the Gantt chart (Fig. 4.16). By adding a progress line, MS Project draws a line which connects the tasks which are in progress and the tasks which have to be started. Sharp angles pointing to the left represent works which are behind schedule, while sharp angles pointing to the right represent works which are ahead of schedule.

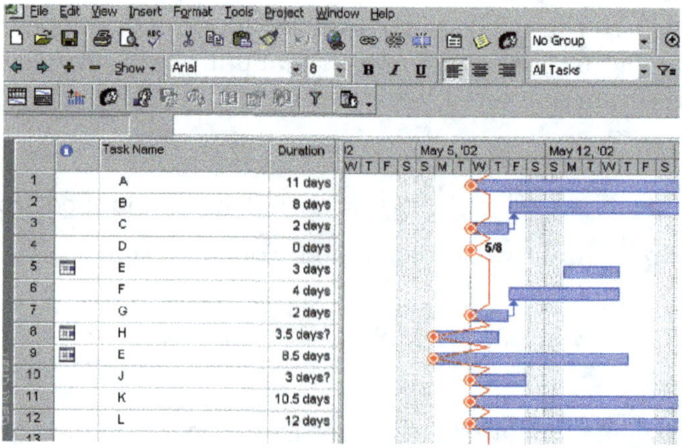


Fig. 4.16 - Progress line which shows where the project is ahead of schedule and where it is behind schedule

### *d.13 Analysis of project execution based on the analysis of financial realization*

Within the analysis of financial realization, three values are used which are dependent on the tasks in order to calculate deviations and relations. Those values are:

- BCWS – Budgeted Cost of Work Scheduled,
- BCWP – Budgeted Cost of Work Performed,

– ACWP – Actual Cost of Work Performed.

The BCWS and BCWP are calculated from the initial project schedule, so it is necessary to first record the initial schedule in order to draft the analysis of financial realization. Recognizing the difference between these three values is made easier through the following formulas:

$$\text{BCWS} = \text{initially planned price} \cdot \text{initially planned hours} \tag{4.6}$$

$$\text{BCWP} = \text{initially planned price} \cdot \text{actually consumed hours} \tag{4.7}$$

$$\text{ACWP} = \text{actual price of a working hour} \cdot \text{actually consumed hours} \tag{4.8}$$

By using these three values two types of deviations and two types of relations can be determined:

– *Schedule Variance* (SV) – compares the BCWS with the BCWP and displays the difference in costs which has occurred only due to the difference between the planned and actually completed work:

$$\text{SV} = \text{BCWP} - \text{BCWS} \tag{4.9}$$

– *Cost Variance* (CV) – compares BCWP with ACWP and shows the difference in costs which has occurred only due to the difference in the price of work of resources:

$$\text{CV} = \text{BCWP} - \text{ACWP} \tag{4.10}$$

The variances are expressed in dollars.

– Schedule Performance Index – SPI which represents the actually completed work with the planned work:

$$\text{SPI} = \text{BCWP}/\text{BCWS} \tag{4.11}$$

– Cost Performance Index – CPI which represents the relation of costs envisaged by the initial plan with the actual costs which were incurred for a certain task:

$$\text{CPI} = \text{BCWP}/\text{ACWP} \tag{4.12}$$

MS Project enables the calculation of the following cost estimates:

– Estimate At Completion – EAC which encompasses the fixed costs, actual costs and remaining costs for a certain task; this estimate is accurate only if the remainder of the task is completed as planned in the budget;
– Budget At Completion – BAC which represents the sum of the fixed costs and costs for resources according to the initial plan: BAC encompasses overtime work according to the overtime hourly rate and standard work according to the standard hourly rate;
– Variance at Completion – VAC which represents the difference between the BAC and EAC.

The variances and deviations may be reviewed in the task display of MS Project. Variances from the plan and variances of costs may be positive or negative. The larger the number, the larger the variance. A positive variance from the plan means that the task is going ahead of schedule. When the variance of costs is positive, the costs for the project are less than the budget of envisaged costs.

The financial realization indexes are small numbers. If there is no variance from the plan, the SPI amounts to 1.0. If there is no variance of costs the CPI has a value of 1.0. Indexes above 1.0 mean that the project is developing well. Indexes below 1.0 indicate that the project is behind schedule or that the planned budget has been exceeded.

If the VAC is positive, the project will be completed within budget. A negative VAC value means that the project will exceed the budget.

Analysis of financial realization may also be applied to the resources and engagements.

*e) Example of application of MS Project*

An illustrated example of the application of MS Project for forming a Gantt chart for a technological development project is provided below [19]. The tasks and engagement of participants are described in Section 4.3.1. A display of the tasks and engagement of participants is provided in Table 4.2. Task number 10 represents a milestone point of the project (MP).

Brief instructions are provided below for forming the Gantt chart which is displayed in Fig. 4.17. Forming a Gantt chart is carried out using the following steps:

Table 4.2 – Tasks and engagement of participants

| Task | Participants | | | | | | | |
|---|---|---|---|---|---|---|---|---|
| | A | B | C | D | E | F | G | H |
| 1. | x | x | x | x | | | | |
| 2. | | | | | x | x | x | x |
| 3. | x | x | x | x | x | x | x | x |
| 4. | x | x | x | x | | | | |
| 5. | | | | | x | x | x | x |
| 6. | x | x | x | x | x | x | x | x |
| 7. | x | x | x | x | | | | |
| 8. | | | | | x | x | x | x |
| 9. | x | x | x | x | x | x | x | x |
| 10. | | | | | | | | |

*e.1 Defining the start date of the project*

The start date of the project is determined by the contract and is set using the option Project / Project Information / Start date.

*e.2 Working parameters of the project – currency symbol*

The general parameters of the project are selected from the menu Tools / Options. The currency symbol is entered in using the option View / Symbol, and the position of the currency symbol in relation to the amount is defined in the Placement field.

*e.3 General project characteristics*

From within MS Project, the File menu is selected, followed by the option Properties / Summary, whereby a series of fields appears containing the general characteristics of the project. The Title field refers to the name of the project. The Subject field serves as the place to enter the registration number of the project. The fields Author and Manager contain the name of the contact person and name of the project manager. The fields Category and Keywords are intended for defining the scientific-technological field and the list of keywords by choice of the project manager. Finally, the field Comments is envisaged for comments from the project manager. Comments normally contain an estimation of the fulfillment of the project plan (in percents).

*e.4 Project calendar*

Types of project calendars are discussed in Section 4.3.2.2.a. Defining the days in which tasks begin, or days in which tasks are completed, is conducted from the menu Tools / Change Working Time.

*e.5 Project Gantt chart*

The basic display of the project is the Gantt chart. It is selected from the menu View / Gantt Chart. A simple Gantt chart encompasses the tasks which last exactly one calendar month, begin on the first working day in the month and are completed on the final working day in the month.

*e.6 Entry of project tasks*

A project represents a collection of tasks. The Task Name column serves for defining tasks. A collection of tasks makes up a project activity.

*e.7 Data about a task*

Entering in data about a task is performed by double-clicking on the number in the column left of the task name. The beginning and end of a task are set by selecting the option Project / Task Information / General / Dates / Start / Finish. The percentage of accomplishment is defined by using the option Project / Task Information / General / Percent Complete. It is important to set the duration of a task to not change with a change of assigned researchers. This is achieved by ensuring that the option Project / Task Information / Advanced / Effort Driven is not selected.

By selecting the option Project / Task Information / Notes it is possible to enter in notes. This is intended to be a short text about the task which contains information about the planned and executed activities and achieved results.

*e.8 Linking tasks*

Tasks which make up a logical whole (project activity) are linked together. Select the first task by clicking on it, press and hold the shift key while pressing the cursor arrow the necessary number of times in order to select the tasks which make up the activity, and then from the Edit menu select the option Link Tasks. If the tasks to be linked are not listed one after the other, use the Ctrl key instead of the Shift key.

*e.9 Project milestone point (MP)*

A project milestone point is a task with a duration of zero days. This task is entered into the Task Name column and zero is provided as the duration of the task in the Duration column. The milestone point has a date (option Project / Task Information / General / Dates / Start) and notes (options Project / Task Information / Notes). The purpose of a milestone point is to emphasize a date when a report is submitted or an import result is obtained.

*e.10 Project executors*

Project executors may be entered in selecting the menu View / Resource Sheet / View / Table Entry. The name and last name of an executor is given in the Resource Name column, while the monthly amount is entered into the column Cost / Use.

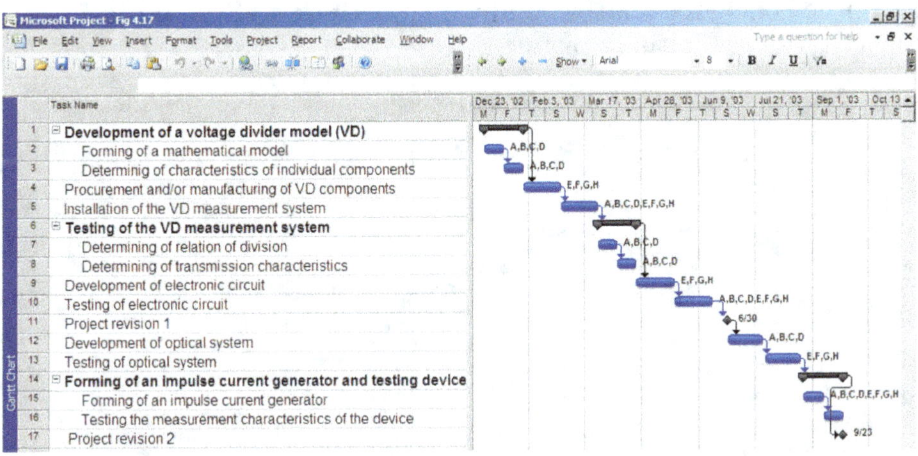


LEGEND:

| | |
|---|---|
| TASK | COLLECTIVE |
| PROGRESS | MILESTONE |
| INTERPHASES | ACHIEVING THE TASK |
| COLLECTIVE PROJECT | ACHIEVING PROGRESS |
| EXTERNAL TASK | ACHIEVING MILESTONE |

Fig. 4.17 - Gantt chart formed using MS Project for the technological development project [19]

*e.11 Assigning task executors*

When entering tasks and resources, each task is assigned executors. First, the task is selected which is to be assigned an executor (option Tools / Resources / Assign Resources). Then the name of the executor is highlighted, the Assign option is selected and the percentage of engagement is entered.

*e.12 Project reports*

For the purpose of recognizing and controlling entered data, select the option View / Reports / Costs in order to form the appropriate reports.

Activities 1, 4 and 9 in Fig. 4.17 are shown as collective activities which are composed of individual sub-activities.

Fig. 4.18 shows the appearance of the Gantt chart following the first revision conducted on 30.06.2003. Based on the displayed chart, it can be concluded that the individual activities of the project lasted longer than was initially planned.

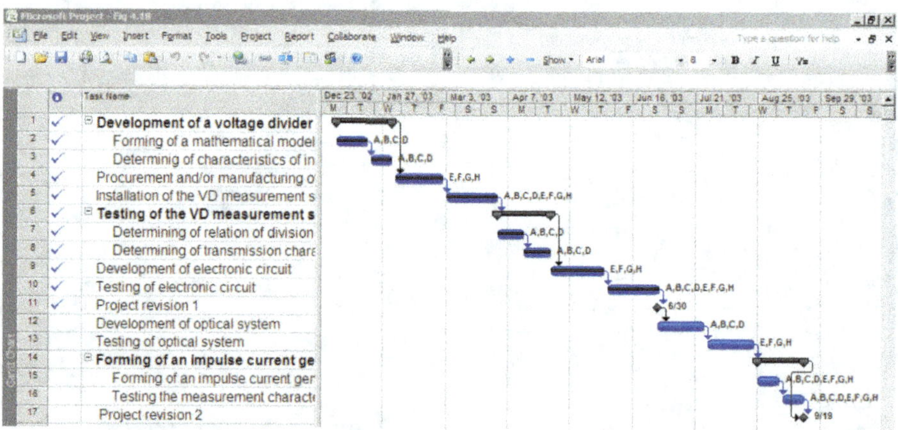


Fig. 4.18 – Appearance of the Gantt chart following the first revision

Forming a critical path is the most important way for managing the total duration of a project. Fig. 4.19 shows a graph for identifying critical points of a project. Critical tasks are highlighted in red. Every change to the planning of critical tasks will affect the project completion date.

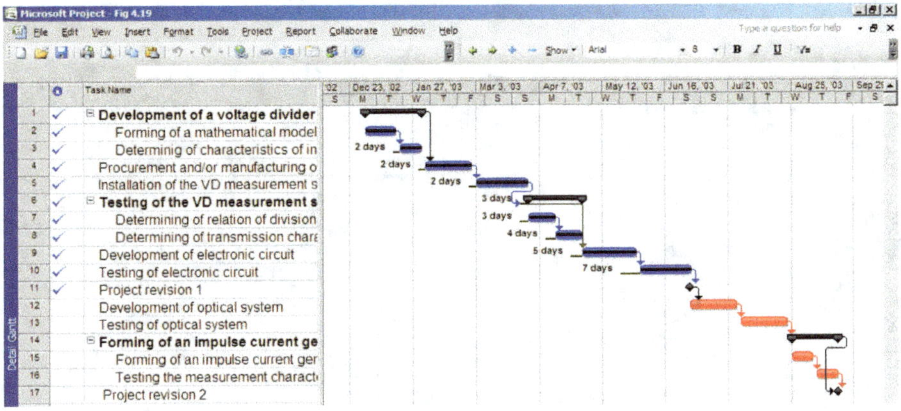


Fig. 4.19 – Graph for identifying the critical points of a project

Due to the second revision of tasks was planned to be the earliest on 30.09.2003, it is possible to change task number 17 within the list of project tasks from As Soon As Possible to Start No Earlier Than. After conducting the changes not a single tasks remains critical, but rather the time reserve of individual tasks is increased. The graph for identifying critical points of the project after conducting the changes appears as in Fig. 4.20.

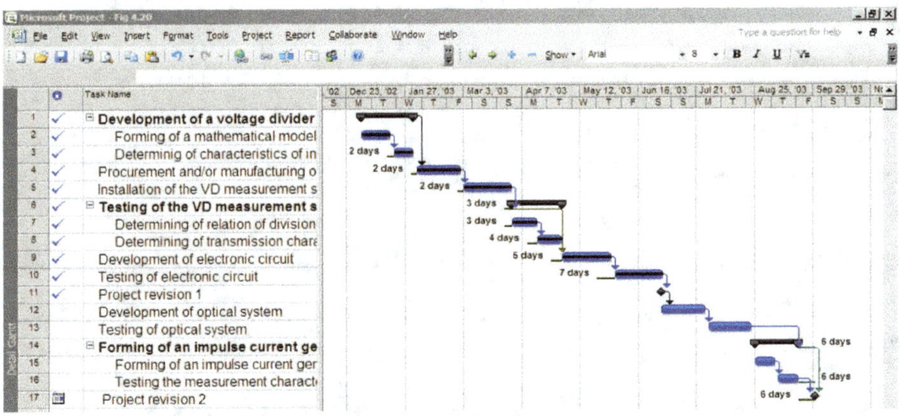


Fig. 4.20 - Graph for identifying the critical points of a project following the conducted changes

A comparative display of the basic and realized project plan is presented in Fig. 4.21.

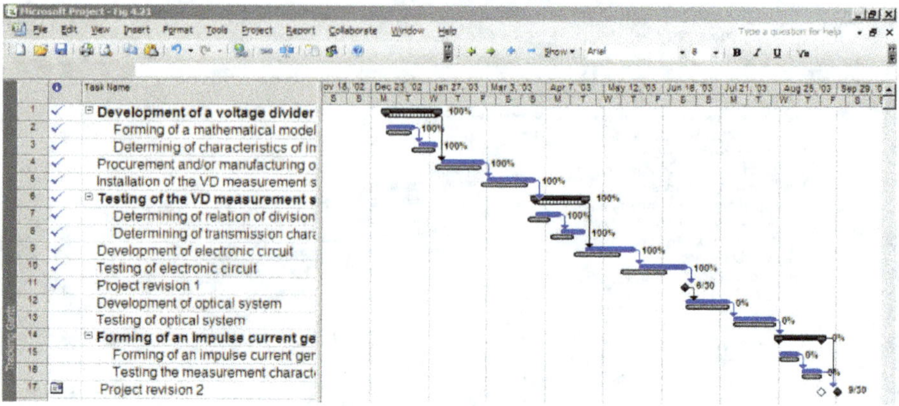


Fig. 4.21 – Comparative display of the basic and realized project plans

The starting costs of tasks, planned costs, actual costs and deviations in costs for the considered technological development project are displayed in Fig. 4.22.

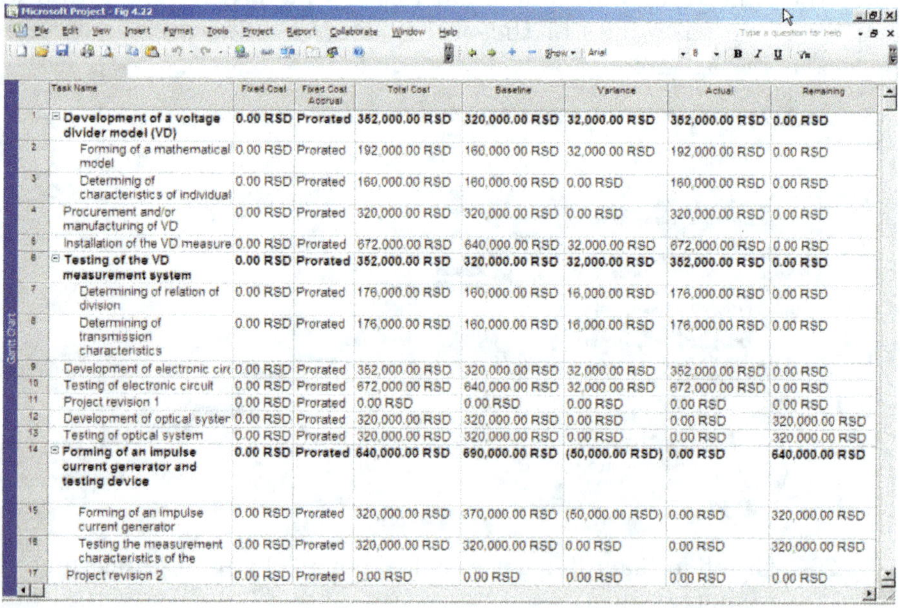

| | Task Name | Fixed Cost | Fixed Cost Accrual | Total Cost | Baseline | Variance | Actual | Remaining |
|---|---|---|---|---|---|---|---|---|
| 1 | **Development of a voltage divider model (VD)** | **0.00 RSD** | **Prorated** | **352,000.00 RSD** | **320,000.00 RSD** | **32,000.00 RSD** | **352,000.00 RSD** | **0.00 RSD** |
| 2 | Forming of a mathematical model | 0.00 RSD | Prorated | 192,000.00 RSD | 160,000.00 RSD | 32,000.00 RSD | 192,000.00 RSD | 0.00 RSD |
| 3 | Determinig of characteristics of individual | 0.00 RSD | Prorated | 160,000.00 RSD | 160,000.00 RSD | 0.00 RSD | 160,000.00 RSD | 0.00 RSD |
| 4 | Procurement and/or manufacturing of VD | 0.00 RSD | Prorated | 320,000.00 RSD | 320,000.00 RSD | 0.00 RSD | 320,000.00 RSD | 0.00 RSD |
| 5 | Installation of the VD measure | 0.00 RSD | Prorated | 672,000.00 RSD | 640,000.00 RSD | 32,000.00 RSD | 672,000.00 RSD | 0.00 RSD |
| 6 | **Testing of the VD measurement system** | **0.00 RSD** | **Prorated** | **352,000.00 RSD** | **320,000.00 RSD** | **32,000.00 RSD** | **352,000.00 RSD** | **0.00 RSD** |
| 7 | Determining of relation of division | 0.00 RSD | Prorated | 176,000.00 RSD | 160,000.00 RSD | 16,000.00 RSD | 176,000.00 RSD | 0.00 RSD |
| 8 | Determining of transmission characteristics | 0.00 RSD | Prorated | 176,000.00 RSD | 160,000.00 RSD | 16,000.00 RSD | 176,000.00 RSD | 0.00 RSD |
| 9 | Development of electronic circ | 0.00 RSD | Prorated | 352,000.00 RSD | 320,000.00 RSD | 32,000.00 RSD | 352,000.00 RSD | 0.00 RSD |
| 10 | Testing of electronic circuit | 0.00 RSD | Prorated | 672,000.00 RSD | 640,000.00 RSD | 32,000.00 RSD | 672,000.00 RSD | 0.00 RSD |
| 11 | Project revision 1 | 0.00 RSD | Prorated | 0.00 RSD | 0.00 RSD | 0.00 RSD | 0.00 RSD | 0.00 RSD |
| 12 | Development of optical syster | 0.00 RSD | Prorated | 320,000.00 RSD | 320,000.00 RSD | 0.00 RSD | 0.00 RSD | 320,000.00 RSD |
| 13 | Testing of optical system | 0.00 RSD | Prorated | 320,000.00 RSD | 320,000.00 RSD | 0.00 RSD | 0.00 RSD | 320,000.00 RSD |
| 14 | **Forming of an impulse current generator and testing device** | **0.00 RSD** | **Prorated** | **640,000.00 RSD** | **690,000.00 RSD** | **(50,000.00 RSD)** | **0.00 RSD** | **640,000.00 RSD** |
| 15 | Forming of an impulse current generator | 0.00 RSD | Prorated | 320,000.00 RSD | 370,000.00 RSD | (50,000.00 RSD) | 0.00 RSD | 320,000.00 RSD |
| 16 | Testing the measurement characteristics of the | 0.00 RSD | Prorated | 320,000.00 RSD | 320,000.00 RSD | 0.00 RSD | 0.00 RSD | 320,000.00 RSD |
| 17 | Project revision 2 | 0.00 RSD | Prorated | 0.00 RSD | 0.00 RSD | 0.00 RSD | 0.00 RSD | 0.00 RSD |

Fig. 4.22 - Starting costs of tasks, planned costs, actual costs and deviations in costs for the considered technological development project

Fig. 4.23 shows the earned value schedule indicators. Fig. 4.24 refers to the earned value cost indicators.

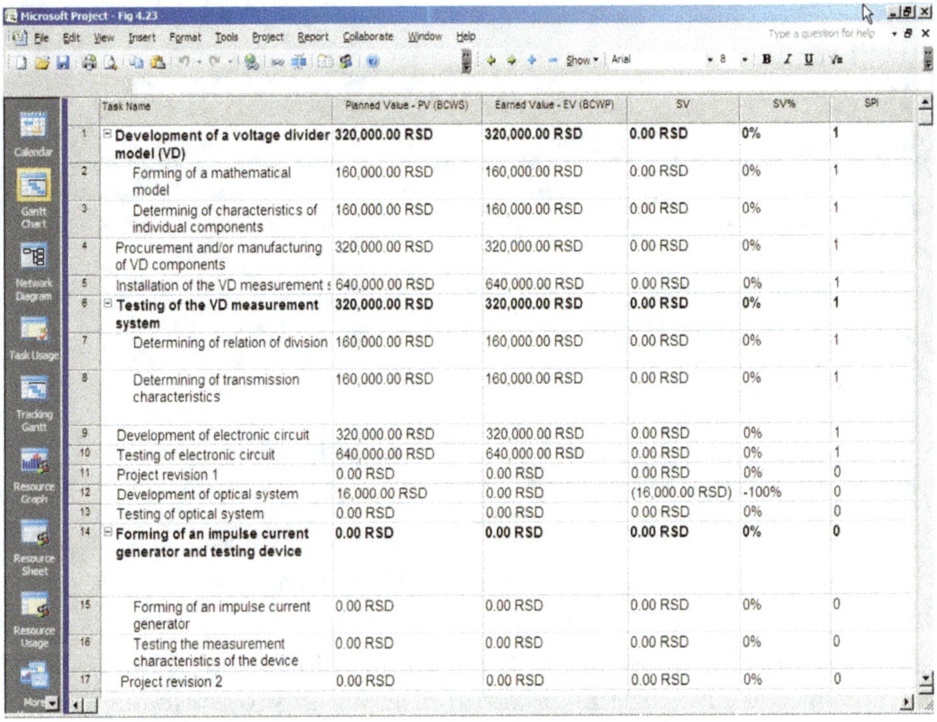

| | Task Name | Planned Value - PV (BCWS) | Earned Value - EV (BCWP) | SV | SV% | SPI |
|---|---|---|---|---|---|---|
| 1 | **Development of a voltage divider model (VD)** | **320,000.00 RSD** | **320,000.00 RSD** | **0.00 RSD** | **0%** | **1** |
| 2 | Forming of a mathematical model | 160,000.00 RSD | 160,000.00 RSD | 0.00 RSD | 0% | 1 |
| 3 | Determinig of characteristics of individual components | 160,000.00 RSD | 160,000.00 RSD | 0.00 RSD | 0% | 1 |
| 4 | Procurement and/or manufacturing of VD components | 320,000.00 RSD | 320,000.00 RSD | 0.00 RSD | 0% | 1 |
| 5 | Installation of the VD measurement : | 640,000.00 RSD | 640,000.00 RSD | 0.00 RSD | 0% | 1 |
| 6 | **Testing of the VD measurement system** | **320,000.00 RSD** | **320,000.00 RSD** | **0.00 RSD** | **0%** | **1** |
| 7 | Determining of relation of division | 160,000.00 RSD | 160,000.00 RSD | 0.00 RSD | 0% | 1 |
| 8 | Determining of transmission characteristics | 160,000.00 RSD | 160,000.00 RSD | 0.00 RSD | 0% | 1 |
| 9 | Development of electronic circuit | 320,000.00 RSD | 320,000.00 RSD | 0.00 RSD | 0% | 1 |
| 10 | Testing of electronic circuit | 640,000.00 RSD | 640,000.00 RSD | 0.00 RSD | 0% | 1 |
| 11 | Project revision 1 | 0.00 RSD | 0.00 RSD | 0.00 RSD | 0% | 0 |
| 12 | Development of optical system | 16,000.00 RSD | 0.00 RSD | (16,000.00 RSD) | -100% | 0 |
| 13 | Testing of optical system | 0.00 RSD | 0.00 RSD | 0.00 RSD | 0% | 0 |
| 14 | **Forming of an impulse current generator and testing device** | **0.00 RSD** | **0.00 RSD** | **0.00 RSD** | **0%** | **0** |
| 15 | Forming of an impulse current generator | 0.00 RSD | 0.00 RSD | 0.00 RSD | 0% | 0 |
| 16 | Testing the measurement characteristics of the device | 0.00 RSD | 0.00 RSD | 0.00 RSD | 0% | 0 |
| 17 | Project revision 2 | 0.00 RSD | 0.00 RSD | 0.00 RSD | 0% | 0 |

Fig. 4.23 – Earned Value Schedule Indicators

The displayed financial indicators represent just a portion of the display which can be generated using the program MS Project for the purpose of efficiently managing a project.

### 4.3.2.3 Closing considerations

MS Project is an efficient program for project management. This program is an extraordinary tool for developing plans, assigning resources to tasks, tracking progress, managing budgets and analyzing workloads. Its real power comes in the form of its abilities to calculate dynamic plans based on linking activities and scheduling resources. Despite being similar to other Microsoft products, MS Project distinguishes itself by allowing the user to simultaneously manage the people,

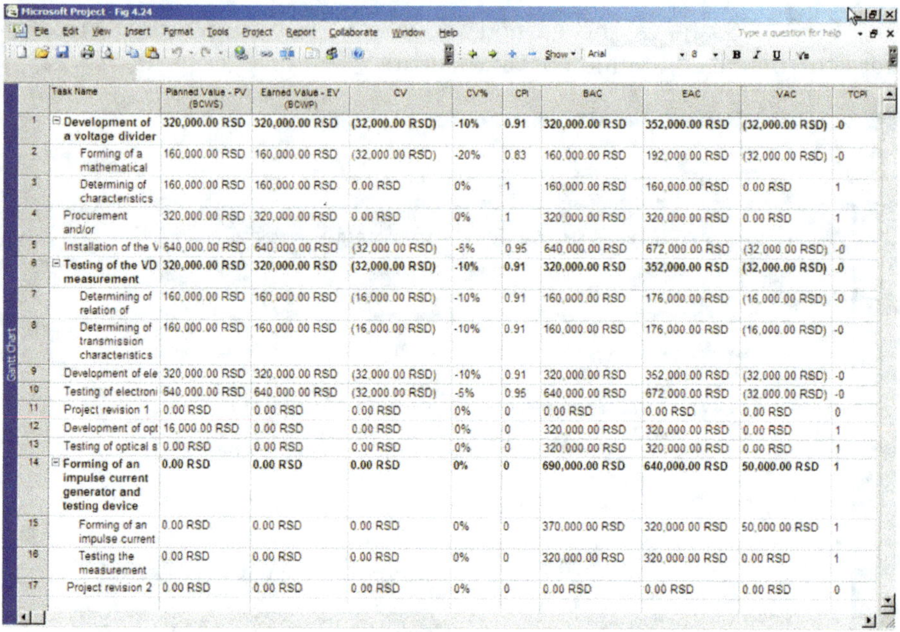

| | Task Name | Planned Value - PV (BCWS) | Earned Value - EV (BCWP) | CV | CV% | CPI | BAC | EAC | VAC | TCPI |
|---|---|---|---|---|---|---|---|---|---|---|
| 1 | **Development of a voltage divider** | **320,000.00 RSD** | **320,000.00 RSD** | **(32,000.00 RSD)** | **-10%** | **0.91** | **320,000.00 RSD** | **352,000.00 RSD** | **(32,000.00 RSD)** | **-0** |
| 2 | Forming of a mathematical | 160,000.00 RSD | 160,000.00 RSD | (32,000.00 RSD) | -20% | 0.83 | 160,000.00 RSD | 192,000.00 RSD | (32,000.00 RSD) | -0 |
| 3 | Determinig of characteristics | 160,000.00 RSD | 160,000.00 RSD | 0.00 RSD | 0% | 1 | 160,000.00 RSD | 160,000.00 RSD | 0.00 RSD | 1 |
| 4 | Procurement and/or | 320,000.00 RSD | 320,000.00 RSD | 0.00 RSD | 0% | 1 | 320,000.00 RSD | 320,000.00 RSD | 0.00 RSD | 1 |
| 5 | Installation of the V | 640,000.00 RSD | 640,000.00 RSD | (32,000.00 RSD) | -5% | 0.95 | 640,000.00 RSD | 672,000.00 RSD | (32,000.00 RSD) | -0 |
| 6 | **Testing of the VD measurement** | **320,000.00 RSD** | **320,000.00 RSD** | **(32,000.00 RSD)** | **-10%** | **0.91** | **320,000.00 RSD** | **352,000.00 RSD** | **(32,000.00 RSD)** | **-0** |
| 7 | Determining of relation of | 160,000.00 RSD | 160,000.00 RSD | (16,000.00 RSD) | -10% | 0.91 | 160,000.00 RSD | 176,000.00 RSD | (16,000.00 RSD) | -0 |
| 8 | Determining of transmission characteristics | 160,000.00 RSD | 160,000.00 RSD | (16,000.00 RSD) | -10% | 0.91 | 160,000.00 RSD | 176,000.00 RSD | (16,000.00 RSD) | -0 |
| 9 | Development of ele | 320,000.00 RSD | 320,000.00 RSD | (32,000.00 RSD) | -10% | 0.91 | 320,000.00 RSD | 352,000.00 RSD | (32,000.00 RSD) | -0 |
| 10 | Testing of electroni | 640,000.00 RSD | 640,000.00 RSD | (32,000.00 RSD) | -5% | 0.95 | 640,000.00 RSD | 672,000.00 RSD | (32,000.00 RSD) | -0 |
| 11 | Project revision 1 | 0.00 RSD | 0.00 RSD | 0.00 RSD | 0% | 0 | 0.00 RSD | 0.00 RSD | 0.00 RSD | 0 |
| 12 | Development of opt | 16,000.00 RSD | 0.00 RSD | 0.00 RSD | 0% | 0 | 320,000.00 RSD | 320,000.00 RSD | 0.00 RSD | 1 |
| 13 | Testing of optical s | 0.00 RSD | 0.00 RSD | 0.00 RSD | 0% | 0 | 320,000.00 RSD | 320,000.00 RSD | 0.00 RSD | 1 |
| 14 | **Forming of an impulse current generator and testing device** | **0.00 RSD** | **0.00 RSD** | **0.00 RSD** | **0%** | **0** | **690,000.00 RSD** | **640,000.00 RSD** | **50,000.00 RSD** | **1** |
| 15 | Forming of an impulse current | 0.00 RSD | 0.00 RSD | 0.00 RSD | 0% | 0 | 370,000.00 RSD | 320,000.00 RSD | 50,000.00 RSD | 1 |
| 16 | Testing the measurement | 0.00 RSD | 0.00 RSD | 0.00 RSD | 0% | 0 | 320,000.00 RSD | 320,000.00 RSD | 0.00 RSD | 1 |
| 17 | Project revision 2 | 0.00 RSD | 0.00 RSD | 0.00 RSD | 0% | 0 | 0.00 RSD | 0.00 RSD | 0.00 RSD | 0 |

Fig. 4.24 – Earned Value Cost Indicators

materials and costs of a project. The aforementioned information shows the advantages in using this powerful program in project management.

# LITERATURE


[1] B. Ivković, Ž. Popović: Project Management in Civil Engineering, Faculty of Civil Engineering, Belgrade, Nauka, 1995.

[2] Law on Planning and Construction of Buildings, Official Gazette of the Republic of Serbia, No. 47/2003 and 34/2006.

[3] G. Dotlić: Power Engineering Through Standards, Laws, Rulebooks and Technical Recommendations, revised and amended 4th edition, SMEITS, 2006.

[4] FIDIC – International Federation of Consulting Engineers, http://www/fidic.org

[5] UN ECE – United Nations – Economic Commission for Europe, http://www.unece.org.

[6] Law on Public Procurement, Republic of Serbia, Dec. 2008.

[7] Rulebook on the Conditions and Procedure for Issuance and Revocation of Licenses for Authorized Urbanists, Designers, Contractors and Authorized Planners, Official Gazette of the Republic of Serbia, No. 116/2004.

[8] Rulebook on Amendments to the Rulebook on the Conditions and Procedure for Issuance and Revocation of Licenses for Authorized Urbanists, Designers, Contractors and Authorized Planners, Official Gazette of the Republic of Serbia, No. 69/2006.

[9] Law on Safety and Health at the Workplace, Official Gazette of the Republic of Serbia, No. 101/2005.

[10] Rulebook on the Minimum Technical Documentation for Issuance of Approval for Construction of High-Rise Investment Structures, Official Gazette of the Republic of Serbia, No. 39/1980.

[11] Main Project of the Adaptation of the 110 kV/ 35 kV "Vreoci" Transformer Station and Transmission Lines 110 kV of the Substation of the TPP "Kolubara", Faculty of Electrical Engineering, Belgrade, Orderer Electrical Utility of Serbia, Mining Kolubara – Lazarevac, 2005-2007, (Chief of project Z. Stojković).

[12] M. Bogner, A. Petrović: Evaluation of Services in Structural Engineering (2nd amended edition), SMEITS, Belgrade, 1999.

[13] D. Stašić: Norms for Electrical Works in Civil Engineering, Kultura, Belgrade, 2005.

[14] B. Ljutić, M. Stamatović: Project Management for Managers – from Idea to Realization, Panda Graf, Belgrade, 1995.

[15] A Guide to the Project Management Body of Knowledge (PMBOK Guide), Project Management Institute, Pennsylvania USA, 2000 Edition.

[16] A. J. Schwab: Managementwissen für Ingenieure, 2. neubearbeitete Auflage, Springer, Berlin, Heidelberg, New York, 1999.

[17] Code of Practice for Project Management for Construction and Development, third edition, Blackwell Publishing, 2003.

[18] P. Blattner, L. Urlich, K. Cook, T. Dysk : Microsoft Excel 2000 Guide (special edition), CET, Belgrade, 2000.

[19] Proof Tests of the Power Equipment Insulation, Project No. ETR. 6.04.0033.B, Faculty of Electrical Engineering, Belgrade, Orderer Ministry of Science, Technology and Development of the Republic of Serbia, Belgrade, 2002 - 2004, (project manager Z. Stojković).

[20] Realization of Multifunctional Optoelectronic Devise for Control and Monitoring of High-Voltage Equipment with Applications in the Commercial Sector, Project TR – 6622B, Faculty of Electrical Engineering, Belgrade, Orderer Ministry of Science and Environmental Protection of the Republic of Serbia, Belgrade, 2005-2007, (project manager Z. Stojković).

[21] Development of a Magnetic Field Spatial Scanner for Diagnostics of Equipment in Power Systems and Environmental Protection, Project TR – 17031, Faculty of Electrical Engineering, Belgrade, Orderer Ministry of Science and Technological Development of the Republic of Serbia, Belgrade, 2008-2011, (project manager Z. Stojković).

[22] P. Mihailović, S. Petričević , Z. Stojković, J. Radunović : Development of Portable Fiber Optic Current Sensor for Power Systems Monitoring, IEEE Transactions on Instrumentation and Measurement, Vol. 51, No. 1, Feb 2004, pp. 24-30.

[23] S. Petričević, Z. Stojković, J. Radunović: Practical Application of Fiber Optic Current Sensor in Power System Harmonics Measurement; IEEE Transactions on Instrumentation and Measurement, Vol. 55, No. 3, June 2006, pp. 923-930.

[24] S. Petričević, Z. Stojković, P. Mihailović , J. Radunović : Development of a Fiber Optic Impulse Current Sensor for High Voltage Equip-

ment Tests, International Journal of Electrical Engineering Education (IJEEE), Vol. 45, No. 1, January 2008, pp. 1-16.

[25] G. Courter, A. Marquis: Project 2000, Computer Library, 2001.

[26] C. Chatfield, T. Johnson: Microsoft Office Project 2003 Step by Step, Microsoft Press, 2004.

[27] E. Marmel: Microsoft Project 2007 Bible, Wiley Publishing, Inc, 2007.

# LIST OF ABBREVIATIONS

ACWP – Actual Cost of Work Performed
ANSI – American National Standards Institute
ATP – Alternative Transient Program
BAC – Baseline Cost or Budget At Completion
BCA – Benefit / Cost Analysis
BCC – Budget Control Charts
BCWP – Budgeted Cost of Work Performed
BCWS – Budgeted Cost of Work Scheduled
CAD – Computer-Aided Design
CAE – Computer-Aided Engineering
CB – Circuit Breaker
CC – Corner Cubicle
CIGRE – International Council on Large Electric Systems
CM – Construction Management
CO – Commisioning and Operating
CP – Cost Plus
CPI – Cost Performance Index
CPIC – Control Point Identification Charts
CPM – Critical Path Method
CS – Counselling Services
CV – Cost Variance
DBMS – Database Management System
DD – Detail Design
DIN – Deutsches Institut fur Normung
DSSCW – Design and Supervision Services for Construction of Works
EAC – Estimate At Completion
EM – Electrogeometric Model
EMTP – Electromagnetic Transient Program
ES – Evaluation Study
ESELC – Early Streamer Emission Lightning Conductor
ET – Estimated Time
FACTS – Flexible AC Transmission Systems

Z. Stojković, *Computer-Aided Design in Power Engineering*,
DOI: 10.1007/978-3-642-30206-0, © Springer-Verlag Berlin Heidelberg 2012

FD – Feasibility and Design
FF – Finish to Finish
FIDIC – International Federation of Consulting Engineers
FOCS – Fiber Optic Current Sensor
FOICS – Fiber Optic Impulse Current Sensor
FORTRAN – FORmula TRANslating system
FP – Fixed Price
FS – Feasibility Stage
FS – Feasibility Study
FS – Fnish to Start
GERT – Graphical Evaluation and Review Technique
GIC – Grounding Impulse Characteristics
GIS – Gas Insulated Switchgears
GM – Generic Model
GPS – General Position System
GUI – Graphical User Interface
HLA – High Level Architecture
HPGL – Hewlett-Packard Graphics Language
HPP – Hydroelectric Power Plant
IEC – International Electrotechnical Commision
IEEE – Institute of Electrical and Electronics Engineers
IH – In-House
IS – Inception Stage
JDE – Jet Database Engine
LISP – LISt Processing language
LS – Lump Sum
MATLAB – MATrix LABoratory
MC – Milestone Charts
MCA – Monte Carlo Analysis
MDM – Master Data Management
MP – Milestone Point
MPT – Most Probable Time
MSDE – Microsoft Data Engine
MTBF – Mean Time Before Failure
MTBF – Mean Time Between Failure
OLAP – Online Analytical Processing
OS – Opportunity Study

OSI – Open Systems Interconnection
OT – Optimistic Time
PAM – Protective Angle Method
PCC – Project Control Charts
PERT – Program Evaluation and Review Technique
PFS – Prefeasibility Study
PIS – Pre-Investment Studies
PLC – Programmable Logic Controller
PM – Project Management
PMIS – Project Management Information System
PT – Pesimistic Time
RSM – Rolling Sphere Method
SA – Sensitivity Analysis
SCADA – Supervisory Control And Data Acquisition
SD – Scheme Design
SDDS – Specialized Design and Development Services
SF – Start to Finish
SP – Sealing Price
SPI – Schedule Performance Index
SPLCAD – Software Power Line Computer-Aided Design
SQL – Structured Query Language
SS – Start to Start
SS – Support Study
STATCOM – Static Synchronous Compensator
SV – Schedule Variance
TACS – Transient Analysis of Control System
TGPR – Transient Ground Potential Rise
THD – Total Harmonic Distortion
TK – Turn Key
TKC – Turn Key Contract
TPP – Thermal Power Plant
TS – Transformer Station
UCS – User Coordinate System
UN ECE – United Nations Economic Commission for Europe
UP – Unit Prices
UTM – Universal Transverse Mercator
VAC – Variance At Completion

VB – Visual Basic
VBA – Visual Basic for Applications
VBE – Visual Basic Environment
VD – Voltage Divider
VFT – Very Fast Transients
VLISP – Visual LISP
VRLA – Valve Regulated Lead Acid
WCS – World Coordinate System

# INDEX OF KEY WORDS


Dielectric constant 29, 46, 89
Documentation of the program 70
Effective length of a grounding grid 100
Equivalent
circuit 12, 25, 32, 40, 45
Graph 31
Integration 49, 189, 241
Interpolation 49, 94
Selection of computerized methods 49
Classifications of systems 20
Configuration 37, 76, 77, 87
Integration step 81, 83, 85
Local instability 85
Model 19, 87
Modeling 19, 100
Nonlinear grounding grid model 41, 103
Normal tree 31, 32
Optimization 86, 96, 136, 204
Parameters of an equivalent
grounding grid 44, 105, 125
Program 50, 160, 169
Programming language 23, 56, 304
Design 57, 78, 98, 369
Calculation 135, 160, 167
Computer 21, 50, 54
Substation 77, 160, 192, 219
Series resistance 25, 27, 66
Risk of failure 77, 302
$SF_6$ switchgear 192, 210
Simulation 19, 21, 37, 86
System of differential equations 30, 66
Self-inductance 25, 29, 113
Self and mutual parallel conductances 28
Soil resistivity 29, 81, 113
External inductance 29, 113
Power frequency grounding
resistance 28, 46, 48, 113
Discharge impulse current 35, 87, 113
Program structure 55, 75
Structural diagram 21
Study 23, 190, 372, 387
Software tool 23, 37, 56, 99
Testing the program 74, 297
Test examples 72
Internal inductance 29, 113
Grounding grid 25, 214, 264
Characteristic values of the fault current 160
Short circuit 160, 220, 227
MATLAB® 61, 135
Simulink® 62, 136
SimPowerSystems 63, 136
Synchronous machine model 152
Asynchronous machine model 153
Power transformer model 147
Breaker model 146
SPLCAD 167
Turbine regulator 180



Z. Stojković, *Computer-Aided Design in Power Engineering*,
DOI: 10.1007/978-3-642-30206-0, © Springer-Verlag Berlin Heidelberg 2012

Arrangement 203, 295, 300
Single-pole diagram 313, 317
Experimental results 77, 208, 297
EMTP/ATP 56, 188, 196
ATPDraw™ 60, 191
Arc model 199
Modeling of GIS components 203
Potential of the metal enclosure 194, 207
Switching operations of the disconnector 193
Very fast transients 193
Transient ground potential rise 194
MS Excel 216, 221
MS Access 216, 224
Electronic databas 227
Macro 238
Designing high-voltage substations 219
Power engineering 321, 334, 369
Relational database 227
Visual Basic for Applications 218
AutoCAD 255
AutoLISP 255
Visual LISP 255
Model of a turbogenerator 255
Lightning protection 77, 263
Designing lightning protection for substations 281
Designing lightning protection for overhead lines 293
Protected zone 264
Lightning conductor 269
Early streamer emission lightning conductor 269
Lightning conductor with circular ring 270
Overhead line shielding zone 300
Electrotechnical documentation 312
Electrotechnical graphic symbols 313
EPLAN 334
SIMARIS SIVACON 339
SIMARIS DESIGN 339
Types of electrical diagrams 315
Principal (block) diagram 315
Multiple-pole diagram 319
Wiring diagram 319
Connection plan 319
Equivalent diagram 320
Logical diagram 320
Low-voltage switchgear 339
Investment project 369
Investor 371
Resource displays 404, 413
3D model 168, 255
Analysis of financial realization 417
Algorithm 51, 266, 271, 297
Sensitivity analysis 38, 87, 396
Analysis 42, 49, 62
Project cycle 396
Phases of project management 397
Gantt charts 397, 399
General contractor 371
General design 370
Main design 370
Principle design 370
Contractor design 371
MS Project 369
As-built design 371
Project 369
Designer 371
Engineering supervisor 371
Realization of the project 371

Content of main design 377
Expert supervisor 375
Participants in
realization of the project 371
Project management 369
Types of designs 369
Types of contracts 388

www.ingramcontent.com/pod-product-compliance
Lightning Source LLC
LaVergne TN
LVHW011246200926
848479LV00003B/70
*9783642302053*